CAMBRIDGE LIBRARY COLLECTION

Books of enduring scholarly value

Physical Sciences

From ancient times, humans have tried to understand the workings of the world around them. The roots of modern physical science go back to the very earliest mechanical devices such as levers and rollers, the mixing of paints and dyes, and the importance of the heavenly bodies in early religious observance and navigation. The physical sciences as we know them today began to emerge as independent academic subjects during the early modern period, in the work of Newton and other 'natural philosophers', and numerous sub-disciplines developed during the centuries that followed. This part of the Cambridge Library Collection is devoted to landmark publications in this area which will be of interest to historians of science concerned with individual scientists, particular discoveries, and advances in scientific method, or with the establishment and development of scientific institutions around the world.

Cosmos

Polymath Alexander von Humboldt (1769-1859), a self-described 'scientific traveller', was one of the most respected scientists of his time. Humboldt's wanderlust led him across Europe and to South America, Mexico, the U.S., and Russia, and his voyages and observations resulted in the discovery of many species previously unknown to Europeans. Originating as lectures delivered in Berlin and Paris (1827–8), his multi-volume *Cosmos: Sketch of a Description of the Universe* (1845–60) represented the culmination of his lifelong interest in understanding the physical world. As Humboldt writes, 'I ever desired to discern physical phenomena in their widest mutual connection, and to comprehend Nature as a whole, animated and moved by inward forces.' Volume 1 (1846) investigates celestial and terrestrial phenomena, from nebulae to the temperature of the earth, as well as 'organic life'. Throughout, he stresses the method of, and limits to, describing the universe's physical nature.

Cambridge University Press has long been a pioneer in the reissuing of out-of-print titles from its own backlist, producing digital reprints of books that are still sought after by scholars and students but could not be reprinted economically using traditional technology. The Cambridge Library Collection extends this activity to a wider range of books which are still of importance to researchers and professionals, either for the source material they contain, or as landmarks in the history of their academic discipline.

Drawing from the world-renowned collections in the Cambridge University Library, and guided by the advice of experts in each subject area, Cambridge University Press is using state-of-the-art scanning machines in its own Printing House to capture the content of each book selected for inclusion. The files are processed to give a consistently clear, crisp image, and the books finished to the high quality standard for which the Press is recognised around the world. The latest print-on-demand technology ensures that the books will remain available indefinitely, and that orders for single or multiple copies can quickly be supplied.

The Cambridge Library Collection will bring back to life books of enduring scholarly value (including out-of-copyright works originally issued by other publishers) across a wide range of disciplines in the humanities and social sciences and in science and technology.

Cosmos

Sketch of a Physical Description of the Universe

VOLUME 1

ALEXANDER VON HUMBOLDT
EDITED BY EDWARD SABINE

CAMBRIDGE UNIVERSITY PRESS

Cambridge, New York, Melbourne, Madrid, Cape Town, Singapore,
São Paolo, Delhi, Dubai, Tokyo

Published in the United States of America by Cambridge University Press, New York

www.cambridge.org
Information on this title: www.cambridge.org/9781108013635

This edition first published 1846
This digitally printed version 2010

ISBN 978-1-108-01363-5 Paperback

COSMOS:

SKETCH

OF A

PHYSICAL DESCRIPTION OF THE UNIVERSE.

BY

ALEXANDER VON HUMBOLDT.

VOL. I.

Naturæ vero rerum vis atque majestas in omnibus momentis fide caret, si quis modo partes ejus ac non totam complectatur animo.—PLIN. H. N. lib. vii. c. 1.

TRANSLATED UNDER THE SUPERINTENDENCE OF

LIEUT.-COL. EDWARD SABINE, R.A., FOR. SEC. R.S.

LONDON:

PRINTED FOR

LONGMAN, BROWN, GREEN, AND LONGMANS,

PATERNOSTER ROW; AND

JOHN MURRAY, ALBEMARLE STREET.

1846.

EDITOR'S PREFACE.

THIS translation of Cosmos was undertaken in compliance with the wish of Baron von Humboldt. The Editor, in common he believes with many others, is indebted to the earlier writings of the Author of Cosmos, for awakening in his mind a taste for pursuits, which have formed a large portion of his interest and added greatly to his enjoyment in life: long cherished feelings of gratitude for this obligation, combined with those of personal regard, have been motives with himself, and with Mrs. Sabine,—by whom the Translation has been made,—to surmount the hesitation which they might otherwise have felt in venturing on a task embracing so extensive a range of subjects. Should this translation be favourably received, it will be a great gratification to them hereafter to reflect, that they have been instrumental in making known to the English reader, the work in which the illustrious Author has embodied the fruits of his active and useful life.

The two introductory discourses, which occupy 48 pages
in the German edition, have been rewritten by M. de
Humboldt himself in the French language, for the French
edition, in which they fill 78 pages.   These were communicated to the Editor in their passage through the press,
and by the Author's desire have been followed in preference
to the corresponding portion of the German text, where
modifications or additions had been introduced.

Short as the interval has been since Cosmos was written,
it has not been unmarked by the progress which has been
made in several branches of scientific knowledge.   In
astronomy it has been distinguished by the discovery of a
new planet, Astrea, making the number of those bodies
belonging to our solar system twelve instead of eleven:
also of the two heads of Biela's comet, a phænomenon previously unknown.   These discoveries, however, in no
respect affect the reasonings in Cosmos.   The optical
means at the command of astronomers have also been
improved, by the construction of a telescope of unparalleled
dimensions by the Earl of Rosse; and the few trials which
have yet been made of its powers, lead to the belief, that the
greater part, if not the whole, of the nebulæ will be resolved
by it into stars: happily the Author of Cosmos will himself
have an opportunity, in the succeeding volumes, of stating
the influence which a discovery of this nature may exercise

CONTENTS.

ORGANIC LIFE.

present work, the whole of which, with the exception of a portion of the Introduction, was written for the first time in the years 1843 and 1844; the discourses in Berlin having been delivered from November 1827 to April 1828, previous to my departure for Northern Asia. A representation of the actual state of our knowledge, in which year by year the acquisition of new observations imperatively demands the modification of previous opinions, must, as it appears to me, gain in unity, freshness, and spirit, by being definitely connected with some one determinate epoch.

The first volume contains a general view of nature, from the remotest nebulæ and revolving double stars to the terrestrial phænomena of the geographical distribution of plants, of animals, and of races of men; preceded by some preliminary considerations on the different degrees of enjoyment offered by the study of nature and the knowledge of her laws; and on the limits and method of a scientific exposition of the physical description of the Universe. I regard this as the most important and essential portion of my undertaking, as manifesting the intimate connection of the general with the special, and as exemplifying in form and style of composition, and in the selection of the results taken from the mass of our experimental knowledge, the spirit of the method in which I have proposed to myself to conduct the whole work. In the two succeeding volumes

I design to consider some of the particular incitements to
the study of Nature,—to treat of the history of the contemplation of the physical universe, or the gradual development
of the idea of the concurrent action of natural forces
co-operating in all that presents itself to our observation,—
and lastly, to notice the specialities of the several branches
of science, of which the mutual connection is indicated in
the general view of nature in the present volume.   References
to authorities, together with details of observation, have been
placed at the close of each volume, in the form of Notes.
In the few instances in which I have introduced extracts
from the works of my friends, they are indicated by marks
of quotation; and I have preferred the practice of giving the
identical words to any paraphrase or abridgment.   The delicate and often contested questions of discovery and priority,
so dangerous to introduce in an uncontroversial work, are
rarely touched upon : the occasional references to classical
antiquity, and to that highly favoured transition period
marked by the great geographical discoveries of the fifteenth
and sixteenth centuries, have had for their principal motive,
the wish, which is occasionally felt when dwelling on general
views of nature, to escape from the more severe and dogma-
tical restraint of modern opinions, into the free and imaginative domain of earlier presentiments.

  It has sometimes been regarded as a discouraging consi-

simplicity of the earlier ages of the world; to them the undisturbed succession of the planetary movements, and the progressive development of animal and vegetable life, were pledges of an order yet undiscovered in other relations, but of which they instinctively divined the existence. To us in an advanced civilization belongs the enjoyment of the precise knowledge of phænomena. From the time when man in interrogating nature began to experiment, or to produce phenomena under definite conditions, and to collect and record the fruits of experience, so that investigation might no longer be restricted by the short limits of a single life, the *philosophy of nature* laid aside the vague and poetic forms with which she had at first been clothed, and has adopted a more severe character:—she now weighs the value of observations, and no longer divines, but combines and reasons. Exploded errors may survive partially among the uneducated, aided in some instances by an obscure and mystic phraseology: they have also left behind them many expressions by which our nomenclature is more or less disfigured; while a few of happier, though figurative origin, have gradually received more accurate definition, and have been found worthy of preservation in our scientific language.

The aspect of external nature, as it presents itself in its generality to thoughtful contemplation, is that of unity in diversity, and of connection, resemblance and order, among created things most dissimilar in their form;—one fair harmonious whole. To seize this unity and this harmony, amid such an immense assemblage of objects and forces,—to embrace alike the discoveries of the earliest ages and those of our own time,—and to analyse the details of phenomena without sinking under their mass,—are efforts of human reason, in the path wherein it is given to man to press towards the full comprehension of nature, to unveil a portion of her secrets, and, by the force of thought, to subject, so to speak, to his intellectual dominion, the rough materials which he collects by observation.

If we attempt to analyse the different gradations of enjoyment derived from the contemplation of nature, we find, first, an impression which is altogether independent of any knowledge of the mode of action of physical forces, and which does not even depend on the particular character of the objects contemplated. When we behold a plain bounded by the horizon, and clothed by a uniform covering of any of the social plants (heaths, grasses, or cistusses),—when we gaze on the sea, where its waves, gently washing the shore, leave behind them long undulating lines of weeds,—then while the heart expands at the free aspect of nature, there is at the same time revealed to the mind an impression of the existence of comprehensive and permanent laws governing the phenomena of the universe. The mere contact with nature, the issuing forth into the open air,—that which by an expression of deep meaning my native language terms *in das Freie*,—exercises a soothing and a calming influence on the sorrows and on the passions of men, whatever may be the region they inhabit, or the degree of intellectual culture which they enjoy. That which is grave and solemn in these impressions is derived from the presentiment of order and of law, unconsciously awakened by the simple contact with external nature; it is derived from the contrast of the narrow limits of our being with that image of infinity, which every where reveals itself in the starry heavens, in the boundless plain, or in the indistinct horizon of the ocean.

Other impressions, better defined, affording more vivid enjoyment and more congenial to some states of the mind, depend more on the peculiar character and physiognomy of the scene contemplated, and of the particular region of the earth to which it belongs. They may be excited by views the most varied; either by the strife of nature, or by the barren monotony of the steppes of Northern Asia, or by the happier aspect of the wild fertility of nature reclaimed to the use of man, fields waving with golden harvests, and peaceful dwellings rising by the side of the foaming torrent: for I regard here less the force of the emotion excited, than the relation of the sensations and ideas awakened to that peculiar character of the scene which gives them form and permanence. If I might yield here to the charm of memory, I would dwell on scenes deeply imprinted on my own recollection—on the calm of the tropic nights, when the stars, not sparkling, as in our climates, but shining with a steady beam, shed on the gently heaving ocean a mild and planetary radiance;—or I would recal those deep wooded valleys of the Cordilleras, where the palms shoot through the leafy roof formed by the thick foliage of other trees, above which their lofty and slender stems appear in lengthened colonnades "a forest above a forest(1) ;"—or the Peak of Teneriffe, when a horizontal layer of clouds has separated the cone of cinders from the world beneath, and suddenly the ascending current of the heated air pierces the veil, so that the traveller, standing on the very edge of the crater, sees through the opening the vine-covered slopes of Orotava, and the orange gardens and bananas of the coast. In such scenes it is no longer alone the peaceful charm, of which the face of nature is never wholly destitute, which speaks to our minds, but the peculiar character of the landscape, the new and beautiful forms of vegetable life, the grouping of the clouds, and the vague uncertainty with which they mingle with the neighbouring islands, and the distant horizon half visible through the morning mist. All that the senses but partially comprehend, and whatever is most grand and awful in such romantic scenes, open fresh sources of delight. That which sense grasps but imperfectly offers a free field to creative fancy; the outward impressions change with the changing phases of the mind; and this without destroying the illusion, by which we imagine ourselves to receive from external nature that with which we have ourselves unconsciously invested her.

When far from our native country, after a long sea voyage, we tread for the first time the lands of the tropics, we experience an impression of agreeable surprise in recognising, in the cliffs and rocks around, the same forms and substances, similar inclined strata of schistose rocks, the same columnar basalts, which we had left in Europe : this identity, in latitudes so different, reminds us that the solidification of the crust of the earth has been independent of differences of climate. But these schists and these basalts are covered with vegetable forms of new and strange aspect. Amid the luxuriance of this exotic flora, surrounded by colossal forms of unfamiliar grandeur and beauty, we experience (thanks to the marvellous flexibility of our nature) how easily the mind opens to the combination of impressions connected with each other by unperceived links of secret analogy. The imagination recognises in these strange forms nobler developments of those which surrounded our childhood; the colonist loves to give to the plants of his new home names borrowed from his native land, and these strong untaught impressions lead, however vaguely, to the same end as that laborious and extended comparison of facts, by which the philosopher arrives at an intimate persuasion of one indissoluble chain of affinity binding together all nature.

It may seem a rash attempt to endeavour to analyse into its separate elements the enchantment which the great scenes of nature exert over our minds, for this effect depends especially on the combination and unity of the various emotions and ideas excited; and yet if we would trace back this power to the objective diversity of the phænomena, we must take a nearer and more discriminating view of individual forms and variously acting forces. The richest and most diversified materials for such an analysis present themselves to ˙the traveller in the landscapes of Southern Asia, in the great Indian Archipelago, and, above all, in those parts of the new continent where the highest summits of the Cordilleras approach the upper surface of the aerial ocean by which our globe is enveloped, and where the subterranean forces which elevated those lofty chains still shake their foundations.

Graphic descriptions of nature, arranged under the guidance of leading ideas, are calculated not merely to please the imagination, but also to indicate to us the gradation of those impressions to which I have already alluded, from the uniformity of the sea beach, or of the steppes of Siberia, to the rich luxuriance of the torrid zone. If we represent to ourselves Mount Pilatus placed on the Shreckhorn (²), or the Schneekoppe of Silesia on the summit of Mont Blanc, we shall not yet have attained to the height of one of the colossi of the Andes, the Chimborazo, whose height is twice that of Etna; and we must pile the Rigi or Mount Athos on the Chimborazo, to have an image of the highest summit of

the Himalaya, the Dhavalagiri. But although the mountains of India far surpass in their astonishing elevation (long
disputed, but now confirmed by authentic measurements)
the Cordilleras of South America, they cannot, from their
geographical position, offer that inexhaustible variety of
phenomena by which the latter are characterized. The impression produced by the grandest scenes of nature does not
depend exclusively on height. The chain of the Himalaya
is situated far without the torrid zone. Scarcely is a single
palm tree (³) found so far north as the beautiful valleys
of Kumaoon and Nepaul. In 28° and 34° of latitude,
on the southern slope of the ancient Paropamisus, nature
no longer displays that abundance of tree ferns, of arborescent grasses, of Heliconias, and of Orchideous plants,
which, within the tropics, ascend towards the higher plateaux
of the mountains. On the slopes of the Himalaya, under
the shade of the Deodar and the large-leaved oak peculiar
to these Indian Alps, the rocks of granite and of mica schist
are clothed with forms closely resembling those which characterize Europe and Northern Asia; the species indeed are
not identical, but they are similar in their aspect and physiognomy, comprising junipers, alpine birches, gentians, par-
nassias, and prickly species of Ribes (⁴). The chain of the
Himalaya is also wanting in those imposing volcanic phenomena, which, in the Andes and in the Indian Archipelago,
often reveal to the inhabitants, in characters of terror, the
existence of forces residing in the interior of our planet.
Moreover, on the southern declivity of the Himalaya, where
the vapour-loaded atmosphere of Hindostan deposits its
moisture, the region of perpetual snow descends to a zone
of not more than 11000 or 12000 (11700 or 12800 Eng.)

feet of elevation: thus the region of organic life ceases at a limit nearly three thousand feet (⁵) below that which it reaches in the equinoctial portion of the Cordilleras.

But the mountainous regions which are situated near the equator possess another advantage, to which attention has not been hitherto sufficiently directed. They are that part of our planet in which the contemplation of nature offers in the least space the greatest possible variety of impressions. In the Andes of Cundinamarca, of Quito, and of Peru, furrowed by deep barrancas, it is permitted to man to contemplate all the families of plants and all the stars of the firmament. There, at a single glance, the beholder sees lofty feathered palms, humid forests of bamboos, and all the beautiful family of Musaceæ; and, above these tropic forms, oaks, medlars, wild roses, and umbelliferous plants, as in our European homes; there, too, both the celestial hemispheres are open to his view, and, when night arrives, he sees displayed together the constellation of the Southern Cross, the Magellanic clouds, and the guiding stars of the Bear which circle round the Arctic pole. There, the different climates of the earth, and the vegetable forms of which they determine the succession, are placed one over another, stage above stage; and the laws of the decrement of heat are indelibly written on the rocky walls and the rapid slopes of the Cordilleras, in characters easily legible to the intelligent observer. Not to weary the reader with details of phenomena which I long since attempted (⁶) to represent graphically, I will here retrace only a few of the more comprehensive features which, in their combination, form those pictures of the torrid zone. That which, in impressions received solely by our senses, partakes of an uncertainty, similar to the effect of the misty atmosphere, which, in mountain scenery, renders at times every outline dim and indistinct,—when scrutinized by reasoning on the cause of the phænomena, may be clearly viewed and correctly resolved into separate elements, to each of which its own individual character is assigned; and thus, in the study of nature, as well as in its more poetic description, the picture gains in vividness and in objective truth by the well and sharply-marked lines which define individual features.

Not only is the torrid zone, through the abundance and luxuriance of its organic forms, most rich in powerful impressions, —it has also another advantage, even greater in reference to the chain of ideas here pursued, in the uniform regularity which characterises the succession both of meteorological and of organic changes. The well-marked lines of elevation which separate the different forms of egetable life, seem there to offer to our view the invariability of the laws which govern the celestial movements, reflected as it were in terrestrial phænomena. Let us dwell for a few moments on the evidences of this regularity, which is such, that it can even be measured by scale and number.

In the burning plains which rise but little above the level of the sea, reign the families of Bananas, of Cycadeæ, and of Palms, of which the number of species included in our floras of the tropical regions has been so wonderfully augmented in our own days by the labours of botanic travellers. To these succeed, on the slopes of the Cordilleras, in mountain valleys, and in humid and shaded clefts of the rocks, tree ferns raising their thick cylindrical stems, and expanding their delicate foliage, whose lace-like indentations are seen against the deep azure of the sky. There, too, flourishes the Cinchona, whose fever-healing bark is deemed the more salutary the more often the trees are bathed and refreshed by the light mists which form the upper surface of the lowest stratum of clouds. Immediately above the region of forests the ground is covered with wide bands of flowering social plants, small Aralias, Thibaudias, and myrtle-leaved Andromedas. The Alp rose of the Andes, the magnificent Befaria, forms a purple girdle round the spiry peaks. On reaching the cold and stormy regions of the Paramos, shrubs and herbaceous plants, bearing large and richly-coloured blossoms, gradually disappear, and are succeeded by a uniform mantle of monocotyledonous plants. This is the grassy zone, where vast savannahs (on which graze lamas and cattle descended from those brought from the old world) clothe the high table lands and the wide slopes of the Cordilleras, whence they reflect afar a yellow hue. Trachytic rocks, which pierce the turf, and rise high into those strata of the atmosphere which are supposed to contain a smaller quantity of carbonic acid, support only plants of inferior organization—Lichens, Lecideas, and the many-coloured dust of the Lepraria, forming small round patches on the surface of the stone. Scattered islets of fresh-fallen snow arrest the last feeble traces of vegetation, and are succeeded by the region of perpetual snow, of which the lower limit is distinctly marked, and undergoes extremely little change. The elastic subterranean forces strive, for the most part in vain, to break through the snow-clad domes which crown the ridges of the Cordilleras;—but even where these forces have actually opened a permanent channel of communication with the outer air, either through crevices or circular craters, they rarely send forth currents of lava, more

often erupting ignited scoriæ, jets of carbonic acid gas and
sulphuretted hydrogen, and hot steam. The contemplation of
this grand and imposing spectacle appears to have produced
on the minds of the earlier inhabitants of those countries
only vague feelings of astonishment and awe. It might
have been imagined, that, as we have before said, the well-
marked periodic return of the same phænomena, and the
uniform manner in which they group themselves in ascending
zones, would have rendered easier a knowledge of the laws of
nature; but so far as history and tradition enable us to
trace, we do not find that the advantages possessed by those
favoured regions have been so improved. Recent researches
have rendered it very doubtful whether the primitive seat of
Hindoo civilisation, one of the most wonderful phases of the
rapid progress of mankind, were really within the tropics.
Airyana Vaedjo, the ancient cradle of the Zend, was to the
north-west of the Upper Indus; and after the separation of
the Iraunians from the Brahminical institution, it was in a
country bounded by the Himalaya and the small Vindhya
chain, that the language which had previously been common
to the Iraunians and Hindoos, assumed among the latter
(together with manners, customs, and the social state), an
individual form in the Magadha, or Madhya Desa (7). The
farthest extension of the Sanscrit language and civilisation
towards the south-east, to far within the torrid zone, has
been developed by my brother, Wilhelm von Humboldt, in
his great work on the Kavi, and other languages of kindred
structure (8).

Notwithstanding the greater difficulties with which in
more northern climates, the discovery of general laws was
surrounded, by the excessive complication of phænomena, and

the perpetual local variations, both in the movements of the atmosphere and in the distribution of organic forms, it was to the inhabitants of the temperate zone that a rational knowledge of physical forces first revealed itself. It is from this northern zone, which has shown itself favourable to the progress of reason, to the softening of manners, and to public liberty, that the germs of civilisation have been imported into the torrid zone, either by the great movements of the migration of races, or by the establishment of colonies, very different in their institution in modern times from those of the Greeks and Phœnicians.

In considering the influences which the order and succession of phenomena may have exercised on the greater or less facility of recognising their producing causes, I have indicated that important point in the contact of the human mind with the external world, at which there is added to the charm attendant on the simple contemplation of nature, the enjoyment springing from a knowledge of the laws which govern the order and mutual relations of phænomena. Thenceforth the persuasion of the existence of an harmonious system of fixed laws, which was long the object of a vague intuition, gradually acquires the certainty of a rational truth, and man, as our immortal Schiller has said—" Amid ceaseless change, seeks the unchanging pole (9)."

In order to reascend to the first germ of this more thoughtful enjoyment, we need only cast a rapid glance on the earliest glimpses of the *Philosophy of Nature*, or of the ancient doctrine of the *Cosmos*. We find amongst the most savage nations (and my own travels have confirmed the truth of this assertion), a secret and terror-mingled presentiment of the unity of natural forces, blending with the dim perception of an invisible and spiritual essence manifesting itself through these forces, whether in unfolding the flower and perfecting the fruit of the food-bearing tree, or in the subterranean movements which shake the ground, and the tempests which agitate the atmosphere. A bond connecting the outward world of sense with the inward world of thought may be here perceived; the two become unconsciously confounded, and the first germ of a philosophy of nature is developed in the mind of man without the firm support of observation. Amongst nations least advanced in civilization, the imagination delights itself in strange and fantastic creations. A predilection for the figurative influences both ideas and language. Instead of examining, men content themselves with conjecturing, dogmatising, and interpreting supposed facts which have never been observed. The world of ideas and of sentiments does not reflect back the image of the external world in its primitive purity. That which in some regions of the earth, and among a small number of individuals gifted with superior intelligence, manifests itself as the rudiment of natural philosophy, appears in other regions and among other races of mankind as the result of mystic tendencies and instinctive intuitions. It is in the intimate communion with external nature, and the deep emotions which it inspires, that we may also trace, in part, the first impulses to the deification and worship of the destroying and preserving powers of nature. At a later epoch of human civilization, when man, having passed through different stages of intellectual development, has arrived at the free enjoyment of the regulating power of reflection, and has learned, as it were, by a progressive enfranchisement, to separate the world of ideas from that of the perceptions of sense, a vague presentiment of the unity of natural forces no longer suffices him. The exercise of thought then begins to accomplish its noble task, and, by observation and reasoning combined, the students of nature strive with ardour to ascend to the causes of phæ-nomena.

The history of science teaches us how difficult it has been for this active curiosity always to produce sound fruits. Inexact and incomplete observations have led, through false inductions, to that great number of erroneous physical views which have been perpetuated as popular prejudices among all classes of society. Thus, by the side of a solid and scien-tific knowledge of phænomena, there has been preserved a system of pretended results of observation, the more diffi-cult to shake because it takes no account of any of the facts by which it is overturned. This empiricism—melan-choly inheritance of earlier times—invariably maintains whatever axioms it has laid down; it is arrogant, as is every thing that is narrow-minded; whilst true physical philosophy, founded on science, doubts because it seeks to investigate thoroughly,—distinguishes between that which is certain and that which is simply probable,—and labours incessantly to bring its theories nearer to perfection by extending the circle of observation. This assemblage of incomplete dogmas bequeathed from one century to another,—this system of physics made up of popular prejudices,—is not only injurious because it perpetuates error with all the obstinacy of the supposed evidence of ill-observed facts, but also because it hinders the understanding from rising to the level of great views of nature. Instead of seeking to discover the *mean* state, around which, in the midst of

apparent independence and irregularity, the phenomena really
and invariably oscillate, this false science delights in multi-
plying apparent exceptions to the dominion of fixed laws;
and seeks, in organic forms and in the phænomena of nature,
other marvels than those presented by internal progressive
development, and by regular order and succession. Ever
disinclined to recognise in the present the analogy of the
past, it is always disposed to believe the order of nature
suspended by perturbations, of which it places the seat, as if
by chance, sometimes in the interior of the earth, sometimes
in the remote regions of space.

It is the special object of this work to combat these
errors, which, originating in vicious empiricism and de-
fective induction, have survived even amongst the higher
classes of society (often by the side of much literary
cultivation), and thus to augment and ennoble the enjoyments which nature affords, by imparting a deeper view into
her inner being. Such enjoyment (as our Carl Ritter has
well shewn) is highest, when the whole mass of facts collected from different regions of the earth is comprehended
in one glance, and placed under the dominion of intellectual combination. Increased mental cultivation, in all
classes of society, has been accompanied by an increased
desire for the embellishment of life through the augmentation of the mass of ideas, and of the means of generalising
those already received. Nor is such a desire unworthy of
notice in reference to vague accusations, which represent the
minds of men, in this our age, as occupied almost exclusively
by the material interests of life.

I touch, almost with regret, on a fear which seems to me
to arise either from a too limited view, or from a certain

feeble sentimentality of character; I mean the fear that
nature may lose part of her charms, and part of the magic of
her power over our minds, when we begin to penetrate her
secrets,—to comprehend the mechanism of the movements
of the heavenly bodies,—and to estimate numerically the intensity of forces. It is true that, properly speaking, the
forces of nature can only exert over us a magical power, by
their action being to our minds enveloped in obscurity, and
beyond the conditions of our experience. Even supposing
that they would thus be the better fitted to excite our imagination, that assuredly is not the faculty which we should
prefer to evoke, whilst engaged in those laborious subsidiary
observations, which have for their ultimate object the knowledge of the grandest and most admirable laws of the universe. The astronomer occupied in determining, by the aid
of the heliometer, or of the doubly refracting prism ([10]), the
diameter of planetary bodies; or patiently engaged for years
in measuring the meridian altitudes of certain stars and
their distances apart,—or, searching for a telescopic comet
among a crowded group of nebulæ, does not feel his imagination more excited, (and this is the very warrant of the
accuracy of his work,) than the botanist who is intent on
counting the divisions of the calix, the number of sta-
mens, or the sometimes connected, and sometimes independent, teeth of the capsule of a moss. And yet it is these
precise angular measurements, and minute organic relations,
which prepare and open the way to the higher knowledge of
nature and of the laws of the universe. The physical philosopher (as Thomas Young, Arago, and Fresnel,) measures
with admirable sagacity the waves of light of unequal length,
which by their interferences reinforce or destroy each other,

even in respect to their chemical action; the astronomer, armed with powerful telescopes, penetrates space, and contemplates the satellites of Uranus at the extreme confines of our solar system, or (like Herschel, South, and Struve) decomposes faintly sparkling points into double stars, differing in colour and revolving round a common centre ot gravity; the botanist discovers the constancy of the gyratory motion of the chara in the greater number of vegetable cells, and recognises the intimate relations of organic forms in genera, and in natural families. Surely the vault of heaven studded with stars and nebulæ, and the rich vegetable covering which mantles the earth in the climate of palms, can scarcely fail to produce on these laborious observers impressions more imposing, and more worthy of the majesty of creation, than on minds unaccustomed to lay hold of the great mutual relations of phænomena. I cannot therefore agree with Burke when he says, that our ignorance of natural things is the principal source of our admiration and of the feeling of the sublime. The illusion of the senses, for example, would have nailed the stars to the crystalline dome of the sky; but astronomy has assigned to space an indefinite extent; and if she has set limits to the great nebula to which our solar system belongs, it has been to shew us further and further beyond its bounds, (as our optic powers are increased,) island after island of scattered nebulæ. The feeling of the sublime, so far as it arises from the contemplation of physical extent, reflects itself in the feeling of the infinite which belongs to another sphere of ideas. That which it offers of solemn and imposing it owes to the connexion just indicated; and hence the analogy of the emotions and of the pleasure excited in us in the midst of the wide sea;

or on some lonely mountain summit, surrounded by semi-transparent vaporous clouds; or, when placed before one of those powerful telescopes which resolve the remoter nebulæ into stars, the imagination soars into the boundless regions of universal space.

The mere accumulation of unconnected observations of details, without generalization of ideas, may no doubt have conduced to the deeply-rooted prejudice, that the study of the exact sciences must necessarily tend to chill the feelings, and to diminish the nobler enjoyments attendant on the contemplation of nature. Those who in the present day cherish such an error in the midst of rapid progress and new vistas of knowledge, fail in appreciating the value of every enlargement of the sphere of intellect, and of the tendency to rise from separate facts to results of a higher and more general character. To this fear of sacrificing, under the influence of scientific reasoning, something of the free enjoyment of nature, is often added another fear, namely, that the extent of the field of natural knowledge forbids to the greater part of mankind access to its enjoyments. It is true that in the midst of the universal fluctuation of forces, and of the seemingly inextricable network of organic life, alternately developed and destroyed, every step in the more intimate knowledge of nature leads to the entrance of new labyrinths; but to those engaged in the pursuit the very multiplicity of paths presenting themselves, the exciting effort of divining the true one, the presentiment of fresh mysteries to be unveiled, are all full of enjoyment. The discovery of each separate law indicates, even if it does not reveal, to the intelligent observer the existence of some other higher and more general law. Nature, according

to the definition of a celebrated physiologist ([11]), and as
the word itself indicated with the Greeks and Romans, is
"that which is in perpetual growth and progress, and which
subsists in continual change of form and internal development." The series of organic types, as presented to our
view, gradually gains enlargement and completeness as previously unknown regions are penetrated and surveyed,—as
living organic forms are compared with those which have
disappeared in the great revolutions which our planet has
undergone,—as microscopes have been rendered more perfect, and have been more extensively employed. Amid this
immense variety of animal and vegetable forms and their
transformations, we see, as it were, incessantly renewed the
primordial mystery of all organic and vital development,
the problem of metamorphosis, so happily treated by
Goethe,—a solution corresponding to our intuitive desire
to arrange all the varied forms of life under a small
number of fundamental types. As observation continually
increasing reveals yet more and more of the treasures of
nature, man becomes imbued with the intimate conviction that, whether we regard the surface or the interior of the earth, the depths of the ocean, or the
celestial spaces, the scientific conqueror will never complain
with the Macedonian, that there are no fresh worlds to subject to his dominion ([12]). General considerations, whether
relating to matter agglomerated in the celestial bodies, or to
the distribution of organic life on the surface of the earth, are
not only in themselves more attractive than special studies,
but they also offer peculiar advantages to the greater number
of men who can devote but little time to such occupations.
The different branches of the study of natural history are

only accessible in certain positions of social life; nor do they present the same charm in all seasons and in all climates. If our interest is fixed exclusively upon one class of objects, the most animated accounts of travellers from distant regions will have no attraction for us, unless they happen to touch on the chosen subjects of our studies.

As the history of nations, if it were possible that it could always successfully trace back events to their true causes, would no doubt solve to us the ever-recurring enigma of the alternately impeded and accelerated progress of human society; so, likewise, the physical description of the universe, the science of the Cosmos, if grasped by a powerful intellect, and based on the knowledge of all that has been discovered up to a given epoch, would remove many of those apparent contradictions, which the complication of phænomena, caused by a multitude of simultaneous perturbations, presents at the first glance.

The knowledge of laws, whether revealing themselves in the ebb and flow of the ocean, in the paths of comets, or in the mutual attractions of multiple stars, renders us more conscious of the " calm of nature :" and one might say that " the discord of the elements,"—that long-cherished phantasm of the human mind in its earlier and more intuitive contemplations—is gradually dispelled as science extends its empire. General views lead us habitually to regard each organic form as a definite part of the entire creation, and to recognize in the particular plant or animal, not an isolated species, but a form linked in the chain of being to other forms living or extinct. They assist us in comprehending the relations which exist between the most recent discoveries, and those which have prepared the way for them. They enlarge the bounds of our intellectual existence, and while we ourselves may be living in retirement they place us in communication with the whole globe. Under their guidance we follow with eager interest the investigations of travellers and observers in every variety of climate. We accompany, in thought, the bold navigators of the polar seas; and, amidst the realm of perpetual ice, view with them that volcano of the antarctic pole, whose fires are seen from afar, even at the season when no night favours their brightness. The intellectual objects, both of these adventurous voyages, and of those stations of observation recently established in almost every latitude, are not strange to us; for we can comprehend some of the wonders of terrestrial magnetism, and general views lend an irresistible attraction to the consideration of those *magnetic storms*, which embrace the whole circumference of the earth at the same instant of time.

Let me be permitted to elucidate the preceding considerations, by touching on a few of those discoveries whose importance cannot be justly appreciated without some general knowledge of physical science. For this purpose I will select instances which have recently attracted much attention. Who, without some general knowledge of the ordinary paths of comets, could perceive how fruitful in consequences was Encke's discovery, by which a comet, that in its elliptic orbit never passes out of our planetary system, reveals the existence of an ethereal fluid obstructing its tangential force? A rapidly-spreading half-knowledge brings scientific results ill understood into the conversation of the day, and the supposed danger of collision between two heavenly bodies, or of a deterioration of climate from cosmical causes, are again brought forward in a new and more

deceptive form.   Clear views of nature, even if merely historical, are sufficient preservatives against these dogmatizing
fancies.   The history of the atmosphere and of the annual
variations of its temperature, extends already sufficiently far
back to shew that these consist in repeated small oscillations
around the mean temperature of a station, thereby dispelling
the exaggerated fear of a general and progressive deterioration of the climates of Europe.   Encke's comet, which is
one of the three interior comets, completing its course
in 1200 days, must, from its position and the form of its
path, be as harmless to the inhabitants of our globe as
Halley's great comet of 1759 and 1835, which has a period
of seventy-six years.   The path of another comet of short
period, Biela's, which completes its course in six years,
does, indeed, intersect the earth's path, but it can only
approach us when its perihelion coincides with our winter
solstice.

    The quantity of heat received by a planetary body (the
unequal distribution of which determines the great meteorological processes of our atmosphere) depends conjointly on
the light-evolving power of the sun (*i. e.* the nature of its
surface, or the state of its gaseous covering), and on the relative positions of the sun and planet.   There are, indeed,
periodical variations, which the form of the earth's orbit and
the obliquity of the ecliptic undergo in obedience to the
universal law of gravitation; but these changes are so slow,
and restricted within such small limits, that their thermic
effects would hardly be appreciable by our present thermometric instruments in many thousands of years.   Supposed
cosmical causes of diminished temperature or moisture, or
of epidemic diseases,—of which the idea has been enter-

tained in modern times, as well as in the middle ages,—are, therefore, wholly out of the range of our actual experimental knowledge.

I may also borrow from physical astronomy other examples, of which the grandeur and the interest cannot be felt without some general knowledge of the forces which animate the universe, and may adduce the elliptic revolutions of many thousands of double stars, or suns, around each other, or rather around their common centre of gravity, revealing the existence of the Newtonian attraction in those distant worlds ;—the periodical abundance or paucity of spots on the sun (openings in the opaque but luminous envelope of the solid nucleus) ;—and the periodic appearance, observed for some years past about the 12th or 13th of November and the 10th or 11th of August, of countless multitudes of shooting stars, moving with planetary swiftness, which probably form a belt of asteroids intersecting the orbit of the earth.

Descending from the skies to the earth, we may notice how the oscillations of a pendulum in air (the theory of which has been perfected by Bessel's acuteness) have thrown light on the internal density, I might say on the degree of solidification, of our planet ; they have also served, in a certain sense, to sound terrestrial depths, conveying information respecting the geological nature of strata otherwise inaccessible. In this manner, as well as in others, we are enabled to trace a striking analogy between the production of granular rocks in lava currents which have flowed down the slopes of active volcanoes, and those granites, porphyries, and serpentines, which, issuing from the interior of the earth, have broken, as eruptive rocks, through the secondary strata, modifying them by their contact, either hardening them by the introduction of silex, or changing them into dolomite, or causing in them the formation of crystals of various kinds. The elevation of sporadic islets, of domes of trachyte, and of cones of basalt, by the elastic forces which emanate from the fluid interior of our planet, conducted the first geologist of our age, Leopold von Buch, to the theory of the elevation of continents and of mountain chains generally. This action of subterranean forces, in breaking through and elevating sedimentary rocks, of which the coast of Chili has offered a recent example, shows us how the oceanic shells which M. Bonpland and myself found on the ridge of the Andes, at an elevation of more than 4600 mètres (about 15100 English feet), may have been conveyed there, not by a rise of the ocean, but by volcanic agencies elevating into ridges the heat-softened crust of the globe.

I use the term volcanic agency in its most general sense, applying it, whether on the earth or on her satellite the moon, to that reaction of the interior of a planet on its crust, which on our globe at least has been very different at different epochs. Those who are unacquainted with the experiments which show the increase of internal heat at increasing depths in the earth to be so rapid, that granite[13] is supposed to be in a state of fusion about twenty geographical miles below the surface, cannot have a clear comprehension of the causes of the simultaneity of volcanic eruptions occurring at great distances apart,—of the extent and intersection of circles of commotion in earthquakes,—or of the constancy of temperature and of chemical composition in thermal springs during many years of observation, and of the difference of temperature of Artesian wells of unequal depth. And yet the knowledge of the internal terrestrial heat throws a faint light on the early history of our planet, by showing the possibility of a generally prevalent tropical climate, arising from the heat issuing from crevices in the recently oxydized crust of the globe; a state of things in which the temperature of the atmosphere would depend far more on the reaction of the interior of the planet upon its crust, than on its relative position in respect to the central body or sun.

The cold zones of the earth present to the researches of the geologist many buried products of a tropical climate :— in the coal formations, upright stems of palms, coniferae, and tree ferns, goniatites, and fishes with rhomboidal enamelled scales (14);—in the Jura limestone, colossal skeletons of crocodiles and long-necked Plesiosauri, Planulites, and stems of Cycadeæ ;—in the chalk small Polythalamia and Bryozoa, in part identical with some of our living marine animals ;—in tripoli or polishing slate, in semi-opal, and in the substance called mountain meal, agglomerated masses of fossil infusoria, such as Ehrenberg's all-animating microscope has disclosed to us ;—and lastly, in transported soils and in caves, bones of hyenas, lions, and elephantine pachydermata. An enlarged knowledge of other natural phænomena renders these objects no longer an occasion for mere barren curiosity and wonder, but for intelligent study and deep consideration.

The multiplicity of diverse objects which I have here purposely crowded together, leads directly to the question, whether general views of nature can posess a sufficient degree of clearness, without a deep and earnest application to separate studies, whether of descriptive natural history, of geology, of physics, or of mathematical astronomy ? In attempting a reply, we must discriminate carefully between the teacher who undertakes the selection, combination, and presentation of the results, and the person who receives them, when thus presented, as something not sought out by himself, but communicated to him by another. To the first, some exact knowledge of the special is indispensably necessary; before proceeding to the generalisation of ideas, he should have wandered long in the domains of the separate sciences, and have himself observed, experimented, and measured. I cannot deny that where positive knowledge is wanting in the reader, general results, which in their mutual connection lend so great a charm to the contemplation of nature, are not susceptible of being always developed with equal clearness; but, nevertheless, I permit myself the pleasure of thinking, that in the work which I am preparing, the greater number of the truths presented will admit of being exhibited without the necessity of always reascending to fundamental principles and ideas. The picture of nature thus drawn, even though some parts of its outlines may be less sharply defined, will still possess truth and beauty, and will still be suited to enrich the intellect, to enlarge the sphere of ideas, and to nourish and vivify the imagination.

Our scientific literature has been reproached, and perhaps not without justice, with not sufficiently separating the general from the special — the view of that already gained from the long recital of the means which have led to it. This reproach even led the greatest poet of our age ([15]) to exclaim with impatience, that "the Germans have the gift of rendering the sciences inaccessible." Whilst the scaffolding stands, it obscures the effect of the finished building. Who can

doubt that the uniformity of figure observed in the distribution of our continental masses, by which they taper towards
the south, and spread out in breadth towards the north
(a fact or law on which the distribution of climates, the
prevailing direction of atmospheric and oceanic currents, and
the great extension of tropical forms into the southern temperate zone so materially depend), may be fully apprehended,
together with its consequences, without any acquaintance
with those geodesical and astronomical determinations, by
means of which the precise forms and dimensions of the
continents have been delineated in our maps?   Thus, too,
the physical description of the earth teaches us that the
length of the equatorial axis of our planet exceeds that of
its polar axis by a certain number of miles, and informs us
of the mean equality of the compression of the northern
and southern hemispheres, without the necessity of relating
in detail the measurements of degrees and the pendulum
experiments, by means of which we have arrived at the
knowledge that the true figure of the earth is that of an irregular ellipsoid of revolution, and is reflected in the irregularity of the movements of the earth's satellite, the moon.

Enlarged views of physical geography have been essentially advanced by the appearance of the admirable work
("Erdkunde im Verhältniss zur Natur und zur Geschichte
des Menschen, oder allgemeine vergleichende Geographie")
in which Carl Ritter has characterized so powerfully the physiognomy of our globe, and has shewn the influence of its
external configuration, both on the physical phænomena which
take place on its surface, and on the migration of nations,
their laws, their manners, and their history.

France possesses an immortal work, Laplace's " Exposi-

tion du Système du Monde," in which the results of the
highest mathematical and astronomical labours of all preceding ages are presented detached from all details of de-
monstration.   In this work the structure of the heavens is
reduced to the simple solution of a great problem in mechanics; and yet, assuredly, it has never been accused of in-
completeness, or want of profoundness.   The separation of
the general from the special not only renders it possible to
embrace at one view, with greater clearness, a wider field of
knowledge, but it also lends to the treatment of natural
science a character of greater elevation and grandeur.   By
the suppression of details the masses are better seen, and
the reasoning faculty is enabled to grasp that which might
otherwise escape our limited powers of comprehension.

The high degree of improvement which the last half century has witnessed in the study of all the separate branches
of natural science, but especially in those of chemistry, general physics, geology, and descriptive natural history, is emi-
nently favourable to the presentation of general results. When
first looked at singly and superficially all phænomena appear
unconnected; as observations multiply and are combined by
reflection, and as a deeper insight into natural forces is
obtained, more and more points of contact and links of
mutual relation are discovered, and it becomes more and
more possible to develope general truths with conciseness,
without superficiality.   In an age of such rapid and brilliant progress as the present, it is a sure criterion of the
number and value of the discoveries to be hoped for in any
particular science, if, though studied with great assiduity
and sagacity, its facts still appear for the most part unconnected, with little mutual relation, or even in some instances

in seeming contradiction with each other. Such is the kind
of expectation at present excited by meteorology, by many
parts of optics, and, since the admirable labours of Melloni
and Faraday, by the study of radiant heat and of electro-
magnetism. The circle of brilliant discoveries has here still
to be run through; although the Voltaic pile already reveals
the wondrous connection of electrical, magnetical, and chemical phenomena. Who will venture to affirm, that we yet
know with precision that part of the atmosphere which is
not oxygen, or that thousands of gaseous substances affecting our organs may not be mixed with the nitrogen? or
who will say that we already know even the whole number
of the forces which pervade the universe?

It is not the purpose of this work to attempt to reduce
all sensible phænomena to a small number of abstract principles, having their foundation in pure reason only. The
physical cosmography of which I attempt the exposition
does not aspire to the perilous elevation of a purely rational
science of nature. Leaving to others, who may perhaps
adventure on them with more success, these depths of a
purely speculative philosophy, my essay on the Cosmos consists of *physical geography,* joined with the *description
of the heavenly bodies in space:* its aim is to present
a view of the material universe, which may rest on the
empirical foundation of the facts registered by science,
compared and combined by the operations of the intellect. It is within these limits alone that the undertaking can harmonise with the wholly objective tendency of
my mental disposition, and with the labours which have
occupied my long scientific career. The unity which I seek
to attain in the development of the great phænomena of

nature, is similar in kind to that which historical compositions may offer. All that belongs to the specialities of the
actual,—to its individualities, variabilities, and accidents,
whether in the form and connection of natural objects and
phænomena, or in the struggle of man with the elements, or
of nations with each other,—does not admit of being *ration-
ally constructed*, that is to say, of being deduced from
ideas alone. I venture to think that a like degree of
empiricism attaches to the description of the material universe, and to civil history; but in reflecting on physical
phænomena and historical events, and in reasoning back-
ward to their causes, we recognise more and more the
grounds of that ancient belief, that the forces inherent in
matter, and those which regulate the moral world, exert their
action under the government of a primordial necessity, and
in recurring courses of greater or less period. It is this
necessity, this occult but permanent connection, this periodical recurrence in the progressive development of forms, of
phænomena and of events, which constitute nature obedient
to the first-imparted impulse of the Creator. Physical science, as the name imports, limits itself to the explanation
of the phænomena of the material world by the properties
of matter. All beyond this belongs not to the domain of
the physics of the universe, but to a higher class of ideas.
The discovery of laws, and their progressive generalisation, are the objects of the experimental sciences. Kant,
who has never been deemed an irreligious philosopher, has
traced with rare sagacity the limits of physical explanations,
in his celebrated "Essay on the Theory and Structure of
the Heavens," published at Königsberg, in 1755.

The study of a science which promises to lead us over the

wide range of creation, may be likened to a journey in a
distant country. Before undertaking it we are inclined to
measure, perhaps not without mistrust, both our own
strength and that of the guide who offers to conduct us.
But our fears may be lessened by remembering how in our
days, an increasing knowledge of the mutual relation of
phænomena, leading to the attainment of general results, has
more than kept pace with the vast increase of separate
observations. The chasms which divide facts from each
other are rapidly filling up; and it has often happened that
facts observed at a distance have thrown a new and unexpected light on others nearer home, which had long seemed
to resist all efforts at explanation. Plants and animals which
had long appeared insulated, become connected with others
by the discovery of intermediate forms before unknown; and
the geography of beings endowed with organic life receives
completeness, as we behold species, genera, and whole
families, peculiar to one continent, reflected, so to speak,
in analogous forms, or, as it were, in *equivalents* in
the opposite continent. These transitions may be traced,
in the sometimes fuller, sometimes more rudimentary,
development of particular parts, or in their different relative
importance in the balance of forces, or in the junction of
distinct organs, or sometimes in resemblances to intermediate
forms, not permanent, but only characteristic of particular
phases of a normal development. Passing to the consideration of inorganic bodies, and to examples which characterise
strongly the advances of modern geology, we see how,
according to the grand views of Elie de Beaumont, chains
of mountains, dividing different climates, floras, and nations,
reveal to us their *relative* age, by the nature of the sedimen-

tary rocks uplifted by them, and by the directions which
they follow over the long crevices produced by the action
of the forces which have elevated in ridges portions of
the crust of the globe.   Relations of superposition of
trachyte and of syenitic porphyry, of diorite and of serpentine, which remain doubtful, if studied in the aurife-
rous soils of Hungary, in the platinum district of the
Oural, or on the South Western slope of the Siberian
Altai, are clearly made out by the aid of observations on
the high table-lands of Mexico and Antioquia, and in
the unhealthy ravines of the Choco. The most important
of the materials, which in modern times have afforded a solid
basis for physical geography, have not been accumulated by
chance.   In conformity with its characteristic tendencies,
our age has recognised, that facts obtained by observations
in different regions of the earth, can only be expected to
prove fruitful in results, when the traveller is previously
acquainted with the state and wants of the science which he
seeks to advance, and when his researches are conducted
under the guidance of sound ideas, and some insight into
the character and connection of natural phænomena.

By means of the happy, though often too easily satisfied tendency towards general conceptions, a tendency dangerous only
in its abuse, a considerable portion of the results of natural
knowledge may become the common property of all educated
persons, producing a sound information very different both
in substance and in form from those superficial compilations,
which contained the sum of what, up to the close of the
last century, was complacently designated by the unsuitable
term of *popular scientific knowledge*. I take pleasure in
persuading myself that it is possible for scientific subjects to

be presented in language, grave, dignified, and yet animated ; and that those who are able to escape occasionally from the restricted circle of the ordinary duties of civil life, and regret to find that they have so long remained strangers to nature, may thus have opened to them access to one of the noblest enjoyments, which the activity of the rational faculties can afford to man. The study of general natural knowledge awakens in us, as it were, new perceptions which had long lain dormant; we enter into a more intimate communion with the external world, and no longer remain without interest or sympathy for that which at once promotes the industrial progress and intellectual ennoblement of man.

The clearer our insight into the connection of phænomena, the more easily shall we emancipate ourselves from the error of those, who do not perceive that for the intellectual cultivation, and for the prosperity of nations, all branches of natural knowledge are alike important ; whether the measuring and describing portion, or the examination of chemical constituents, or the investigation of the physical forces by which all matter is pervaded. It has not been uncommon presumptuously to depreciate investigations arbitrarily characterized as " purely theoretic," forgetting that in the observation of a phænomenon which shall, at first sight, appear isolated, may lie concealed the germ of a great discovery. When Galvani first stimulated the nervous fibre by the contact of two dissimilar metals, his immediate contemporaries could not have foreseen that the voltaic pile would discover to us in the alkalis, metals of a silvery lustre, easily inflammable, and so light as to float in water ; that it would become the most important instrument of chemical analysis, and at the same time a thermoscope and a magnet. When

Huyghens first applied himself, in 1678, to the enigma of the phænomena of polarisation of light exhibited in doubly-refracting spar, and observed the difference between the two portions into which a beam of light divides itself in passing through such a crystal, it was not foreseen that through the admirable sagacity of a physical philosopher of the present day ([16]), the phænomena of *chromatic polarisation* would lead us to discern, by means of a minute fragment of Iceland spar, whether the light of the sun proceeds from a solid nucleus, or from a gaseous covering; whether comets are self-luminous, or reflect borrowed light.

An equal appreciation of all parts of natural knowledge is an especial requirement of the present epoch, in which the material wealth and the increasing prosperity of nations are in great measure based on a more enlightened employment of natural products and forces. The most superficial glance at the present condition of European states shews, that those which linger in the race cannot hope to escape the partial diminution, and perhaps the final annihilation, of their resources. It is with nations as with nature, which, according to a happy expression of Goethe ([17]), knows no pause in unceasing movement, development, and production, and has attached a curse to standing still. The danger to which I have alluded must be averted by the earnest cultivation of natural knowledge. Man can only act upon nature, and appropriate her forces to his use, by comprehending her laws, and knowing those forces in relative value and measure. Bacon has said that, in human societies, knowledge is power,—both must rise or sink together. Knowledge and thought are at once the delight and the prerogative of man; and they are also a part of the wealth of

nations, and often afford to them an abundant indemnifica-
tion for the more sparing bestowal of natural riches. Those
states which remain behind in general industrial activity, in
the selection and preparation of natural substances, in the application of mechanics and chemistry,—and where a due appreciation of such activity fails to pervade all classes,—must see
their prosperity diminish ; and that the more rapidly as neighbouring states are meanwhile advancing, both in science and
in the industrial arts, with, as it were, renewed and youthful
vigour.

The improvement of agriculture in the hands of freemen,
and on properties of moderate extent,—the flourishing state
of the mechanical arts freed from the trammels of the spirit
of corporation,—commerce augmented and animated by the
multiplied contact of nations with each other,—are brilliant
results of the general progress of intelligence, and of the
amelioration of political and civil institutions in which that
progress is reflected. The picture presented by modern history ought to convince those who seem tardy in apprehending
the instruction which it is fitted to convey. Nor let it be
feared that the predilection for industrial progress, and for
those branches of natural science most immediately connected
with it, which characterize the age in which we live, has any
necessary tendency to check intellectual exertion in the fair
fields of classical antiquity, history, and philosophy; or to
deprive of the life-giving breath of imagination, the arts and
the literature which embellish life. Where all the blossoms
of civilization unfold themselves with vigour under the
shelter of wise laws and free institutions, there is no danger
of the development of the human mind in any one direction
proving prejudicial to it in others. Each offers to the nation

precious fruits,—those which furnish necessary subsistence and comfort, and are the foundation of material wealth,— and those fruits of creative fancy which, far more enduring than that wealth, transmit the glory of the nation to the remotest posterity. The Spartans, in spite of the Doric severity of their mode of thought, " prayed the Gods to grant them the beautiful with the good ([18])."

As in that higher sphere of thought and feeling to which I have just alluded, in philosophy, poetry, and the fine arts, the primary aim of every study ought to be an inward one, that of enlarging and fertilising the intellect; so the direct aim of science should ever be the discovery of laws, and of the principles of unity, order, and connection, which every where reveal themselves in the universal life of nature. But by that happy connection, whereby the useful is ever linked with the true, the exalted, and the beautiful, science thus followed for her own sake will pour forth abundant, over-flowing streams, to enrich and fertilise that industrial pros-perity, which is a conquest of the intelligence of man over matter.

The influence of mathematical and physical knowledge on national prosperity, and on the present condition of Europe, requires here only a passing allusion : the well-nigh boundless course which we have to travel over, warns me that it would ill become me to digress more widely from the leading object of our undertaking,—the contemplation of nature as a whole. Accustomed to distant excursions, I have perhaps fallen into the error of describing the path before us as more smooth and pleasant than it will be really found, as those are wont to do who love to guide others to the summit of lofty moun-tains : they praise the view, even when great part of the dis-

tant prospect is hidden by the clouds; knowing, indeed, that
this half transparent misty veil is itself not altogether without
a secret charm for the imagination. I too ought to fear, that
from the height to which this physical description of the
universe aspires, many parts of the wide horizon may appear
dimly lighted and imperfectly defined,—that much of the
prospect may remain vague and obscure, and this not only
by reason of the want of connection arising from the imperfect state of some branches of science, but also still more,
(and how, in so comprehensive a work, should I not will-
ingly own it?) because of the deficiencies of the guide
who has imprudently adventured to attempt to scale these
lofty summits.

The object of this introductory discourse has been less to
represent the importance of natural knowledge, which is
admitted by all, and may well dispense with any eulogium,
than to show how, without prejudice to the thorough and
fundamental study of separate branches, a higher point of
view may be indicated, from whence all the forms and the
forces of nature may be contemplated in intimate and living
connection.

The idea of physical geography, extended so as to embrace
all that we know of the material creation in space as well
as on our own globe, passes into that of physical cosmography; the one term is moulded upon the other. But the
science of the Cosmos, as I understand it, is not the mere
encyclopædic aggregation of the most general and important results, extracted from separate works on natural history,
physics, and astronomy. Such results are only to be used
as materials, and in so far as they illustrate the concurrent
action of the various forces in the universe, and the manner

in which they reciprocally call forth or limit each other. The distribution of organic types in different regions and climates (*i. e.* the geography of plants and animals,) differs as widely from descriptive botany and zoology, as does a geological knowledge of the globe from mineralogy properly so called. The physical description of the universe is not therefore to be confounded with *encyclopædias of the natural sciences*. In the work before us it is proposed to consider partial facts only in their relation to the whole. The higher the point of view here indicated, the more the study requires a peculiar mode of treatment, and to be presented in animated and picturesque language.

But thought and language are of old intimately allied : if the language employed lends to the presentation grace and clearness; if by its organic structure, its richness, and happy flexibility, it favours the attempt to delineate the phænomena of nature, it at the same time reacts almost insensibly on thought itself, and breathes over it an animating influence. *Words*, therefore, are more than signs and forms; and their mysterious and beneficent influence is there most powerfully manifested, where the language has sprung spontaneously from the minds of the people, and on its own native soil. Proud of my country, whose intellectual unity is the firm foundation of every manifestation of her power, I look with joy to these privileges of my native land. Highly favoured indeed is he, who, in attempting an animated representation of the phænomena of the universe, is permitted to draw from the depths of a language, which through the elevation and free exercise of powerful thought, in the domain of creative fancy no less than in that of searching reason, has for centuries exerted so powerful an influence over the minds and the destinies of men.

LIMITS AND METHOD OF EXPOSITION OF THE PHYSICAL

DESCRIPTION OF THE UNIVERSE.

In the preceding discourse, I have sought to make manifest, and to illustrate by examples, how greatly the enjoyment of nature, varying as it does in the inward sources from which it springs, may be heightened by a clear insight into the connection of phenomena, and of the laws by which they are regulated. I have now to examine more particularly the spirit of the method of exposition, and to indicate the limits of the science of physical cosmography, such as I have conceived it, and have now endeavoured to display it, after many years of preparatory studies in many regions of the earth. Would, that in so doing, I might flatter myself with the hope of thereby justifying the bold title of my work, and freeing it from the reproach of presumption!

Before entering on the view of nature, which forms the larger portion of the present volume, I would touch on some general considerations intimately connected with each other, with the nature of our knowledge of the external world, and with the relations which this knowledge presents, at different epochs of history, to the different phases of the intellectual cultivation of nations. These considerations will have for their objects :—

1. The idea and the limits of physical cosmography as a distinct and separate science.

2. A rapid review of the known phenomena of the universe, under the form of a general view of nature.

3. The influence of the external world on the imagination and feelings. This, in modern times, has acted as a powerful incitement to the study of the natural sciences, through the instrumentality of animated descriptions of distant regions, descriptive poetry (a branch of modern literature), of landscape painting when it seizes the characteristic physiognomy of vegetable or of geological forms, and by the cultivation and arrangement of exotic plants, in well-contrasted groups.

4. The history of the contemplation of nature, or the progressive development of the idea of the Cosmos, with the exposition of the historical and geographical facts which have led to the systematic connection of the phænomena as they are thus presented.

The higher the point of view from which all the phænomena are to be regarded in this study, the more necessary it is to circumscribe it within its just limits, and to distinguish it from all analogous and auxiliary ones. The physical description of the universe is founded on the contemplation of all the material creation (whether substances or forces) co-existing in space. For man, as an inhabitant of the earth, it may be ranged under two leading divisions; the telluric and the celestial. I will pause a few moments on the first of these, (or on that portion of the science of the Cosmos which concerns the Earth,) in order to illustrate the independence of the study, and the nature of its relation to general physics, descriptive natural history, geology, and comparative geography. An encyclopædic aggregation of these would no more constitute the telluric portion of the Cosmos, than a mere dry enumeration of the philosophical opinions prevailing in different ages, would deserve to be called the history of philosophy.

The confusion between the boundaries of closely allied branches of study has been the greater, because for centuries different portions of our empirical knowledge have been designated by terms which are either too comprehensive, or too restricted, for the notions they were intended to convey; and which have besides the disadvantage of having borne a very different sense in the languages of classical antiquity from which they are borrowed. The terms of physics, physiology, natural history, geology, and geography, arose and grew into general use long before clear ideas were entertained of the diversity of the objects which those sciences ought to embrace, and consequently of their respective limits.—Such is the influence of long habit upon language, that in one of the nations of Europe most advanced in civilisation, the word "physic" is applied to medicine; and in a Society of justly deserved and universal renown, writings on technical chemistry, geology, and astronomy empirically treated, (all branches of purely experimental science), are classed under the general title of "Philosophical Transactions." The attempt has often been made, and almost always in vain, to substitute new and more appropriate names for those ancient terms,—vague, it is true, but which, however, are now generally understood. These changes have been proposed, for the most part, by those who have occupied themselves with the general classification of all branches of human knowledge; from the great Encylopædia (Margarita Philosophica) of Gregory Reisch ([19]), Prior of the Chartreuse of Freiburg, towards the end of the fifteenth century, to Lord

Bacon; and from Bacon to D'Alembert; and still more recently to a sagacious physicist of our own time, André-Marie Ampère ([20]). The selection of an inappropriate Greek nomenclature has, perhaps, been even more prejudicial to the last of these attempts, than the abuse of the binary division and the excessive multiplication of groups.

The physical description of the universe as an object of external sense, does indeed require the aid of general physics and of descriptive natural history; but the consideration of the material creation, all the parts of which are linked together by mutual connection, under the figure of a natural whole animated and moved by inward forces, gives to the science which now occupies us a peculiar character. Physical science dwells on the general properties of matter; it is an abstract representation of the manifestations of physical forces, and in the work in which its earliest foundations were laid, in the eight books of Physics of Aristotle ([21]), all the phænomena of nature are depicted as the moving vital activity of a universal force.

The telluric portion of the physical description of the universe, to which I preserve the old and expressive title of *physical geography*, treats of the distribution of magnetism on our planet in its relations of intensity and direction, but does not teach the laws of magnetic attraction and repulsion, or the means of eliciting powerful electro-magnetic effects, whether transitorily or permanently. Physical geography describes in bold and general outlines the compact or indented configuration of continents, and the distribution of their masses in both hemispheres,—a distribution which powerfully influences the differences of climates and the most important meteorological processes of the atmosphere; it seizes the predominant character of mountain chains, whether parallel, or transverse and intersecting, and whether belonging to the same or to different epochs and systems of elevation; it examines the *mean* height of continents above the present surface of the sea, or the position of the centre of gravity of their volume; the relation of the highest summits of the great chains to the general line of their crests, to the vicinity of the sea, and to the mineral character of the rocks of which they consist. It depicts to us the eruptive rocks as active principles of movement, traversing, uplifting, and inclining at various angles, the passive sedimentary rocks : it considers volcanoes either as isolated or ranged in single or in double series, and extending their sphere of action to various distances, either by means of long narrow bands of erupted rocks, or by earthquakes operating in circles which widen or contract in the course of centuries. It describes the strife of the liquid element with the firm land; it shews the features which are common to all great rivers in the upper and in the lower portion of their course, and how they become subject to bifurcation. It characterizes rivers either as breaking their way through great mountain chains, or following, for a time, a course parallel to them, either close to their foot or at a considerable distance, according to the influence which the elevation of the mountain system may have exercised on the neighbouring plains. It is only the general results of comparative orography and hydrography which belong to the science whose proper limits I am endeavouring to trace, and not the enumeration of our loftiest mountains, active volcanoes, or rivers with the extent of their watershed and the number of their tributaries. All these details belong to geography properly so called, in its more restricted sense. We here consider phænomena only in their mutual connection, and in their relations to the different zones of our planet, and to its general physical constitution. The specialities either of inanimate substances or of organic beings, classed according to analogy of form and composition, do indeed form a highly interesting subject of study, but quite foreign to the present work.

Particular descriptions of countries are, it is true, the most available materials for a general physical geography ; but the most careful successive accumulation of such descriptions would be as far from affording a true picture of the general conformation of the irregular surface of our planet, as a series of all the floras of different regions would be from forming what I should designate by the term of a " Geography of Plants." It is the work of the intellect, by comparing and combining isolated observations, to extract from the specialities of organic formation (morphology and the descriptive natural history of plants and animals,) that which is common to them in regard to their climatic distribution ;—to investigate the numerical laws, or the proportion of certain forms or particular families to the whole number of species ;—to assign the latitude or geographical position of the zone where (in the plains) each of these forms reaches its maximum number of species, and its highest organic development. These considerations will lead us to perceive the manner in which the picturesque character of the landscape in different latitudes, and the impression which it pro-· duces on the mind, depend principally on the laws of the geography of plants, or the relative number and more vigorous growth of those which predominate in the general mass. The

systematically-arranged catalogues, to which the too pompous
name of "Systems of Nature" was formerly given, present to
us an admirable connection and arrangement by analogies of
structure, whether completely developed, or (according to
views of an evolution in spirals) in the different phases passed
through in vegetables, by the leaves, bracteas, calix, blossom,
and fruit, and in animals by their cellular and fibrous tissues,
and their articulations or less perfectly developed parts. But
these ingeniously classified so called "systems of nature,"
do not shew us organic beings as they are grouped over the
surface of our planet, in districts, zones of latitude, or of
elevation, and according to other climatic influences arising
from general and often very distant causes. But, as we
have already said, the final aim of physical geography is to
recognise unity in the vast variety of phenomena, and by
the exercise of thought and the combination of observations,
to discern that which is constant through apparent change.
In the exposition of the terrestrial portion of the Cosmos,
we may sometimes find occasion to descend to very special
facts, but it will only be for the purpose of recalling the
connection existing between the laws of the actual distribution of organic beings over the surface of the globe, and the
laws of the ideal classification by natural families, analogy
of internal organisation, and progressive evolution.

It follows from these discussions on the limits of different
sciences, and particularly from the distinction which it is
necessary to draw between descriptive botany (morphology)
and the geography of plants, that, in the physical description
of the globe, the innumerable multitude of organised bodies,
which form so large a portion of the beauties of creation,
ought to be considered rather with reference to *zones of*

habitation, and to the differently inflected *isothermal curves*, than according to principles of gradation in the development of their internal organisation. But botany and zoology, which are the two branches of the descriptive natural history of organised bodies, are the fruitful sources from whence we draw the materials, without which the study of the relations and connection of phenomena would want a solid foundation.

We will here add an important observation. The first general glance over the vegetation of an extensive portion of a continent, shews us an assemblage of dissimilar forms,— gramineæ, orchideæ, coniferæ, and oaks : we perceive these families and genera, instead of being locally associated, scattered apparently as it were by chance : but this irregular dispersion is only apparent ; and it is the province of physical geography to shew that vegetation every where presents constant numerical relations in the development of its forms and types ; that, in the same climates, species which are wanting in one country, are replaced in a neighbouring one by other species of the same families, according to a law of substitution, which seems to belong to the yet unknown relations of organised beings ; and by which the numerical proportion of particular great families to the whole mass of the phenogamous floras in adjoining countries is maintained. There is thus revealed in the multitude of organic forms by which these regions are peopled a principle of unity, a primitive plan of distribution. There is also discovered in each zone, diversified according to the families of plants, a slow but continuous action on the aerial ocean, an action which depends on the influence of light—that primary and essential condition of all organic vitality on the solid or liquid surface of our planet. It might be said, according to a fine expression of Lavoisier, that the marvel of the ancient mythus of Prometheus is incessantly renewed before our eyes.

When we apply the march which it is proposed to follow in the exposition of the physical description of the earth, to the sidereal part of the science of the Cosmos, or to the description of what is known to us of the regions of space, and of the heavenly bodies which they contain, we shall find our task remarkably simplified. If, according to ancient but inexact forms of nomenclature, we distinguish between *physics*, or the general consideration of matter, its forces, and its movements,—and *chemistry*, or the consideration of the different nature of substances, their elementary composition, and their attractions not depending on relations of mass or the laws of gravitation,—we must of course recognise that the telluric portion of our study embraces both *physical* and *chemical* processes. By the side of the fundamental force of gravitation, we discover around us on the earth the action of other forces, taking effect either when the particles of matter are in contact, or at exceedingly small distances apart ([22]) ; to which forces we give the name of *chemical affinity.* Under various modifications, by electricity, by heat, by condensation in porous bodies, or by the contact of an intermediate substance, these forces are incessantly in action in inorganic matter, and in the tissues of animals and plants. But, in the regions of space, we are only cognizant by direct observation of physical phænomena, and among these (excepting in the case of the small asteroids, which appear to us under the form of aerolites or shooting stars,) we know with certainty only those effects which depend on the quantitative relations of matter or the distribution of masses ; and which may therefore be contemplated as governed by simple dynamic laws.   Effects due to specific differences, or to heterogeneous qualities of matter, do not as yet enter into our calculations of the celestial mechanics.

It is only through the phænomena of light (the propagation of luminous waves) and the effects of gravitation, that the inhabitants of our earth enter into relation with matter in space, whether existing in spheroids, or in a dispersed form.   The fact of a periodical direct action of the sun and moon on the variations of terrestrial magnetism is still highly problematical. The only direct experimental knowledge which we possess of any of the specific properties or qualities of matter not belonging to our planet, is derived from the fall of the aerolites or meteoric stones, already alluded to. Their direction and enormous velocity of projection (a velocity wholly planetary) render it more than probable, that these masses, enveloped in vapours and reaching the earth in a high state of temperature, are small heavenly bodies, which the attraction of our planet has caused to deviate from their previous path. The aspect so familiar to us of these asteroids, and the analogy which their composition presents to the minerals of which the crust of our globe is formed, are indeed very striking.   The inference to which they point appears to me to be, that the planetary and other masses were agglomerated in rings of vapour, and afterwards in spheroids, under the influence of a central body ; and that being originally integral parts of the same system, they consist of substances chemically identical.   Pendulum experiments, and especially those made by Bessel with so high a degree of precision, confirm the Newtonian axiom, that the acceleration occasioned by the attraction of the earth is identical in bodies the most heterogeneous in composition,— viz., water, gold, quartz, granular limestone, and portions of different aerolites. Purely astronomical observations add their testimony to the proofs afforded by the pendulum. The almost identical results found for the mass of Jupiter, from its influence on his own satellites, on Encke's comet of short period, and on the small planets, Vesta, Juno, Ceres, and Pallas, equally teach that, as far as our observations reach, the attraction of gravitation is determined solely by the quantity of matter [23].

This absence of all perception (derived either from observation or from theoretical considerations) of any heterogeneous qualities of matter, gives to celestial mechanics a high degree of simplicity. The study of the immense regions of space being directed by the laws of motion only, the sidereal portion of the Cosmos draws from the pure and abundant sources of mathematical astronomy, as the terrestrial portion does from those of physics, chemistry, and organic morphology. But the domain of the three last-named sciences embraces phænomena so complex, and, to the present time, so little susceptible of the application of rigorously exact methods, that the physical knowledge of the globe cannot boast of the certainty and simplicity in the exposition of facts and of their mutual connection, which characterise the celestial portion of the Cosmos. This difference may be the true reason why, in the early times of the intellectual cultivation of the Greeks, the natural philosophy of the Pythagoreans was directed to the heavenly bodies in space, rather than to the earth and her productions; and became through

Philolaus, and subsequently through the analogous views of Aristarchus of Samos, and Seleucus of Erythrea, of far greater avail towards the knowledge of the true system of the universe, than the natural philosophy of the Ionic school could ever become to the physical knowledge of the earth. Giving less heed to the properties and specific differences of the various kinds of matter, the great *statical* school, in its Doric gravity, preferred to turn its regards towards all that relates to measure, form, and number ([24]); while the Ionic school dwelt on the qualities of matter, its real or supposed transformations, and its relations of origin. It was reserved to the powerful genius, and to the at once profoundly philosophical and practical mind of Aristotle, to enter equally deeply and successfully into the world of abstract ideas, and into that of the rich diversity of material substances, of organised beings, and animated existence.

Several highly esteemed treatises on physical geography have prefixed to them an introductory astronomical section, in which the earth is first considered in its planetary dependence, and in its relation to the solar system. This order of proceeding is opposite to that which I propose to follow. The dignity of the physical description of the universe requires that the sidereal portion, which Kant has called the natural history of the heavens, should not be made subordinate to the terrestrial portion. In the science of the Cosmos, according to the expression of Aristarchus of Samos —that ancient herald of the Copernican doctrine—the sun (together with all his satellites) is viewed but as one of the countless host of stars. It is then with these celestial bodies with which space is peopled, that the physical description of the universe ought to begin. It should commence with such a graphic sketch of the universe (such a true *map of the world*) as was traced by the bold hand of the elder Herschel. If, notwithstanding the smallness of our planet, the telluric portion of the present work occupies the largest space, and is treated with the greatest fulness, this arises only from the unequal amount of our knowledge of that which is within and that which is beyond our reach. The subordination of the celestial to the terrestrial portion is met with, however, in the great geographical work of Bernard Varenius ([25]), written in the middle of the seventeenth century. He distinguishes, with great acuteness, between general and special geography; and subdivides the first into an absolute, or properly *terrestrial* portion (when treating of the surface of the earth in its different zones), and a relative or planetary one, when considering the solar and lunar relations of our planet. It is a permanent glory to Varenius, that his "General and Comparative Geography" was found capable of fixing, in a high degree, the attention of Newton. In the imperfect state, in the time of Varenius, of the auxiliary branches of knowledge from which his resources had to be drawn, it was not possible that the execution of the work should correspond to the greatness of the undertaking. It was reserved to our own time to see comparative geography, in its most extended sense, and even embracing its influence on the history of man, treated in a masterly manner by my own countyman, Carl Ritter ([26]).

The enumeration of the more important results of the astronomical and physical sciences, which in Cosmos radiate towards a common centre, may justify, in some degree, the title which I have ventured to affix to this work, written in the late evening of my life. I might add, that the title is

perhaps more adventurous than the enterprise itself, circum-
scribed within the limits which I have proposed. In all
my previous investigations, I have hitherto avoided, as much
as possible, the introduction of new names to express general ideas. When I have attempted to enlarge our nomenclature, it has been solely in the specialities of descriptive
botany and zoology, when objects observed for the first time
rendered new names necessary. The expression of physical
cosmography, or description of the universe, is formed on
that of physical geography, or description of the earth,
which has long been used. The powerful genius of Descartes has left us some fragments of a great work, which he
intended should appear under the title of "Monde," and
for which he had begun to study special subjects, and even
human anatomy. The little used, but precise expression of
the *science of the Cosmos*, recals to the inhabitant of our
globe that we are treating of a wider horizon, of the assemblage of all the material things with which space is filled,
from the remotest nebulæ, to the climatic distribution of
the thin vegetable tissues of variously coloured lichens,
which clothe the surface of our rocks.

In every language, views entertained in the infancy of
nations have led to the confusion of the ideas of *earth* and
world : the common expressions of "voyages round the
world," "map of the world," "new world," are instances
of this confusion. The more accurate and more noble expressions * of "system of the world," "creation of the

---

* Our language does not possess all the expressions referred to by M. de
Humboldt. We have no direct English equivalents for the expressive German terms "Weltgebäude," "Weltraum," and "Weltkörper."

TRANSLATOR.

world," and others of a similar nature, relate either to the whole of the bodies with which celestial space is filled, or to the origin of the entire universe.

It was natural that, amidst the extreme variability of the phenomena presented by the surface of the earth and the surrounding aerial ocean, men should have been impressed by the aspect of the vault of heaven and the regular and uniform movements of the sun and planets. The word *Cosmos*, which, in its primitive signification, in the Homeric times, expressed the ideas of *ornament* and *order*, was subsequently applied to the order and harmony observed in the movements of the heavenly bodies; then to those bodies generally; and finally, to the entire universe.   It is asserted by Philolaus,—the genuine fragments of whose writings have been commented on with so much sagacity by M. Böckh,— that, according to the general testimony of antiquity [27], "Pythagoras was the first who used the word Cosmos to express the order which reigns in the universe, or the world or universe itself."   From the Italic school of philosophy, the term used in this sense passed into the language of the poets of nature, Parmenides and Empedocles, and thence into that of prose writers.   We need not enter here into the distinction, which, following the Pythagorean views, Philolaus draws between Olympus, Uranus, and Cosmos, or how the latter word, used in the plural, has been applied individually to celestial bodies (the planets) circling round the central "hearth," or focus of the world, or to "world-islands," or groups of stars.   In my work, the word Cosmos is employed as signifying the heavens and the earth, or the whole world of sense, or the material universe; agreeably to general Hellenic usage subsequently to the time of Pythagoras, and in conformity with its definition by the unknown author of the treatise, entitled "De Mundo," which was long erroneously attributed to Aristotle. If scientific names had not long varied from their true linguistic meaning, the present work might properly have been entitled "*Cosmography*," divided into *Uranography* and *Geography*. The desire of imitating the Greeks led the later Romans, in their feebler philosophical essays, to give the signification of universe to the word *mundus*, the primary meaning of which was merely that of *ornament*, without including order or regularity in the arrangement of parts. The introduction of this technical term, in the same double signification as the Greek word Cosmos, was probably due to Ennius ([28]), who was a follower of the Italic school, and translated the writings of Epicharmus, or one of his imitators, on the Pythagorean Philosophy. A physical *history* of the universe, in the extended sense of the word, ought, if materials for writing it existed, to trace the variations to which the Cosmos has been subjected in the course of ages, from those new stars which have suddenly become visible or have disappeared in the firmament, from nebulæ dissolving or condensing towards their centres,—to the first cryptogamic vegetation on the surface of the recently cooled crust of the globe, or that which now clothes the coral reef newly risen above the ocean. On the other hand, the object of the physical *description* of the universe is to present a view of all that co-exists in space, and of the simultaneous action of natural forces, with the resulting phænomena. But if we wish to comprehend existing nature well, we cannot separate entirely and absolutely the consideration of the present state of things, from that of the successive phases through which they have previously passed. The mode of formation, or of production, is often an important element of their character. Nor is it in the organic world only that matter is constantly undergoing change, and dissolving, to be formed into new combinations : the globe on which we live also reveals the knowledge of an earlier state : the strata of sedimentary rocks, which compose a large portion of its crust, present to us earlier forms of organic life, which have now almost entirely disappeared ; and these forms are associated in groups, successively replacing each other. The different superimposed strata thus present to us the buried faunas and floras of different epochs. In this sense the description of nature cannot be separated from its history ; for, in studying the present, the geologist, in tracing the mutual relations of the facts which come before him, is conducted back to ages long past; and this intermixture of past and present is in some respects analogous to that which may be observed in the study of languages, where the etymologist finds traces of successive grammatical developments leading him back to the primitive state of the idiom reflected as it were in forms of speech now in use. In the material world, this reflex of the past is the clearer, from our now seeing similar eruptive and sedimentary rocks in process of formation. The particular forms of domes of trachyte, basaltic cones, bands of amygdaloid with long parallel pores, and white deposits of pounce with black scoriæ intermixed, give to the eye of the geologist a peculiar kind of animation to the landscape, acting on his imagination as traditional monuments of an earlier world. Their form is their history.

The sense in which the Greeks and Romans employed the

SCIENCE OF THE COSMOS.

word *history* shows that they too had the intimate persuasion, that, to form a complete idea of the actual condition of
things, it was necessary to consider them in their succession. It is not, however, in the definition given by Verrius
Flaccus (²⁷), but in the zoological writings of Aristotle, that
the word *history* presents itself as signifying an exposition
of the results of experience and observation. The elder
Pliny's physical description of the world bears the title of
"Natural History;" and in his nephew's letters, the nobler
appellation of "History of Nature." The earlier Greek
historic writers scarcely separated the description of countries
from the relation of events of which they had been the
theatre. In their writings, physical geography and history
were long gracefully and pleasantly interwoven, until the
increasing complexity of political interests, and the agitations of civil life, expelled the geographical element from
the history of nations, and obliged it to become the subject
of a separate study.

It remains to examine, whether we can hope, by the operation of thought, to reduce the immense diversity of phænomena comprehended by the Cosmos, to a unity of principle,
similar to that presented by the evidence of what are
specially called "rational truths." In the present state of
our empirical knowledge at least, we dare not entertain such
a hope. Experimental sciences, founded on observation
of the external world, cannot aspire to completeness; the
nature of things and the imperfection of our organs are alike
opposed to it. We shall never succeed in exhausting the
inexhaustible riches of nature, and no generation of men
will ever be able to boast of having comprehended all phænomena. It is only by distributing them into groups, that

we have been able to discover in some the empire of laws, grand
and simple as Nature herself.   Doubtless, the bounds of
this empire will be enlarged as the physical sciences gradually
enlarge their domain, and become more perfect.   Brilliant
examples of such progress have appeared in our own times, in
the phænomena of electro-magnetism, and in those of the propagation of luminous waves and of radiant heat.   The doctrine of evolution shows us how, in organic development,
all that is formed is sketched out as it were beforehand, and
how the tissues of both vegetable and animal matter are
uniformly produced by the multiplication and transformation of cells.

The generalisation of laws which were first applied to
smaller groups of phænomena advances by successive gradations, and their empire is extended, and their evidence
strengthened, so long as the reasoning process is directed to
really analogous phænomena.   But as soon as dynamic views
no longer suffice, and the specific properties of heterogeneous matter come into play, fear may be entertained
lest, in the too obstinate pursuit of laws, we may arrive at
impassable chasms : the principle of unity fails us, and the
guiding clue breaks, when, in tracing the effects of natural
forces, we come to specific kinds of action.   The law of
equivalents, and of definite numerical proportions in com-
pound substances, so happily recognised by modern chemists,
and proclaimed under the antique form of atomic symbols,
remains hitherto isolated, and unsubjected to the mathematical laws of motion and gravitation.

Those natural productions, which are objects of direct
observation, may be logically distributed in classes, orders,
and families.   Such distribution does no doubt give greater

clearness to descriptive natural history; but the study of
organised bodies, arranged in linear connection, though it
gives greater unity and simplicity to the distribution of
groups, cannot rise to the height of a classification founded
on a single principle of composition and internal organization. As different gradations are presented by natural laws,
according as they embrace narrower or wider circles of phænomena, so there are successive steps in empirical investigation. It begins by single perceptions, which are afterwards classed according to their analogy or dissimilarity.
Observation is succeeded, at a much later epoch, by experiment, in which phænomena are made to arise under conditions previously determined on by the experimentalist, guided
by preliminary hypotheses, or a more or less just intuition
of the true connection of natural objects and forces. The
results obtained by observation and experiment lead by the
path of induction and analogy to the discovery of empirical
laws; and these successive phases in the application of the
human intellect have marked different epochs in the life of
nations. It has been by adhering closely to this inductive
path, that the great mass of facts has been accumulated
which now forms the solid foundation of the natural sciences.

Two forms of abstraction govern the whole of this class
of knowledge; viz. the determination of *quantitative*
relations, according to number and magnitude; and relations
of *quality*, embracing the specific properties of heterogeneous
matter. The first of these forms, more accessible to the
exercise of thought, belongs to the domain of mathematics;
the other, more difficult to seize, and apparently more mysterious, to that of chemistry. In order to submit phænomena to calculation, recourse is had to a hypothetical con-

struction of matter by a combination of molecules and atoms, whose number, form, position, and polarity, determine, modify, and vary the phænomena. The suppositions of imponderable matter, and vital forces peculiar to each mode of organization, have complicated and perplexed the view. Meanwhile, the prodigious mass of empirical knowledge is enlarging with increasing rapidity ; and investigating reason tries at times, with varying success, to break through ancient forms and symbols invented to effect the subjection of rebellious matter to mechanical constructions.

We are yet very far from the time, even supposing it possible that it should ever arrive, when a reasonable hope could be entertained of reducing all that is perceived by our senses to the unity of a single principle. The complication of the problem, and the immeasurable extent of the Cosmos, seem to forbid the expectation of such success in the field of natural philosophy being ever achieved by man ; but the partial solution of the problem—the tendency towards a general comprehension of the phænomena of the universe— does not the less continue to be the high and enduring aim of all natural investigation. For my own part, faithful to the character of my earlier writings, and to that of the labours which have occupied my scientific career, in measurements, experiments, and investigation of facts, I limit myself in the present work to the sphere of empirical conceptions. It is the only ground on which I feel myself able to move without a sense of insecurity. This mode of treating an aggregation of observed facts does not exclude their combination by reasoning, their arrangement under the guidance of leading ideas, their generalisation wherever it can be justly effected ; and the constant tendency to the discovery of laws. A

purely rational conception of the universe, founded on principles of speculative philosophy, would no doubt assign to the
science of the Cosmos a still more elevated aim. I am far
from blaming efforts which I have not myself attempted,
solely because their success hitherto has been extremely
doubtful. Contrary to the wishes and counsels of those
profound and powerful thinkers who have given new life to
speculations belonging to antiquity, systems of a philosophy
of nature have in our country (Germany) turned men's
minds for a time from the graver studies of the mathematical and physical sciences. The intoxication of supposed
conquests already achieved,—a novel and extravagantly symbolical language,—a predilection for formulæ of scholastic
reasoning more contracted than were ever known to the
middle ages,—have, through the youthful abuse of noble
powers, characterised the short saturnalia of a purely ideal
science of nature. I say abuse of powers, for superior
minds, which have embraced both speculative studies and
the experimental sciences, took no part in these saturnalia.
The results obtained by serious investigations in the path of
induction, cannot be at variance with a true philosophy of
nature. If there is contradiction, the fault must be either
in the unsoundness of the speculation, or in the exaggerated
pretensions of empiricism, which thinks that it has proved by
its experiments more than is really deducible from them.

The natural world may be opposed to the intellectual, or nature to art, taking the latter term in its higher
sense as embracing the manifestations of the intellectual
power of man; but these distinctions (which are indicated
in the most cultivated languages) must not be suffered to
lead to such a separation of the domain of physics from that

of the intellect, as would reduce the physics of the universe to a mere assemblage of empirical specialities. Science only begins for man from the moment when his mind lays hold of matter,—when he strives to subject the mass accumulated by experience to rational combinations : science is mind applied to nature. The external world only exists for us so far as we receive it within ourselves, and as it shapes itself within us into the form of a contemplation of nature. As intelligence and language, thought and the signs' of thought, are united by secret and indissoluble links, so in like manner, and almost without our being conscious of it, the external world and our ideas and feelings melt into each other. " External phænomena are translated," as Hegel expresses it, in his Philosophy of History, " in our internal representation of them." The objective world, thought by us, reflected in us, is subjected to the unchanging, necessary, and all-conditioning forms of our intellectual being. The activity of the mind exerts itself on the elements furnished to it by the perceptions of the senses. Thus, in the youth of nations, there manifests itself in the simplest intuition of natural facts, in the first efforts made to comprehend them, the germ of the philosophy of nature. These tendencies vary, and are more or less powerful, according to national individualities of character, turn of mind, and stage of mental culture, and whether attained amidst scenery fitted to excite and charm, or to repress and chill the imagination.

History has preserved the record of the varied and hazardous attempts, which have been made to comprehend all phænomena in a theoretical conception, and to discover in them a single natural force pervading, setting in motion, and transforming all matter. In classical antiquity the

earliest of these attempts are found in the treatises of the
Ionic school on the principles of things; treatises in which
the whole of nature was subjected to rash speculation, with
only an extremely scanty basis of observation. This ardour
for deductively determining the essence of things and their
mutual connection from an ideal construction and purely
rational principles, has gradually subsided, with the increasingly brilliant development of the natural sciences resting
on the firm support of observation. Nearer to our own
time, the mathematical portion of natural philosophy has
received the grandest and most admirable enlargement. The
method, and the instrument (analysis), have both been perfected together. We are of opinion, that what has been
conquered by means so diverse,—by the ingenious application
of atomic suppositions, —by the more general and more intimate study of phænomena,—and by new and improved
apparatus,—is the common property of mankind; and cannot
now, any more than in the times of the ancients, be with-
drawn from the free exercise of speculative thought. It
cannot be denied that the results of experience may have
been sometimes undervalued in the course of such processes;
nor ought we to be too much surprised if, in the perpetual
fluctuations of speculative views, as the author of Giordano
Bruno ([30]) has ingeniously expressed it, " Most men see in
philosophy only a succession of passing meteors; and even
the grander forms under which she has revealed herself
partake in the popular estimation of the fate of comets,
which they regard as belonging not to the class of permanent celestial bodies, but to that of mere passing igneous
vapours." But the abuse of speculative thought, and the
false paths into which it has sometimes strayed, ought not to

lead to a view, dishonouring to intellect, which would regard the world of ideas as essentially a region of phantom-like illusions, and philosophy as a hostile power, by which the accumulated treasures of experimental knowledge are threatened. It is unsuitable to the spirit of the age to reject with distrust any attempted generalisation of views, or investigation, in the path of reasoning and induction. Nor is it consonant with a due estimation of the dignity of the human intellect, and of the relative importance of the faculties with which we are endowed, to condemn, at one time, severe reason applied to the investigation of causes and their connection, and at another, that exercise of the imagination which is often precursive to discoveries,—for the achievement of which the imaginative power is indeed an essential auxiliary.

GENERAL VIEW OF NATURE.

WHEN the mind of man attempts to subject to itself the world of physical phænomena ;—when in meditative contemplation of existing things he strives to penetrate the rich fulness of the life of nature, and the free or restricted operations of natural forces ;—he feels himself raised to a height from whence, as he glances round the far horizon, details disappear, and groups or masses are alone beheld, in which the outlines of individual objects are rendered indistinct as by an effect of aerial perspective. This illustration is purposely selected in order to indicate the point of view from whence we design to consider the material universe, and to present it as the object of contemplation in both its divisions, celestial and terrestrial. I do not blind myself to the boldness of such an undertaking. Under all the forms of exposition to which these pages are devoted, the presentation of a general view of nature is the more difficult, because we must not permit ourselves to be overwhelmed by the development of the manifold and the multiform ; but must dwell only on the consideration of masses, great either by actual magnitude, or by the place which they occupy in the subjective range of

ideas. We strive by classification and due subordination of
phenomena, by penetration into the play of obscure forces,
and by an animated representation in which the visible
spectacle may be reflected back as in a faithful mirror,—to
conceive and to describe the whole creation (το παν) in a
manner befitting the dignity of the word Cosmos in its sense
of *universe, order of the material world,* and beauty or
ornament of that universal order.   May the immeasurable
diversity of the elements which crowd together into the
picture of Nature not be found to impair the harmonious
impression of repose and unity, which is the ultimate aim
of every literary or purely artistic composition !

I propose to begin with the depths of space and the
remotest nebulæ, and thence gradually to descend through
the starry region to which our solar system belongs, to the
consideration of the terrestrial spheroid with its aerial and
liquid coverings, its form, its temperature and magnetic
tension, and the fulness of organic life expanding and
moving over its surface under the vivifying influence of
light.   Such a universal sketch, though drawn with only a
few strokes of the pencil, must comprehend from the unmeasured celestial spaces to those microscopic animal and
vegetable organisations which inhabit our pools of standing
water and the weathered surfaces of our rocks.   All that
can be known by the senses, and all that a persevering
study of nature, in every direction, has revealed up to the
present time, constitute the material from which the representation is to be drawn.   This representation must contain
within itself the evidence of its fidelity and truth.   It does
not require for its completeness the enumeration of all
animated forms or of all natural objects or processes ; on

the contrary, order and harmony must be maintained by carefully resisting the tendency to endless division; thus avoiding the danger to which we are subjected by the very abundance of our empirical riches. Doubtless, a considerable portion of the properties of matter are still unknown to us; entire series of phænomena dependant on forces and qualities of which we are ignorant, remain to be discovered; and were it for this reason only, we must fail of the perfect attainment of unity in totality. By the side of the pleasure derived from knowledge already attained, there subsists, not unmixed with melancholy, the longing of the aspiring spirit, still unsatisfied with the present, after regions yet undiscovered and unopened. Such longing draws still closer the link which, by ancient and deep-seated laws of the world of thought, connects the material with the immaterial, and quickens the interchange between that which the mind receives from without, and that which it gives back from its own depths.

If, then, Nature (comprising in the word all natural objects and phænomena) may be regarded as embracing a range infinite in extent and contents, it also presents to the human intellect a problem which it cannot wholly grasp, and of which it can never hope to reach the solution, because it requires a knowledge of all the forces which act in the universe. Such an acknowledgment is due, when present and prospective phænomena are the objects of that direct investigation which does not venture to quit the empirical path and strictly inductive method. But though the constant effort to embrace the *whole* remain unsatisfied, the "History of the contemplation of the Universe" (which is reserved for a subsequent portion of this work) shows us how, in the course of centuries, mankind have gradually arrived at a partial insight into the relative dependence of phænomena. My duty is to depict that which is known, according to its present measure and limits. In all that is subject to motion and change in space, *mean numerical values* are the ultimate object; they are, indeed, the expression of physical laws; they shew to us the constant amid change, the stable amid the flow of phænomena. The advance of our modern physical science, which proceeds by weight and measure, is specially characterised by the attainment and progressive rectification of the mean values of certain quantities. Thus, the only remaining and widely diffused hieroglyphic characters of our present writing,—*numbers,*—reappear, as once in the Italic school, but now in a more extended sense, as powers of the Cosmos.

The earnest investigator delights in the simplicity of numerical relations, indicating the dimensions of celestial spaces, the magnitudes of heavenly bodies, their periodic disturbances, the threefold elements of terrestrial magnetism, the mean pressure of the atmosphere, and the quantity of heat which the sun dispenses in each year and each portion of the year to the several points of the solid and the liquid surface of our planet. Less satisfied is the poet of nature, and still less the mind of the curious multitude. To both of these, Science appears a blank, now that she answers doubtfully, or rejects as unanswerable, questions to which replies were, in earlier times, unhesitatingly adventured. In her severer form and less ample robes she appears deprived of that seductive grace, with which a dogmatising and symbolising physical philosophy could deceive the reason and occupy the imagination. Long before the discovery of the new world, men dreamed of lands in the West, visible from the Canaries or the Azores; and these illusive images were formed, not by any extraordinary refraction of the rays of light, but by the longing gaze striving to penetrate the distant and the unapproached. The fascination which belongs to such unsubstantial images and illusions was offered abundantly by the natural philosophy of the Greeks, the physics of the middle ages, and even by those of the centuries which succeeded them. At the limits of exact knowledge, as from a lofty island shore, the eye loves to glance towards distant regions. The belief of the unusual and the marvellous lends a definite outline to every creation of fancy; and the realm of imagination, a fairy land of cosmological, geognostical, and magnetical dreams, becomes uncontrollably blended with the domain of reality.

Nature, so manifold in signification—sometimes taken as including all the material creation existing and coming into existence, sometimes as a power of internal development, sometimes as the mysterious prototype of all phænomena—reveals herself to the simple senses and feelings of man, by preference in that which is terrestrial, and closely allied to himself. It is in the animated circle of organic forms that we first feel ourselves peculiarly at home. It is where the bosom of the earth unfolds its flowers, and ripens its fruits, and feeds countless tribes of animals, that the image of nature comes most vividly before our souls. The starry vault, the wide expanse of the heavens, belong to the picture of the universe, in which the magnitude of the masses, and the number of congregated suns, or faintly shining nebulæ, excite indeed our admiration and astonishment, but seem estranged from us by the entire absence of any immediate impression of their being the theatres **of**

organic life.   The earliest physical views separate between,
and oppose to each other, the heavens and the earth—the
above and the below in space.   If, then, our view were
intended to correspond solely to the requirements of
sensuous contemplation, it ought to begin with the
description of our native earth.   It should depict first
the terrestrial spheroid;—its magnitude and form; its increasing density and temperature at increasing depths in its
solid or liquid strata ; the relative configuration of sea and
land, and in both the development of organic life in the
cellular tissues of plants and animals ; the atmospheric
ocean, with its waves and currents, and forest-clad mountain
chains, which rise like reefs and shoals from its bottom.
After thus depicting purely telluric relations, the eye would
be raised to the celestial spaces ; the earth, the well-known
seat of organic development, would now be considered as a
planet taking its place in the series of cosmical bodies
revolving around one of the countless host of self-luminous
stars.   This succession of ideas indicates the path pursued
in the earliest mode of contemplation, or that which derives
purely from the senses ; it almost reminds us of the ancient
" sea-girt disc of Earth supporting the Heavens."   It begins
in perception, and its course is from the known and near to
the unknown and distant.   It corresponds to the method
pursued in our elementary works on astronomy (and which
has much to recommend it in a mathematical point of
view), of proceeding from the apparent to the real or true
movements of the celestial bodies.

In a work which proposes not so much to shew the
grounds of our knowledge as to display that which is known,
whether regarded, in the present state of science, as certain,

or as merely probable in a greater or less degree, a different order of succession is to be preferred. Here, therefore, we do not proceed from the subjective point of view of human interest: the terrestrial is treated only as a part of the whole, and in its due subordination. The view of nature should be general, grand, and free; not narrowed by proximity, sympathy, or relative utility. A physical cosmography, or picture of the universe, should begin, therefore, not with the earth, but with the regions of space. But as the sphere of contemplation contracts in dimension, our perception and knowledge of the richness of details, of the fulness of physical phænomena, and of the qualitative heterogeneity of substances, augment. From the regions in which we recognise only the dominion of the laws of gravitation, we descend to our own planet, and to the intricate play of terrestrial forces. The method thus pursued is the opposite of that which is followed when conclusions are to be established. The one recounts what the other demonstrates.

Our knowledge of the external world is obtained through the medium of the senses. It is by the phænomena of light that the presence of matter in the remote regions of space is revealed to us. The eye is the organ by which we are enabled to contemplate the universe ; and, for the last two centuries and a half, telescopic vision has given to later generations a power of which the limit is yet unattained. The first and most general consideration in the Cosmos is that of the *contents of space*,—the distribution of the material universe.

We see matter existing in space, partly in the form of rotating and revolving spheroids, differing greatly in density and magnitude, and partly in that of self-luminous vapour, dispersed in shining nebulous spots or patches. If we consider first the nebulous spots, or cosmical vapour in definite forms, its state of aggregation appears constantly varying. The nebulæ present themselves to the eye in the form of round or elliptic disks of small apparent magnitude, either single, or in pairs which are sometimes connected by a thread of light : when their diameter is greater their forms vary,—some are elongated, others have several branches, some are fan-shaped, some annular, the ring being well defined and the interior dark. They are supposed to be undergoing various and progressive changes of form, as condensation proceeds around one or more nuclei, in conformity with the laws of gravitation. Between two and three thousand of such unresolvable nebulæ (or, at least, in which no stars have hitherto been discovered by the most powerful telescopes), have already been counted, and their positions determined.

The genetic evolution, or perpetual process of formation, which appears to be going on in this part of space, has led philosophical observers to the analogy of organic phænomena. As we see in our forests, at one time, the same kind of tree in all stages of growth, and receive from this co-existence the impression of progressive devolopment ; so, in the great garden of the universe, we seem to see stars in various stages of progressive formation. The process of condensation, which was part of the doctrine of Anaximenes and of the whole Ionic school, appears to be here going on before our eyes. This subject of conjoint investigation and conjecture has a peculiar charm for the imagination. Throughout the range of animated existence, and of moving forces in the physical universe, there is an especial fascination in the recognition of that which is becoming, or about to be,—even greater than in that which is, though the former be indeed no more than a new condition of matter already existing : for of the act of creation itself, the original calling forth of existence out of non-existence, we have no experience, nor can we form a conception of it.

Besides the comparison of different stages of development in nebulæ which appear more or less condensed towards their centres, observers have believed that they could recognize, by direct observation at different epochs, actual changes of form in particular nebulæ; in the nebula in Andromeda, for example; in the nebula in the constellation of the Ship; and in the filamentous portion of the nebula in Orion. Inequality in the instruments employed, differences in the state of the atmosphere, and other optical circumstances, may indeed invalidate part of these results as true historical elements.

Neither the irregularly-shaped nebulæ (to which the name more especially belongs), the separate parts of which are of unequal brightness, and which may, possibly, as their circumference contracts, become finally concentrated into stars,—nor the planetary nebulæ, whose circular or slightly oval disks show throughout a perfectly equable intensity of faint light, must be confounded with *nebulous stars*. These are not stars accidentally projected upon a distant nebulous ground, but the luminous nebulous matter itself forms one mass with the body which it surrounds. The often considerable magnitude of their apparent diameter, and the remote distance from which their faint light reaches us, show that both the planetary nebulæ and the nebulous stars must be of enormous dimensions. New and highly ingenious considerations([31]) on the very different effect which distance produces on the intensity of light of a disk of appreciable diameter, and of a self-luminous point, render it not improbable that the planetary nebulæ are very remote nebulous stars, in which the difference between the central star and the nebulous envelope is no longer sensible even to our telescopic vision.

The magnificent zones of the southern celestial hemisphere, between 50° and 80°, are especially rich in nebulous stars, and in unresolvable nebulæ. Of the two Magellanic clouds which revolve around the starless and desert southern pole, the larger especially appears, by the most recent researches([32]), as a "collection of clusters of stars, composed of globular clusters and nebulæ of different magnitudes, and of large nebulous spaces not resolvable, which, producing a general brightness of the field of view, form, as it were, the background of the picture." The appearance of these clouds, that of the brilliant constellation of the Ship, the milky way between the Scorpion, the Centaur, and the Southern Cross,—I may say, the graceful and picturesque aspect of the whole southern celestial hemisphere, have left on my mind an ineffaceable impression. The zodiacal light which rises in a pyramidal form, and constantly adorns the tropical nights with its mild radiance, is either a vast rotating nebulous ring between the Earth and Mars, or, less probably, the outermost stratum of the solar atmosphere. Besides the luminous clouds and nebulæ of definite form, exact and always accordant observations indicate the existence and general distribution of an infinitely divided and apparently non-luminous matter, which constitutes a resisting medium, and manifests itself by diminishing the eccentricity, and shortening the time of revolution, of Encke's, and, perhaps, also of Biela's comet. We may conceive of this impeding ethereal and cosmical matter, that it is subject to motion ; that it gravitates, notwithstanding its extreme tenuity ; that it is condensed in the vicinity of the great mass of the sun ; and even that it may be augmented in the course of myriads of years by emanations from the tails of comets.

If we now leave the consideration of the attenuated va- porous matter of the immeasurable regions of space (οὐρανοῦ χορτος) [33], whether existing in a dispersed state, as a cosmical ether without form or limits, or in the shape of nebulæ, and pass to those portions of the universe which are condensed into solid spheres or spheroids, we approach a class of phænomena exclusively designated as stars, or as the sidereal universe. Here, too, we find different degrees of solidity, or density, in the agglomerated matter. Our own solar system presents all gradations of *mean density* (or relation of *volume* to *mass*). If we compare the planets, from Mercury to Mars inclusive, with the Sun and with Jupiter, and the two latter bodies with the still inferior density of Saturn, we pass through a descending scale, in which (taking terrestrial substances for illustration) the gradations correspond respectively to the densities of anti- mony, honey, water, and deal wood. In comets (which, in point of number of individual forms, constitute the largest portion of our solar system), that which we call the *head*, or *nucleus*, allows the light of stars to shine unimpaired through its substance : perhaps in no case does the mass of a comet equal the five-thousandth part of that of the earth,—so various do the processes of formation appear in the original,

and, perhaps, still progressive, agglomeration of matter.
Before we pass from the most general considerations to those
which are less so, it is especially desirable to notice this
diversity, not merely as a possibility, but as actually existing.

The purely speculative conceptions of Wright, Kant,
and Lambert, concerning the general arrangement of the
fabric of the universe, and the distribution of matter in
space, have been confirmed by Sir William Herschel in
the surer path of observation and measurement. This
great man, in whom the inspiration of genius was combined
with a spirit of cautious investigation, was the first to
sound the depths of the celestial spaces, in order to determine the limits and the form of the starry stratum which
we inhabit ; the first to enter on the inquiry of the relations of position and distance between our own region of
the heavens and remote nebulæ.   William Herschel (as the
inscription on his monument at Upton finely says), broke
through the inclosures of the heavens (*cælorum perrupit
claustra*) ; like Columbus, he penetrated into an unknown
ocean, and first beheld coasts and groups of islands, whose
true position remains to be determined by succeeding ages.

Considerations respecting the different intensity of light
in stars, and their relative numbers in equal telescopic fields,
have led to the assumption of unequal distances and distribution in space of the strata in which they may be conceived to
exist. Such assumptions, in so far as we may attempt to trace
by them the limits of separate portions of the fabric of the
universe, cannot, indeed, offer the same degree of mathematical certainty as is attained in all that regards our solar
system, or the revolution of the double stars with unequal
velocity around a common centre of gravity, or the apparent

or true movements of the heavenly bodies. If we commence physical cosmography with the most remote nebulæ, we may feel inclined to compare this portion of our subject with the heroic, or mythical, periods of history. Both begin in twilight obscurity—the one of antiquity, the other of inaccessible distance ; and where reality threatens to elude the grasp, imagination becomes doubly incited to draw from its own fulness, and to give outline and permanence to undefined evanescent objects.

If we compare the regions of space to one of the island-studded seas of our planet, we may imagine we see matter distributed in groups, whether of unresolvable nebulæ of different ages condensed around one or more nuclei, or in clusters of stars, or in stars scattered singly. Our cluster of stars, or the island in space to which we belong, forms a lens-shaped, flattened, and every where detached stratum, whose major axis is estimated at seven or eight hundred, and its minor axis at a hundred and fifty times the distance of Sirius. If we assume that the parallax of Sirius does not exceed that accurately determined for the brightest of the stars in the Centaur $(0''\cdot9128)$, it will follow that light traverses one distance of Sirius in three years, while nine years and a quarter are required for the transmission of the light of 61 Cygni, whose considerable proper motion might lead to the inference of great proximity. I take the parallax of this remarkable star $(0''\cdot3483)$ from Bessel's excellent first memoir [34]. Our cluster of stars is a disk of comparatively small thickness, divided, at about a third of its length, into two branches : we are supposed to be near this division, and nearer to the region of Sirius than to that of the constellation of the Eagle ;

almost in the middle of the starry stratum in the direction of its thickness.

The place of our solar system, and the form of the whole lens, are inferred from a kind of stellar scale, *i. e.*, from the different number of stars seen (as already alluded to) in equal telescopic fields of view. The greater or less number of stars measures the relative depth of the stratum in different directions ; giving, in each case, like the marks on a sounding line, the comparative length of visual ray required to reach the bottom ; or, more properly, as above and below do not here apply, the outer limit of the sidereal stratum. In the direction of the major axis, where the greater number of stars are placed behind each other, the remoter ones appear closely crowded together, and, as it were, united by a milky radiance, and present a zone, or belt, projected on the visible celestial vault, or sky. This narrow belt is divided into branches ; and its beautiful, but not uniform, brightness is interrupted by some dark places. As seen by us on the apparent concave celestial sphere, it deviates only a few degrees from a great circle; we being near the middle of the entire starry cluster, and almost in the plane of the milky way. If our planetary system was far *outside* the cluster, the milky way would appear to telescopic vision as a ring, and, at a still greater distance, as a resolvable disk-shaped nebula.

Among the many self-luminous moving suns (erroneously called *fixed* stars), of which our starry island, or nebula, consists, our own sun is the only one known to us by direct observation as a *central body,* in its relations to spherically-agglomerated matter, revolving around it under the manifold forms of planets, comets, and aerolite-asteroids. In the *multiple* stars (double stars, or double suns), so far as we have yet investigated them, there does not prevail the same planetary dependence in respect to relative motion and illumination as that which characterizes our solar system : two or more self-luminous heavenly bodies (whose planets and their satellites, if they exist, escape the power of our telescopes), do, indeed, revolve around a common centre of gravity; but this centre of gravity falls in a space occupied, possibly, only by unagglomerated matter, *i. e.*, cosmical vapour; whilst, in our system, the centre of gravity is included within the surface of a visible central body. If we choose to consider the sun and the earth, or the earth and the moon, as *double stars*, and our whole planetary solar system as a *multiple* star, or group of stars ; yet the analogy suggested by such denominations fails altogether when we regard illumination, and can only apply to motion in conformity with the laws of gravitation.

In such a generalisation of cosmical views as accords with the plan of the present work, namely, with the sketch of a picture of nature or of the universe, the solar system to which our earth belongs may most properly be considered under a two-fold aspect,—first, in reference to the different classes of the individual bodies which it contains, their magnitudes, forms, densities, and distances apart ; and, second, in its relation to the other parts of the starry cluster of which it forms a portion, and to its motion, or change of place, within the same.

The solar system (*i. e.*, those various material bodies which revolve round the sun) consists, according to our present knowledge, of eleven principal planets, eighteen moons or satellites, and myriads of comets, three of which (called planetary comets) do not pass beyond the orbits of the principal planets. We may, with considerable probability, include within the dominion of our sun, and the immediate sphere of its central force, a rotating ring of finely-divided or nebulous matter, situated, perhaps, between the orbits of Venus and Mars, but certainly extending beyond that of the Earth [35], which is called by us the Zodiacal Light; and a host of extremely small asteroids, the paths of which intersect, or very nearly approach, that of the Earth, and which present to us the phænomena of aerolites, or shooting stars. When we take into our consideration all the varied forms of bodies which revolve round the sun in more or less eccentric paths,—unless we are inclined, with the illustrious author of the " Mécanique Céleste," to view the greater number of comets as nebulous stars, wandering from one central system to another [36],—we must acknowledge that the planetary system, distinctively so-called (*i. e.* the group of bodies which in only slightly eccentric orbits revolve around the sun, with their attendant moons), forms (not, indeed, in mass, but in number) a comparatively small portion of the entire solar system.

It has been proposed to consider the telescopic planets, Vesta, Juno, Ceres, and Pallas, with their more eccentric, intersecting and greatly inclined orbits, as forming a middle zone, or group, in our planetary system ; and if we follow out this view, we shall find that the comparison of the inner group of planets, comprising Mercury, Venus, the Earth, and Mars, with the outer group, consisting of Jupiter, Saturn, and Uranus, presents several striking contrasts [37]. The planets of the inner group, which are nearer the sun, are of more moderate size, are denser, rotate around their respective axes more slowly in nearly equal periods, which differ little from twenty-four hours, are less compressed at the poles, and, with one exception, are without satellites. The external planets, more distant from the sun, are of much greater magnitude, five times less dense, more than twice as rapid in their rotation round their axes, more compressed at their poles, and richer in moons in the proportion of 17 to 1; if Uranus has really the six satellites ascribed to it.

In viewing these general characteristics of the two groups, we must admit, however, that they cannot be strictly applied to each of the planets in particular; nor are there any constant relations between the distances of the planets from the central body round which they revolve, and their absolute magnitudes, densities, times of rotation, eccentricities, and inclinations of orbit and of axis. We know as yet of no inherent necessity, no natural mechanical law, (such as the great law of the proportionality of the squares of the periodic times to the cubes of the mean distances from the sun), connecting the above-named six elements of the planets, and the forms of their orbits, either *inter se*, or with their mean solar distances. We find Mars, though more distant from the Sun than either the Earth or Venus, inferior to them in magnitude; being, indeed, that one of the long known greater planets which most nearly resembles in size Mercury, the nearest planet to the solar orb. Saturn is less than Jupiter, and yet much larger than Uranus. The zone of the telescopic planets, which are so inconsiderable in point of volume, viewed in the series of distances commencing from the Sun, comes next before Jupiter, the greatest in size of all the planetary bodies; and yet the disks of these small planets (whose apparent diameters scarcely admit of measure-

ment) are less than twice the size of France, Madagascar,
or Borneo.   Remarkable as is the small density of all the
colossal planets which are farthest from the sun, yet neither
in this respect can we recognise any regular succession [38].
Uranus, even if we assume as correct the smaller mass of
$\frac{1}{24605}$, assigned by Lamont [39], appears to be denser than
Saturn; and (although the inner group of planets differ but
little from each other in this particular) we find both Venus
and Mars less dense than the Earth, which is situated between
them.   The time of rotation decreases, on the whole, with
increasing solar distance, but yet it is greater in Mars than
in the Earth, and in Saturn than in Jupiter.   Among all
the planets, the elliptic paths of Juno, Pallas, and Mercury,
have the greatest eccentricity; and Venus and the Earth,
which immediately follow each other, have the least : while
Mercury and Venus (which are likewise neighbours) present,
in this respect, the same contrast as do the four smaller
planets, whose paths are so closely interwoven.   The eccentricities of Juno and Pallas are nearly equal, but are each
three times as great as those of Ceres and Vesta.   Nor is
there more regularity in the inclination of the orbits of the
planets towards the plane of projection of the ecliptic, or in
the position of their axes of rotation, relatively to their
orbits; on which latter position the relations of climate,
seasons of the year, and length of the days depend, more
than on the eccentricity.   It is in the planets which
have the most elongated ellipses—Juno, Pallas, and Mercury—that we find, though not in equal proportion, the
greatest inclination of the orbits to the ecliptic.   The path
of Pallas is almost comet-like, its inclination twenty-six
times greater than that of Jupiter; whilst, in the little

Vesta, which is so near Pallas, the corresponding angle of inclination is one-fourth less, or scarcely six times greater than in Jupiter. Neither do we find a regular order of succession in the position of the axes of the few planets (four or five), of the planes of rotation of which we have at present any certain knowledge. Judging by the position of the satellites of Uranus (of two of which, *i. e.*, the second and the fourth, a fresh and certain view has been recently obtained), the axis of this the outermost of all the planets, is inclined barely 11° to the plane of its orbit; and Saturn is placed intermediately between this planet, in which the axis of rotation almost coincides with the plane of its orbit, and Jupiter, whose axis is almost perpendicular to it.

In this enumeration of forms in space, they have been depicted simply as they exist, rather than as objects of intellectual contemplation, or in inherent causal connection. The planetary system, in its relations of absolute magnitude, relative position of the axes, density, time of rotation, and different degrees of eccentricity of the orbits, has, to our apprehension, nothing more of natural necessity, than the relative distribution of land and water on the surface of our globe, the configuration of continents, or the elevation of mountain chains. No general law in these respects is discoverable, either in the regions of space, or in the irregularities of the crust of the earth. They are *facts* in nature, which have arisen out of the conflict of various forces acting under unknown conditions. We apply the term *accidental* to what in the planetary formation we are unable to elucidate genetically. If the planets have been formed out of separate rings of nebulous matter revolving round the sun, the different thickness, unequal density, temperature, and electro-magnetic tension of these rings, may have afforded occasion to the differences of form in the spheroidally condensed matter, as the amount of tangential velocity and small variations in its direction may have done to the diversity of form and inclination in the elliptic orbits. Attractions of mass, and the laws of gravitation, have, no doubt, been influential here, as well as during the changes which have produced the irregularities in the terrestrial surface; but we cannot infer, from present forms, the whole series of conditions which may have been passed through. Even the so called law of the distances of the planets from the sun, (which led Kepler to conjecture the existence of some planet filling up the void between Mars and Jupiter), has been found numerically inexact for the distances between Mercury, Venus, and the Earth, and requires an arbitrary supposition in the first member of the series.

The eleven hitherto discovered primary planets, which revolve round our Sun, are attended certainly by fourteen, and probably by eighteen, secondary planets—moons, or satellites; the primary planets being themselves the central bodies of subordinate systems. We seem to recognise here, in the fabric of the universe, an arrangement somewhat similar to that so often shown to us in the development of organic life, where, in the manifold combinations of groups of plants or of animals, the typical form is repeated in *subordinate circles*. The secondary planets, or satellites, are more frequent in the outer region of our planetary system, situated beyond the intersecting orbits of the telescopic planets; none of the planets of the inner division have satellites, except the Earth, whose moon is of great relative magnitude, its diameter being to that of the earth as one to four, whereas the diameter of the largest of all known satellites—the sixth of Saturn—is supposed to be one-seventeenth, and that of the largest of Jupiter's satellites—the third—only one-twenty-sixth part of the respective diameters of the planets round which they revolve. The planets most rich in satellites are found among those most remote, of greatest magnitude, least density, and greatest compression. According to the most recent measurements of Mädler, Uranus is the planet which has the greatest compression, viz. $\frac{1}{9,92}$. The Earth and her moon are 267200 miles apart, and the differences of mass[40] and diameter in these two bodies are much less than we are accustomed to meet with elsewhere in the solar system, between bodies of different orders, or primary planets and their satellites. The density of the Moon is $\frac{5}{9}$ less than that of the Earth, while the second satellite of Jupiter appears, if we may place sufficient dependence on the determinations of magnitude and of mass, to be even actually denser than the great planet round which it revolves.

Among the fourteen satellites concerning which investigation has arrived at some degree of certainty, the system of the seven satellites of Saturn offers the greatest contrasts, both of absolute magnitude and of distance from the central planet. The sixth satellite is probably but little smaller than Mars (whose diameter is twice that of our moon), while, on the other hand, the two innermost satellites (discovered by the forty-foot telescope of William Herschel in 1789, and seen again by John Herschel at the Cape of Good Hope, by Vico at Rome, and by Lamont at Munich) belong, perhaps, together with the remote moons of Uranus, to the smallest cosmical bodies of our solar system, being

visible only under peculiarly favourable circumstances, and
with the most powerful telescopes.　After the sixth and
seventh of the satellites of Saturn comes, in order of volume,
the third and brightest of Jupiter's.　The diameters of satellites deduced from measurements of the apparent magnitude
of their small disks, are subject to many optical difficulties;
fortunately, the calculations of astronomy, which shew the
movements of the heavenly bodies as they will appear to
us when viewed from the earth, depend much more on
motion and mass than on volume.

The *absolute* distance of any satellite from the planet
round which it revolves is greatest in the case of the outermost (or seventh) of the satellites of Saturn, being above
two millions of geographical miles, or ten times the distance
of our moon from the earth.　The distance of the outermost,
or fourth, satellite of Jupiter from that planet is only
1040000 miles; the distance between Uranus and his
sixth satellite (supposing the latter really to exist) amounts
to 1360000 miles.　If we compare in each subordinate
system the volume of the central planet with the dimensions
of the orbit of its outermost satellite, we obtain a new series
of numerical relations.　The distances of the outermost
satellites of Uranus, Saturn, and Jupiter, expressed in semi-
diameters of the respective central planets, are as 91, 64,
and 27; and in this mode of estimation the outermost
of the satellites of Saturn appears to be only a little ($\frac{1}{15}$)
further from the centre of that planet, than our Moon
is from the Earth.　The satellite, which is nearest to
its central planet, is undoubtedly the first or innermost
of Saturn, and it offers, moreover, the only example of a
period of revolution of less than twenty-four hours: its

distance from the centre of the planet is, according to Mädler and Wilhelm Beer, 2·47 semi-diameters of Saturn, or 80088 miles; from the surface of the planet, therefore, only 47480, and from the outermost edge of the ring, only 4916 miles. The traveller may find pleasure in realising to his imagination the smallness of this amount, by remembering the statement of a distinguished navigator, Captain Beechey, that, in three years, he had sailed over 72800 geographical miles. If we estimate distances, not in absolute measure, but in semi-diameters of the primary planets, we find that the first or nearest of Jupiter's satellites (which, in absolute distance, is 26000 miles further from the centre of that planet than our moon is from the earth) is only six semi-diameters of Jupiter from its centre, while our moon is distant from us fully 60⅓ semi-diameters of the Earth.

In the subordinate systems of satellites, or secondary planets, we see reflected in their relations to their primary planets and to each other, all the laws of gravitation which regulate those primary planets in their revolutions round the sun. The twelve moons attendant on Saturn, Jupiter, and the Earth, all move, as do their primary planets, from west to east, and in elliptical orbits differing little from circles. It is only the Earth's moon, and probably the first or innermost of the satellites of Saturn, which have orbits more elliptic than that of Jupiter. The eccentricity of the sixth satellite of Saturn, which has been so accurately observed by Bessel, is 0·029, and is greater than that of the Earth. Near the extreme limits of the planetary system, where, at a solar distance nineteen times greater than that of the Earth, the centripetal force of the solar orb is considerably diminished, the satellites of Uranus (which have, it is true, been as yet but imperfectly investigated) exhibit some remarkable differences from the movements of other satellites and planets. In all other cases, the orbits are but little inclined to the ecliptic, and the movements are from west to east, including Saturn's rings, which may be regarded as belts formed of an aggregation of satellites; but the satellites of Uranus are almost perpendicular to the ecliptic, and the direction of their movement, as confirmed by Sir John Herschel, after many years of observation, is retrograde, or from east to west. If the primary and secondary planets have been formed by condensation from annular rotating portions of the primitive atmospheres of the sun, and of the principal planets, there must have been, in the rings of vapour which revolved round Uranus, singular and unknown relations of retardation or counteraction, to have occasioned the second and the fourth satellite to revolve in a direction opposite to that of the rotation of the central planet.

It appears highly probable, that the times of rotation of *all* secondary planets, or satellites, are the same as their times of revolution round their primary planets; so that they always present to the latter the same face. In the case of the moon, inequalities, consequent on small variations in the revolution, cause, however, fluctuations of from six to eight degrees, or an apparent libration in longitude as well as in latitude, which renders more than one-half of her surface visible to us at different times, showing us sometimes more of her western and southern limbs, and sometimes more of her eastern and northern. It is this libration ([41]) which enables us to see the annular mountain of Malapert, sometimes concealed from us by the moon's southern pole, the arctic landscape round the crater of

Gioja, and the great grey plain near Endymion, which exceeds, in superficial extent, the Mare Vaporum. Three-sevenths of the moon's surface are at all times concealed from the earth, and must always remain so, unless new and unexpected disturbing forces are brought into action.

These cosmical relations remind us involuntarily of nearly similar ones in the intellectual world, where, in the domain of deep research and of meditation on the mysterious elaborations of nature, and on primeval creation, there are regions similarly turned from our view, and apparently unattainable, of which a narrow margin has for thousands of years seemed to show itself, from time to time, to the human race, glimmering now in true, now in delusive light.

Having thus considered, as products of a single tangential impulse, and closely connected with each other by mutual attraction, the primary planets, their satellites, and the concentric rings which belong to one of the outermost planets, we have still to notice,—among the cosmical bodies which revolve around the sun in paths of their own, and receive light from him,—the unnumbered host of comets. If, assuming an equable distribution of the paths of these bodies, the limits of their perihelia, and the possibility of their remaining invisible to the inhabitants of the earth, we estimate their possible numbers by the rules of the calculus of probabilities, the result will be an amount of myriads astonishing to the imagination. Kepler, with his characteristic liveliness of expression, said, even in his day, "there are more comets in space than fishes in the ocean." As yet, however, we hardly

possess as many as one hundred and fifty calculated paths of
comets; but we have notices more or less precise of the
appearance of six or seven hundred of such bodies, and of
their passage through known constellations.   Whilst the
classic nations of the West, the Greeks and the Romans, do,
indeed, sometimes mention the place in the heavens where a
comet was first seen, but never afford us information respecting its apparent path, the literature of the Chinese (who
observed nature diligently, and carefully recorded everything),
supplies us with circumstantial notices of the comets seen
by them, and of the constellations which each passed
through.   These notices extend back to more than five
centuries before the Christian era, and many of them are
still found useful in astronomy([42]).   Of all planetary bodies,
comets,—though their mean mass is probably much less
than the five-thousandth part of the earth,—are those which
occupy the greatest space, their wide-spreading tails often
extending over many millions of miles.   The cone of light
reflecting vapour which radiates from them has been found
in some instances, as in 1680 and 1811, to equal in length
the distance of the earth from the sun, or that of a line
including the orbits of the two planets, Venus and Mercury.
It is even probable that the vapour of the tails of the comets
of 1819 and 1823 mixed with our atmosphere.

    Comets show such diversities of form, diversities belong-
ing rather to the individual than to the class, that the
description given of one of these "wandering light-clouds"
(as they were called by Xenophanes, and Theon of Alex-
andria, contemporaries of Pappus), can only be applied with
much caution to another.   The fainter telescopic comets are
for the most part without tails, and resemble Herschel's

nebulous stars. Their appearance is that of circular nebulæ shining with faint light concentrated towards the centre. This is the simplest type; but not, on that account, a rudimentary type, for it might equally be the type of a cosmical body grown old and exhausted by exhalation. We can distinguish in the larger comets, the "head," or "nucleus," and the "tail," either single or multiple, which the Chinese astronomers more characteristically denominate by the word "brush" (sui). The nucleus usually presents no definite outline, although in some rare instances it appears like a star of the first or second magnitude; and in the larger comets of 1402, 1532, 1577, 1744, and 1843, has even been clearly seen in bright sunshine ([43]). This latter circumstance appears to indicate, in particular individuals, a denser mass, capable of reflecting light with greater intensity. Even in Herschel's large telescope, only two comets showed well-defined disks ([44]), viz. the comet discovered in Sicily in 1807, and the fine comet of 1811, the one having an angle of 1″ and the other of 0″77, whence he inferred their true diameters to be respectively 538 and 428 miles. The measurement of the less well-defined nuclei of the comets of 1798 and 1805, gave diameters of only 24 to 28 miles. In several comets which were examined with great care, and particularly in the above-named comet of 1811 which was so long in sight, the nucleus and its nebulous envelope were entirely separated from the tail by a darker space. The intensity of light in the nucleus of a comet does not increase in a uniform manner towards the centre, but bright zones alternate with concentric nebulous envelopes. The tail sometimes appears single, more rarely double, and in two instances (in the comets of 1807 and

1843) the two branches were of very different length: in the comet of 1744, the tail had six branches, the two exterior ones forming an angle of 60°. The tails have been sometimes straight, sometimes curved; in the latter case, either concave towards both sides, or, as in 1618, convex towards the direction in which the comet is moving; and sometimes the tail even appears inflected like a flame in motion. The tails are always turned from the sun, and so directed that the prolongation of the axis would pass through the centre of that body; this circumstance was remarked (according to Edouard Biot) by the Chinese astronomers as early as 837, but was first clearly stated in Europe by Fracastoro and Peter Apian in the sixteenth century. We may conceive that these emanations form conoidal envelopes of greater or less thickness, which would furnish a very simple explanation of several of the remarkable optical phenomena above mentioned.

Not only are different comets characterised by such great differences of form,—some being entirely without visible tails, others, as the third comet of 1618, having a tail of 104 degrees in length,—but we also see the same comets pass through successive and rapid variations of form. These changes have been well and most accurately described in the comet of 1744, by Heinsius, at Petersburgh, and in Halley's comet, on its last reappearance, in 1835, by Bessel, at Königsberg. A tuft of rays issued from that part of the nucleus which was turned towards the sun. These rays thus issuing were bent backwards, so as to form part of the tail. "The nucleus of Halley's comet, with its emanations, presented the appearance of a burning rocket, the train of which was deflected sideways by a current of air." The rays issuing from the head of the comet were seen by Arago and myself, at the Observatory at Paris, to assume very different forms ([45]) on successive nights. The great Königsberg astronomer concluded from many measurements and from theoretical considerations, " that the out-streaming cone of light deviated notably both to the right and to the left of the true direction towards the sun, but that it always returned to that direction, and passed beyond it to the opposite side; so that both the cone of light, and the body of the comet from which it issued, were subject to a rotatory, or rather vibratory motion, in the plane of the orbit." He finds "that the ordinary power of the attractive force of the sun on heavy bodies is not sufficient to explain such vibrations, and is of opinion that they would seem to indicate a polar force, which tends to turn one of the semi-diameters of the comet towards the sun, and the opposite semi-diameter from the sun. The magnetic polarity possessed by the earth may present something analogous, and should the sun have the opposite polarity, a resulting influence might manifest itself in the precession of the equinoxes." This is not the place for entering more at length into this subject; but observations so remarkable, and so important with reference to the most wonderful class of cosmical bodies belonging to our solar system, ought not to be entirely passed over in this general view of nature ([46]).

Although, in the greater number of cases, the tails of comets increase in magnitude and brilliancy in the vicinity of the sun, and are directed from that body; yet the comet of 1823 offered the remarkable example of two tails, one turned from and the other nearly towards the sun, forming with each other an angle of 160°. Peculiar modifications of polarity, and its unequal distribution and conduction, may in

this rare case have caused a double continuous current of
nebulous matter ([47]).   Aristotle, in his Natural Philosophy,
brings the phænomena of comets through the medium of
these effusions into a singular connection with the existence
of the Milky Way.   He supposes that the countless multitude of stars which form the galaxy give out a luminous
or incandescent matter.   The nebulous belt which traverses
the vault of the heavens is therefore regarded by the Stagirite as an immense comet incessantly reproducing itself ([48]).

The passage over the fixed stars of the nucleus of a
comet, or of its innermost vaporous envelopes, might throw
light on the physical character of these wonderful bodies, but
we are deficient in observations of the kind in which we
can be assured that the passage was perfectly central ([49]) :
for in the immediate vicinity of the nucleus, as I have
already noticed, dense coatings alternate with others of great
tenuity.   On the other hand, there is no doubt that the light
of a star of the tenth magnitude passed through very dense
nebulous matter on the 29th of September, 1835, at a
distance of $7''\cdot78$ from the centre of the nucleus of Halley's
comet, according to Bessel's most careful measurement,
without experiencing any deflection in its rectilinear course
at any moment of its passage ([50]).   Such an absence of refracting power, if actually extending to the centre of the
nucleus, makes it difficult to regard the substance of comets
as of a gaseous nature ;—or, is the absence of refracting
power a consequence of the almost infinite rarity of a
fluid of that description ?—or, does the comet consist of
" detached particles," forming a *cosmical cloud*, which no
more affects the ray of light passing through it, than do the
clouds of our atmosphere, which in like manner have no

influence on the zenith distances of the heavenly bodies ? In the passage of a comet over a star, there has often been noticed a greater or less diminution of the light of the star, but this has been justly ascribed to the brightness of the ground, on which, during the coincidence, the star is seen.

We are indebted to Arago's polarization experiments, for the most important and decisive observations on the nature of the light of comets. His polariscope instructs us concerning the physical constitution of the sun, as well as that of the comets; it informs us whether a luminous ray, which reaches us from a distance of many millions of miles, is a direct, or a reflected or refracted ray; and, if direct, whether the source of light is a solid, a liquid, or a gaseous body. The light of Capella, and that of the great comet of 1819, were examined at the Paris Observatory with the same apparatus. The comet showed polarised, and therefore reflected light; whilst, as was to be expected, the fixed star was proved to be a self-luminous sun ([51]). The existence of polarised cometary light announced itself not only by the inequality of the images, but was shown with still greater certainty, at the reappearance of Halley's comet in 1835, by the more striking contrast of complementary colours, in accordance with the laws of chromatic polarization discovered by Arago in 1811. These fine experiments leave it however still undecided, whether, besides this reflected solar light, comets may not have a proper light of their own. Even in planets, in Venus for example, an evolution of independent light appears very probable.

The variable intensity of the light of comets is not always to be explained by their place in their orbit, and

their distance from the sun.   In particular individuals, it
certainly indicates internal processes of condensation, and
increased or diminished capability of reflecting light.   In
the comet of 1618, as in that which has a period
of revolution of three years, Hevelius saw the nucleus
lessen at the perihelion, and enlarge at the aphelion of the
comet : this remarkable phenomenon, which had long remained unheeded, has since been observed by the distinguished astronomer, Valz, at Nismes.   The regularity of the
alteration of the volume according to the distance from the
sun, appeared exceedingly striking ; the physical explanation
of the phenomenon cannot well be sought in the greater
condensation of the cosmical ether in the vicinity of the
sun, for it is difficult to imagine the nebulous envelope of
the nucleus of the comet to be, like a vesicle, impervious to
the ether ([52]).

The very dissimilar excentricities of the elliptical paths of
different comets, has led in modern times (1819) to a brilliant accession to our knowledge of the solar system.   Encke
has discovered the existence of a comet having so short a
period of revolution, that it always remains within our planetary system, and even reaches its aphelion, or greatest distance from the sun, between the orbits of Jupiter and of the
small planets.   The excentricity of its orbit is 0·845 ; that
of Juno, which has the greatest excentricity amongst the
planets, being 0·255.   This comet has been more than once
seen, though with difficulty, by the naked eye ; in Europe,
in 1819, and, according to Rümker, in New Holland, in
1822.   Its period of revolution is about 3·3 years ; but from
the most careful comparison of the epochs of its return to
its perihelion, the remarkable fact has been discovered

that this period has regularly decreased in every successive revolution from 1786 to 1838; the diminution amounting in an interval of 52 years to 1·8 days. After a careful consideration of all the planetary perturbations, this remarkable phænomenon has led to the adoption, for the purpose of bringing observation and calculation into harmony, of the not improbable supposition of the existence of a fluid of extreme rarity, or ether, dispersed through space, forming a resisting medium ; the resistance lessens the tangential force, and with it the major axis of the comet's orbit. The value of the constant of resistance appears to be somewhat different before and after the perihelion; and this may perhaps be ascribed to the change of form of the small nucleus, or to inequality in the density of the ether in the vicinity of the sun ([53]). These facts, and the investigations to which they have given rise, are amongst the most interesting results of modern astronomy. Encke's comet has also led to a more rigorous examination of the mass of Jupiter, so important an element in all calculations of perturbations ; and has still more recently obtained for us the first, although it is only an approximate, determination of a smaller mass for the planet Mercury.

To this first discovered comet of short period, there was soon added a second planetary comet, having its aphelion beyond the orbit of Jupiter, but within that of Saturn. Biela's comet, discovered in 1826, has a period of revolution of 6·75 years, and its light is still fainter than that of Encke's. The motion of both these comets is direct, or the same as that of the planets, whereas Halley's is opposite or retrograde. Biela's comet presents the first certain instance of the orbit of a comet intersecting that of the earth; its path

may therefore suggest the possibility of a catastrophe,
if we may apply that term to the extraordinary phænomenon of a collision, which has not occurred within
historical times, and of which we cannot with certainty
predict the consequences. Granting that small masses
possessed of enormous volocity may exert a notable force,
Laplace, after showing that the mass of the comet of 1770
is probably less than $\frac{1}{5000}$ of that of the earth, has assigned
with a certain degree of probability, for the average mass
of comets, a quantity far below even $\frac{1}{100000}$ of the
earth's mass, or only about $\frac{1}{1200}$ of that of the moon (54).
The passage of Biela's comet across the earth's orbit must
not of course be confounded with its proximity to, or
encounter with, our globe itself. When this passage took
place on the 29th of October, 1832, the earth was a full
month in time from the point of intersection of the two
paths. The orbits of Encke's and Biela's comets also intersect each other; and it has been justly remarked, that in the
course of the many perturbations which such small bodies
suffer from the greater planetary masses, there is a possibility
of their meeting (55) ; and that should this take place about
the middle of the month of October, the inhabitants of the
Earth might behold the extraordinary spectacle of an encounter between two cosmical bodies, and possibly of their mutual
penetration and amalgamation, or of their destruction by
exhausting emanations. Such events, the consequences either
of deflection produced by disturbing masses, or of originally
intersecting orbits, may have taken place frequently in
the course of millions of years, and in the vast extent of
immeasurable space ; they would, however, be isolated occurrences, having as little general influence on other cosmical

forms, as the breaking forth or extinction of a single volcano has in our less extensive sphere.

A third planetary comet of short period was discovered on the 22d of November, 1843, at the Paris Observatory, by M. Faye. Its elliptic path approaches much more nearly to a circle than that of any of the comets previously known to us, and is included between the orbits of Mars and Saturn; passing, according to Goldschmidt, beyond the orbit of Jupiter. This comet is therefore one of the very few whose perihelia are beyond the orbit of Mars : its period of revolution is 7·29 years, and the present form of its path may perhaps be due to the near approach which it made to the great mass of Jupiter at the end of the year 1839.

If we consider all comets moving in elliptic orbits as members of our solar system, and class them by the lengths of their major axes, the amount of their excentricities, and the duration of their periods of revolution, it seems probable that in the last-named respect, the three comets of Encke, Biela, and Faye, are most nearly approached by the comet which Messier discovered in 1766 (regarded by Clausen as identical with the third comet of 1819), and by the fourth comet seen in the same year (1819), discovered by Blanpain, and thought by Clausen to be identical with the comet of 1743. The orbits of the last-mentioned comet, as well as that of Lexell's, seem to have undergone great alteration by the proximity and attraction of Jupiter; their periods of revolution appear to be only from five to six years, and their aphelia take place near the orbit of Jupiter. Among comets whose periods range from seventy to seventy-six years, we should name, first, Halley's (which has been

of so much importance in theoretical and physical astronomy, and whose last appearance in 1835 was less
brilliant than might have been expected from preceding
ones); the comet of Olbers (of the 6th of March, 1815);
and the one discovered by Pons in 1812, the elliptic
orbit of which was determined by Encke. The two latter
comets were invisible to the naked eye. We now know,
with certainty, of nine returns of Halley's comet, for Laugier's [56] calculations have recently demonstrated the identity
of its orbit with that of the comet of 1378, mentioned in
the Chinese tables of comets, for the knowledge of which we
are indebted to Edouard Biot. During the interval between
its first and last recorded appearances, in 1378 and 1835,
its periods of revolution have fluctuated between 74·91
years and 77·58 years, the mean being 76 1.

Contrasted with the cosmical bodies of which we have
been speaking are a group of comets requiring many thousand years to perform their revolutions, of which the
periods can only be determined with great difficulty and
uncertainty. Argelander assigns a period of 3065 to the
fine comet of 1811, and Encke a period of upwards of
8800 years to the awfully grand one of 1680. According to such views, these bodies recede respectively to distances from the Sun twenty-one and forty-four times greater
than that of Uranus, or to 33600 and 70400 millions
of miles. At these enormous distances, the attractive force
of the Sun still subsists; but whilst the motion of the comet
of 1680 at its perihelion is 212 miles in a second, being
thirteen times greater than that of the Earth, its velocity at
its aphelion is scarcely ten feet in a second, being only
three times greater than that of our most sluggish European

rivers, and but half that which I found in the Cassiquiare, an arm of the Orinoco. Amongst the countless host of uncalculated or still undiscovered comets, it is highly probable that there are many, the major axes of whose orbits may far exceed even that of the comet of 1680. In order to afford through the medium of figures some idea,—I do not say of the extent of the sphere of attraction, but,—of the distance in space of a fixed star or other sun from the aphelion of the comet of 1680, (the one of the bodies of our solar system which, according to our present knowledge, attains the greatest remoteness,) I would here remind the reader, that, according to the most recent determinations of parallax, even the nearest fixed star is at least 250 times more distant from our Sun than this comet at its aphelion. The comet's distance is only 44 times that of Uranus, whilst that of α Centauri is 11000, and of 61 Cygni, according to Bessel's determination, 31000 times that of Uranus.

Having thus considered the greatest known distances of comets from the central body, we may proceed to notice instances of the greatest proximity hitherto measured. The smallest distance between a comet and the Earth occurred in the case of Lexell and Burkhardt's comet of 1770, which has acquired so much celebrity from the perturbations it underwent from the mass of Jupiter. On the 28th of June, 1770, this comet's distance from the Earth was only six times that of the Moon from the Earth. The same comet passed twice, in 1767 and in 1779, through the system of the four satellites of Jupiter, without causing the slightest sensible derangement in these small bodies, whose movements are so well known. The great comet of 1680, when at its perihelion on the 17th of December, was only one-sixth of the sun's diameter from the surface of that body; an approach eight or nine time nearer than Lexell's comet made to the Earth, and equal to only seven-tenths of the Moon's distance from the Earth. Owing to the feebleness of the light of distant comets, perihelia which take place beyond the orbit of Mars can very seldom be observed by the inhabitants of the earth; and of all the comets which have been computed hitherto, that of 1729 is the only one which has its perihelion between the orbits of Pallas and Jupiter; it was even observed beyond the latter planet.

Since a degree of scientific knowledge, sometimes sound, but oftener vague and partial, has extended into wider circles of social life, the fears of possible evils threatened by comets, have perhaps rather increased in weight as their direction has become more definite. The certainty that within the known planetary orbits there are comets which visit our regions at short intervals,—the considerable perturbations which their paths undergo by the attractions of Jupiter and Saturn, whereby apparently harmless bodies might be converted into dangerous ones,—the intersection of the Earth's orbit by that of Biela's comet,—the recognition of the existence of a cosmical ether, which as a retarding medium tends to contract all orbits,—the differences between individual comets, which allow of the supposition of considerable diversity in the mass of the nucleus, —are motives of alarm which, by their number and variety, are fully equivalent to the vague fears which prevailed in former centuries of " fiery swords," and " long haired blazing stars," threatening universal conflagration. The tranquillising considerations which, on the other hand, have been derived from the calculus of probabilities, being addressed to the understanding rather than to the imagination, modern science has been accused, with some degree of justice, of only endeavouring to allay fears which she has herself contributed to excite. That the unexpected and extraordinary should oftener excite fear than joy or hope [57], springs from a source more deeply seated in our common nature. The strange aspect of a large comet, its faint nebulous gleam, its sudden appearance in the vault of heaven, have almost always, and in all regions of the earth, been viewed as the portentous heralds of impending change in the established order of things. The phænomenon itself being of short duration, its reflection is the more naturally looked for in cotemporaneous or immediately succeeding events, and it is seldom difficult to fix on some incident which may be interpreted as the calamity foreshewn. In our own time, however, the popular mind has taken another and more cheerful, though singular, direction in respect to comets. Among the German vineyards, in the beautiful valleys of the Rhine and the Moselle, a favourable influence on the ripening of the grape and on the quality of the wine has been ascribed to these bodies, long regarded as so ill-omened : nor has experience of a contrary kind, which has not been wanting in these days, when comets have been so often seen, been able to shake the belief in this meteorological fable of wandering heat-imparting stars.

I now proceed from the consideration of comets to that of another and still more enigmatical class of bodies ; namely, to those minute asteroids, which, when they arrive in a fragmentary state within our atmosphere, we designate by the names of " aerolites," or " meteoric stones." If I dwell at greater length on these phænomena as well as on comets, and notice particular cases more than might appear suitable in a general view of nature, it is not without a purpose. The great diversity of character in different comets has been already mentioned. The little knowledge which we yet possess of the physical qualities of these bodies, renders it difficult to separate the essential from the accidental in phenomena recurring at intervals, and which have been observed with very unequal degrees of accuracy. It is only the measuring and computing parts of the astronomy of comets, which, in modern times, have made such admirable progress: in the present imperfect state of our knowledge, therefore, a scientific consideration must restrict itself to physiognomical diversity in the nucleus and tail,—to instances of remarkable approximation to other cosmical bodies,—and to extreme cases either of dimensions of orbit, or of periods of revolution. In these phænomena, as well as in those which will next be treated of, fidelity to nature can only be sought in the careful description of individual instances, and by such an animated and graphic mode of expression, as may serve to bring the reality vividly before the mind.

Shooting stars, fire-balls, and meteoric stones, are with great probability regarded as small masses moving with planetary velocities in space, and revolving in conic sections round the Sun, in accordance with the laws of universal gravitation. These masses approach the Earth in their path, are attracted by her mass, and enter our atmosphere, becoming luminous at its limits; when they frequently let fall stony fragments, heated in a greater or less degree, and covered with a shining black crust. A careful investigation of what has been observed at the epochs when periodic

showers of shooting stars fell in Cumana in 1799, and in
North America in 1833 and 1834, shews, that balls of fire
and shooting stars are not only often contemporaneous and
intermingled, but that they pass gradually one into the
other, whether we compare the magnitude of their disks, or
the trains which accompany them, or the velocities of their
movement. While there are exploding and smoke-emitting
balls of fire, which are luminous even in the bright sunshine
of a tropic day ([58]), and sometimes exceed in size the apparent
diameter of the Moon,—there are, on the other hand, shooting
stars which fall in immense numbers, and are of such small
dimensions, that they exhibit themselves only as moving
points, or as *phosphorescent lines* ([59]). Whether among the
many luminous bodies which shoot across the sky, there may
not be some of a different nature from others, still remains uncertain. In the equinoctial zone, I received the impression
that, both on the low burning plains, and at elevations of
twelve or fifteen thousand feet, falling stars were there more
frequent,—of brighter colours,—and more often accompanied
by long brilliant trains of light, than in the colder latitudes ;
but doubtless this impression was occasioned solely by the
exceeding transparency of the tropical atmosphere, which
enables the eye to penetrate farther into its depths ([60]). Sir
Alexander Burnes extols, as a consequence of the serenity
and clearness of the air and sky in Bokhara, the brilliant
and frequently recurring spectacle of variously coloured
meteors.

The connection of meteoric stones with the more splendid
phænomenon of fire-balls, and the fact that meteoric stones
sometimes fall from fire-balls with a force which causes them
to sink to a depth of from ten to fifteen feet into the earth,

have been shewn, among many other instances, by the falls
of aerolites observed at Barbotan, in the Département des
Landes, in France (24th July 1790), at Sienna (16th
June, 1794), at Weston, in Connecticut (14th December,
1807), and at Juvenas, in the Département de l'Ardèche (15th
June, 1821). In other instances, a small and very dark
cloud forms suddenly in a perfectly clear sky, and the
stones are hurled from it with a noise resembling repeated
discharges of cannon. Such a cloud, moving over a whole
district of country, has sometimes covered it with thousands
of fragments, very various in size, but similar in quality.
A phænomenon of still more rare occurrence took place on
the 16th September, 1843, when a large aerolite fell at
Kleinwenden, not far from Muhlhausen, accompanied by a
thundering noise, but with a clear sky, in which no cloud
was formed. As further evidence of the affinity between
fire-balls and shooting stars, it should be noticed that fire-
balls, from which meteoric stones have descended, have sometimes been seen, as at Angers, on the 9th of June, 1822, of
a diameter hardly equal to that of the small Roman candles
in our fire-works.

We have as yet scarcely any knowledge in regard to the
physical and chemical processes which contribute to the
formation of these phænomena. Whether the particles, of
which the compact meteoric masses are composed, exist
originally in a fluid form (as in comets), and only begin to
condense within the fire-ball at the moment when it becomes
luminous to our sight,—or what takes place within the
bosom of the dark cloud from which sounds resembling
thunder are sometimes heard for minutes before the stones
are precipitated from it,—or whether, in the case of smaller

shooting stars, there falls any compact substance, or only a *meteoric dust* ([61]), containing iron and nickel,—are questions still wrapped in great obscurity. We know, by measurement, the astonishing and wholly planetary velocity of shooting stars, fire-balls, and meteoric stones; in this respect, therefore, we are able to recognise what is "general" and "uniform" in the phænomena; but the genetic and cosmical premises, the successive transformations undergone, are not known to us. If meteoric stones circulate in space in already consolidated and dense masses ([62]) (less dense, however, than the mean density of the earth), we must suppose that they form very small nuclei, which, surrounded by inflammable vapours or gases, constitute fire-balls of from 500 to 2600 feet in actual diameter, as inferred in some of the largest amongst them from observations of their height and apparent diameter. The largest meteoric masses yet known to us are those of Bahia in Brazil, and of Otumpa, described by Rubin de Celis: these are seven and seven and a half feet in length. The meteoric stone of Ægos Potamos, celebrated in antiquity, and mentioned in the Chronicle of the Parian Marbles, and which fell about the year of the birth of Socrates, has been described as being of the size of two millstones, and equal in weight to a full waggon load. Notwithstanding the failure of the efforts of the African traveller, Browne, I have not given up the hope that this Thracian meteoric mass, which must have been so difficult to destroy, may be found, after the lapse of more than 2300 years, by some of the European visitors to countries which have now become so easy of access. We learn by a document lately discovered by Pertz, that the enormous aerolite which, in the beginning of the tenth century, fell into

the river near Narni, projected nearly four feet above the surface of the water.   It must be remarked that these meteoric
stones, whether ancient or modern, cannot be regarded as
more than principal fragments of the mass which exploded
in the fire-ball, or descended from the dark cloud.

When we consider the mathematically-proved enormous
velocity with which meteoric stones arrive at the earth from
the extreme limits of the atmosphere, and with which balls
of fire move in a more lengthened course through its denser
strata, it appears to me highly improbable that these metalliferous masses of stone, with their imbedded and perfectly-
formed crystals of olivine, labradorite, and pyroxene, should
have condensed from a gaseous state into a solid nucleus in
so short an interval of time.   Even when the fallen pieces
differ from each other in chemical composition, they almost
always show the peculiar characters of a fragment, having
often a prismatic, or truncated pyramidal form, with slightly
curved faces and rounded angles.   But whence this fragmentary character (first recognised by Schreibers) in a
rotating planetary body ?   Here, as in the sphere of organic
life, there is obscurity in all that belongs to the history of
development.   Meteoric masses kindle and become luminous
at elevations which must be supposed to be almost
entirely deprived of air.   Biot's recent investigations
on the important phænomena of twilight ([63]) even reduce
considerably the height of the line which has been
usually, but somewhat hazardously, termed the "limit" of
the atmosphere; but luminous processes may take place
without the presence of oxygen, and Poisson has imagined
that the ignition of aerolites occurs far beyond the range of
our atmosphere.   In treating of meteoric stones, as well as

of the larger cosmical bodies of our solar system, it is only
in what is subject to calculation and to geometric measurement that we feel ourselves on safe or solid ground. As
early as 1686, Halley pronounced the great ball of fire seen
in that year, the movement of which was opposite to that of
the Earth in her orbit ([64]), a cosmical phænomenon; but it was
not until 1794, that Chladni, with remarkable acuteness,
recognised the general connection between fire-balls and those
stones which had been known to fall through the air, and
the motion of the former bodies in space ([65]). A brilliant confirmation of this view of the cosmical origin of these phænomena has since been furnished by Denison Olmsted, of Newhaven in Massachusetts, who, from the concurrent testimony
of all the observers of the celebrated fall of meteors, or shooting
stars, which took place on the 12th or 13th of November,
1833, has shewn,—that all these bodies, whether fire-balls or
shooting stars, proceeded from the same quarter of the
heavens, i. e. from a point near the star γ Leonis,—and that
this continued to be the point, although in the time during
which the phænomenon was observed, the star materially
altered both its apparent altitude and azimuth. This independence of the earth's rotation showed that the luminous
bodies came from *without, i. e.* that they entered our atmosphere from the external regions of space. From Encke's
computation ([66]) of the whole of the observations which were
made in the United States of North America, between the
latitudes of 35° and 42°, it follows that these meteors
all proceeded from the point in space towards which the
motion of the Earth was then directed. In the subsequent
great falls of shooting stars in the month of November,
observed in 1834 and 1837 in North America, and in 1838

in Bremen, the same general parallelism of the paths of the meteors, and the same direction from the constellation of the Lion, were recognised. The periodical falls of shooting stars which take place at other parts of the year, are also supposed to show greater parallelism of direction than is the case with those which appear sporadically at other seasons. A periodical recurrence, similar to that of November, has been noticed in the month of August; in which month, in 1839, it was observed that the meteors came from a point in the heavens situated between Perseus and Taurus, towards the latter of which constellations the Earth was then moving. This peculiarity of the phænomenon (*viz.*, the retrograde direction, both in November and in August), is especially deserving of being confirmed or refuted by very exact and careful observation on future occasions.

The heights of shooting stars, *i. e.* the heights at which they first become visible, and at which their visibility ceases, are exceedingly various, fluctuating from 16 to 140 miles. This important result, and the enormous velocity of these problematical asteroids, were first shown by Benzenberg and Brandes, from simultaneous observations and determinations of parallax, at the two extremities of a base line of 46000 (49020 English) feet in length ([67]). The relative velocity of their motion was from eighteen to thirty-six miles in a second; and similar, therefore, to that of the planets ([68]). This planetary velocity, and the retrograde or opposite direction of the paths of the meteors to that of the Earth, are the principal grounds which are considered subversive of that hypothesis which attributes the origin of aerolites to the supposed active volcanoes of the Moon. All numerical hypotheses of a greater or less volcanic force on a small cosmical body not surrounded by an atmosphere, must be in their nature exceedingly arbitrary. The reaction of the interior of such a body against its crust may, indeed, be imagined to be ten times, or even a hundred times, more powerful than that of our present terrestrial volcanoes. The direction of masses discharged from a satellite revolving from west to east, might also appear retrograde, in consequence of the Earth arriving later at the point in her orbit where those masses fall. If, however, we duly weigh all the circumstances, which I have thought it necessary to recount lest I should seem to make assertions without having sufficient ground for their support, we shall find the lunar origin ([69]) of meteoric stones to be dependent on a number of different conditions whose concurrence would be requisite in order to change the simply possible into the actual. It appears more in analogy with our other views of the formation of the solar system, to admit the separate existence of small planetary masses circulating independently in space.

It is very probable that a large portion of these minute cosmical bodies may continue their course round the sun undestroyed by the vicinity of our atmosphere, and suffering only an alteration in the eccentricity of their orbits by the attraction of the Earth. It is possible, therefore, that they first become visible to us after many revolutions. The supposed ascent of shooting stars and balls of fire which Chladni attempted, not very happily, to explain by the reaction of the air strongly compressed in their descent, seems, at first sight, a consequence of some unexplained tangential force tending to throw off the meteors from the earth : but Bessel has shown, on theoretical grounds, the improbability of the supposed facts; and this has been since confirmed by Feldt's careful calculations, that, for want of perfect simultaneity in the observed disappearances, an upward movement cannot be regarded as a result of observation ([70]). Future researches must decide whether, as Olbers supposes, the explosion of shooting stars, and the ignition of fire-balls, may not occasionally impel the meteors upwards, or otherwise influence the direction of their paths.

Shooting stars fall either singly or sporadically, or in groups of many thousands which are compared by Arabian writers to flights of locusts. The latter cases are periodical, and the meteors are then seen in streams, moving, for the most part, in parallel directions. Of the periodic groups, those hitherto best known are the phænomena of the 12th to the 14th of November, and of the 10th of August or the day of St. Lawrence, whose "fiery tears" ([71]) were long since recognised in England as a recurring meteorological phæ-nomenon, and are mentioned in an old Church Calendar, as well as in legendary traditions. Although a mixed shower of shooting stars and fire-balls had been seen on the night of the 12th and 13th of November, in 1823, by Klöden, at Potsdam,—and in 1832 throughout Europe, from Ports-mouth, in England, to Orenburg, on the Oural river, and even in the southern hemisphere, in the Isle of France,—still the idea of the periodicity of the phænomenon, and of great showers of falling stars being connected with particular days, was first inferred on the occasion of the fall observed by Olmsted and Palmer in North America, on the 12th and 13th of November, 1833, when the shooting stars seemed, in one part of the sky, to fall as thickly as snow-flakes; in the course of nine hours there fell at least 240000.

Palmer recalled to recollection the fall of meteors in 1799, at Newhaven, which was first described by Ellicott and myself ([72]), and which was shown, by the observations which I brought together, to have extended simultaneously over the new continent, from the Equator to New Herrnhut in Greenland, in lat. 64° 14′, and from 46° to 82° of west longitude from Paris. The identity of the two epochs was perceived with astonishment. The stream which was seen over the whole sky, on the 12th and 13th of November, 1833, from Jamaica to Boston, recurred on the nights of the 13th and 14th of November, 1834, in the United States, but the display was less brilliant.

A second equally regular periodic shower, that of the Feast of St. Lawrence, takes place between the 9th and 14th of August. Muschenbrock ([73]) had called attention, in the middle of the last century, to the frequency of meteors in the month of August; but their regular return about the epoch of St. Lawrence's Day was first made out by Quetelet, Olbers, and Benzenberg. No doubt other periodically recurring showers ([74]) will in time be discovered, possibly about the 22nd and 25th of April, the 6th and 12th of December, and—judging from the falls of aerolites recounted by Capocchi—perhaps also from the 27th to the 29th of November, or on the 17th of July.

Independent as all the occurrences of this kind have hitherto seemed of local circumstances, such as latitude, temperature, and climatic relations, there is an accompanying phænomenon of which it would be wrong to omit the notice, although the coincidence may, perhaps, have been purely accidental. The Aurora Borealis showed itself with great intensity during the occurrence of the most magnificent display of meteors yet observed, viz., that described by Olmsted, on the 12th and 13th of November, 1833. The Aurora was also seen during the periodical phænomenon in 1838, at Bremen, where, however, the fall of meteors was much less striking than at Richmond, near London. I have noticed elsewhere the remarkable observation of Admiral Wrangel [75], which he has repeatedly confirmed to me verbally, viz., that during the appearance of the Aurora on the Siberian coast of the Polar Sea, he frequently saw portions of the sky, which were not previously luminous, but which seemed to kindle when a falling star shot across them, and continued bright for some time afterwards.

It is probable that the different streams of meteors, each consisting of myriads of small cosmical bodies, intersect the orbit of the Earth in the same way that Biela's comet does. According to this view, we may imagine that they form a continuous ring, each pursuing its course in a common direction. The small planets between Mars and Jupiter present, with the exception of Pallas, an analogous arrangement in their closely connected orbits. We cannot yet determine whether the variations in the epochs at which the stream becomes visible to us, and the retardations of the phænomena, to which I long ago called attention, indicate a regular progression or an oscillation of the nodes (*i. e.*, the points of intersection of the ring with the Earth's orbit) ; or whether they are to be explained by the irregular grouping and very unequal distances apart of these very small bodies ; and by the supposition that the zone formed by them has a width which the Earth requires several days to traverse. The system of the satellites of Saturn shows us a group of intimately connected cosmical bodies, occupying a zone of prodigious breadth. In this group the orbit of the outermost, that is the seventh, satellite, has a diameter so considerable, that the earth, in her course round the sun, requires three days to pass through an equal space. If in one of the continuous rings, which we have imagined as formed by the paths of the periodical streams, we suppose the distribution of the asteroids to be such that there are only a few groups so closely congregated as to occasion the appearance of showers, we may conceive how such brilliant phænomena as those of November 1799 and 1833 may be of exceedingly rare occurrence. The highly ingenious Olbers was inclined to think, that the next return of the great phænomenon of fire-balls and shooting stars falling like flakes of snow, would be witnessed from the 12th to the 14th of November, 1867.

Sometimes the stream of the November asteroids has been visible over a small portion only of the earth's surface : for example, in the year 1837, it was seen with great magnificence as a meteoric shower in England, whilst, on the same night, which was uninterruptedly clear, a very attentive and practised observer at Braunsberg, in Prussia, saw only a few sporadic shooting stars between 7 P.M. and sunrise the following morning. Hence, Bessel inferred ([76]) that a small group of the great ring occupied by these bodies approached the earth in England only, while the countries to the eastward passed through a part of the meteoric ring which was comparatively void. Should increased probability be given to the supposition of a regular progression of the line of nodes, or of its oscillation in consequence of perturbations, the discovery of older observations of these phænomena will acquire a special interest. The Chinese annals, which contain notices both of the appearance of comets, and of great showers of falling stars, go back beyond the time of Tyrtæus, or of the second Messenian war: they describe two streams occurring in the month of March, one of which was observed 687 years before the Christian era. Edouard Biot has remarked, that in fifty-two appearances of numerous shooting stars recorded in the Chinese annals, the periods which recur most frequently are from the 20th to the 22d July, old style. This stream may therefore be the same as that which we now observe about St. Lawrence's day (10th of August), supposing it to have somewhat advanced ([77]). If the shower of falling stars of the 21st October, 1366, old style, of which Boguslawski, (the younger), has found a notice in Benessius de Horowic's Chronicon Ecclesiæ Pragensis,—be our present November phænomenon, seen on that occasion in bright daylight,—we should learn from the progression which has taken place in the interval of 477 years, that the centre of gravity of this system of shooting stars describes a retrograde path round the sun. It also follows from the views which have been here developed, that if years pass by in which neither of the streams which have been hitherto indicated (that is, those of November and August) are observed at any part of the earth, the cause may be sought, either in interruptions in the ring by vacant spaces or gaps between the groups of asteroids, or, as Poisson thinks, in the influence which the larger planets ([78]) may exercise on the form and position of the ring.

The solid masses which reach the earth,—whether they have been seen to fall at night from balls of fire, or in the daytime from a small dark cloud usually in a clear sky, and with a loud noise—though considerably heated, are not incandescent. They exhibit, on the whole, a general unmistakeable resemblance to one another in their external form, in the nature of their crust, and in the chemical composition of their principal constituents; and this resemblance is traceable when and wherever they have been collected, at all periods of time, and in all parts of the earth. But this remarkable and early recognised similarity of general character in solid meteoric masses, suffers many exceptions in detail. How different are the very malleable masses of iron from Hradschina in the district of Agram, or those from the banks of the Sisim in the Jeniseisk government, mentioned by Pallas, or those which I brought from Mexico (79), all of which contain 96 per cent. of iron, from the aerolite of Sienna, which hardly contains two per cent. of iron, from the earthy meteoric stone of Alais in the Département du Gard, which falls to pieces when immersed in water, and from those of Jonzac and Juvenas, which are without any metallic iron, and are composed of various crystalline ingredients. These diversities have led to the division of the cosmical masses under consideration into two classes; nickeliferous meteoric iron, and fine or coarse-grained meteoric stones. The crust of these masses, which is only a few tenths of a line in thickness, is very characteristic; it has often a pitchy lustre (80), and is sometimes veined. The only instance which I know of the absence of this crust is in the meteoric stone of Chantonnay in La Vendée, which is marked by another circumstance equally rare, viz. the presence of pores and vesicular cavities, like the meteoric stone of Juvenas. The separation of the black crust from the light gray

mass beneath is always as sharply defined as is that of the
dark leaden-coloured crust of the white granite blocks[81]
which I brought from the cataracts of the Orinoco, and
which are also found by the side of many cataracts in other
parts of the world, as those of the Nile and the Congo.
The greatest heat of our porcelain furnaces can produce
nothing similar to the crust of the aerolites, so distinctly
and sharply separated from the unaltered mass beneath.
Appearances which might seem to indicate a softening of
the fragments have been occasionally recognised, but, in
general, the condition of the greater part of the mass,—
the absence of any flattening from the effect of the fall,—and
the moderate degree of heat perceived on touching the
newly fallen aerolite,—are far from indicating a state of internal fusion during its rapid passage from the limits of the
atmosphere to the earth.

The chemical elements of which meteoric masses consist
have been well analysed by Berzelius, and are the same which
we find dispersed in the crust of the earth; they include iron,
nickel, cobalt, manganese, chrome, copper, arsenic, tin,
potash, soda, sulphur, phosphorus, and carbon; being in all
about one-third of the number of elementary substances with
which we are at present acquainted. Notwithstanding the
identity of their ultimate constituents with those into which
inorganic bodies are chemically decomposable, yet the manner in which these constituents are combined occasions the
general aspect of meteoric masses to be peculiar, and
unlike terrestrial productions. The presence of native iron,
which is found in almost all aerolites, gives them a
specific character, but it is one not necessarily lunar;

for tnere may be other cosmical bodies besides the moon in which water may be entirely wanting, and processes of oxidation may be rare.

The cosmical gelatinous vesicles, the organic masses, re. sembling the *Tremella nostoc*, which, since the middle ages, have been supposed to belong to shooting stars, and the pyrites of Sterlitamak, west of the Oural, which are supposed to have formed the inside of hailstones [82], belong to the fables of meteorology. The aerolites which possess a fine-grained texture, and are composed of olivine, augite, and labradorite [83], are, as Gustav Rose has shewn, the only ones which have a telluric appearance; for example, the aerolite resembling dolorite, found at Juvenas, in the Dé-partement de l'Ardèche. They contain, in fact, crystalline substances quite similar to those of the crust of our earth ; and in the Siberian mass of meteoric iron, the olivine is only distinguished by the absence of nickel, which is there replaced by oxide of tin [84]. As meteoric olivine, like our basalts, contains from 47 to 49 per cent. of magnesia, and as, according to Berzelius, olivine forms one-half of the earthy constituents in meteoric stones, there is no reason to be surprised at the large proportion of silicate of magnesia which we find in these cosmical masses. Since the aerolite of Juvenas contains distinct crystals of augite and labradorite, the numerical proportions of the constituents render it at least probable that the meteoric masses of Château Rénard are examples of diorite composed of horn-blende and albite, and that those of Blansko and Chatonnay are a combination of hornblende and labradorite. The proof which has been supposed to be furnished from the minera-logical resemblances just alluded to, of a telluric or atmo-

spheric origin of aerolites, do not appear to me to have
much force. I would here refer to a remarkable conversation
which took place between Newton and Conduit at Ken-
sington([85]), and ask, why should not the substances belonging to one group of cosmical bodies, or to one planetary
system, be for the most part the same ? Why should it
not be so, if we permit ourselves to surmise that the
planets, and all the spheroidal masses which revolve
around the sun, have been formed by separation from the
once more extended solar atmosphere, as from rings of vapour revolving round the central body ? We are, it seems
to me, no more entitled to call nickel and iron, olivine and
augite, which we find in meteoric stones, exclusively terrestrial substances, than I should be to call plants which grow
wild in Germany, and which I might also meet with beyond the
Oby, "European species of the flora of Northern Asia."
If, in a group of cosmical bodies, the elementary substances are
the same, why should they not form determinate compounds
in accordance with their mutual affinities and attractions, as
in the polar regions of Mars resplendent domes of snow and
ice, and in other smaller cosmical masses, mineral aggregations, containing crystals of olivine, augite, and labradorite?
Even in the field of what must necessarily be conjecture, its
course must not be arbitrary, or irrespective of induction.

Extraordinary obscurations of the sun's disk have occasionally taken place, so that stars have been seen even at
midday. A phænomenon of this nature, not to be explained
by a cloud of volcanic ashes, or by a fog of unusual elevation,
occurred in 1547, at the period of the eventful battle near
Mühlberg, and continued during three entire days. They were
attributed by Kepler, at one time to a " materia cometica,"

and at another to a black cloud produced by sooty exhalations from the solar orb. Obscurations of less duration, which took place in 1090 and 1203, and lasted, one for three, and the other for six hours, were ascribed by Chladni and Schnurrer to the intervention of meteoric masses. Since the common direction of their paths has led us to regard the streams of shooting stars as forming a continuous ring, the epochs of yet unexplained celestial phænomena have been brought into remarkable connection with the regularly recurring epochs of the meteoric displays. Adolph Erman, with much acuteness, and after a careful analysis of all the facts hitherto collected, has called attention in this respect to the times of conjunction with the sun of the August asteroids (the 7th of February), and of the November asteroids (the 12th of May); and has pointed out a remarkable coincidence between the conjunction of the November asteroids, and the celebrated *cold days* of the Saints Mamertus, Pancratius, and Servatius ([86]).

The Greek natural philosophers, generally little disposed to observation, but most persevering and inexhaustible in conjectural interpretations of half-perceived facts, have left on record speculations respecting shooting stars and meteoric stones, which greatly resemble some of the views of the cosmical nature of these phænomena now commonly received. "Shooting stars," says Plutarch ([87]), in the Life of Lysander, "according to some naturalists, are not emanations from the ethereal fire, which become extinguished in the air immediately after being kindled; nor are they an ignition and combustion of air which may have been dissolved in quantity in the upper region; they are rather celestial bodies which fall in consequence of an interruption of the general force of rotation, and are precipitated not only upon inhabited countries, but also beyond them in the ocean, so that they are not found." Diogenes of Apollonia ([88]) expresses himself still more clearly. According to him, "together with the visible stars there move other invisible ones, which are therefore without names. These not unfrequently fall to the earth and become extinguished, like the star of stone which fell in flames at Ægos Potamos." The Apollonian, who regarded all other stars (*i. e.* the luminous ones) as pumice-like bodies, probably founded his opinion respecting shooting stars and meteoric stones on the doctrine of Anaxagoras of Clazomene, who imagined all celestial bodies to be mineral masses, which the fiery ether in its impetuous course had torn from the earth, inflamed, and converted into stars. The Ionic school, therefore, with Diogenes of Apollonia, placed aerolites, and stars or heavenly bodies, in one and the same class: both, indeed, were alike regarded as of telluric origin, but this was in the view of all having been once formed from the earth, and having taken their places round her as a central body ([89]); precisely as, according to modern ideas, the planets of a system are conceived to have been formed around the central body or sun, and from its once extended atmosphere. This view is not, therefore, to be confounded with that usually implied in what is called the telluric or atmospheric origin of meteoric stones; or with the extraordinary notion of Aristotle, who supposed the enormous mass of Ægos Potamos to have been carried up by a tempestuous wind.

A presumptuous scepticism, which rejects facts without examination of their truth, is in some respects even more injurious than an unquestioning credulity; it is the tendency of both to impede accurate investigation. Although for upwards of two thousand years the annals of different nations had told of falls of stones, which in many instances had been placed beyond doubt by the testimony of irreproachable witnesses; although the Bætylia formed an important part of the meteor worship of the ancients, and the companions of Cortes saw at Cholula the aerolite which had fallen on the neighbouring pyramid; although Caliphs and Mongolian princes had had swords forged of fresh fallen meteoric stones; and even although human beings had been killed by the falling stones (namely, a friar at Crema, on the 4th of September, 1511 ; a monk at Milan, 1650 ; and two Swedish sailors on board a ship in 1674) ; yet, until the time of Chladni, who had already earned for himself imperishable renown in physics by the discovery of his figure representations of sound, this great cosmical phænomenon remained almost unheeded, and its intimate connection with the rest of the planetary system unknown. Those who are persuaded of this connection, if susceptible of emotions of awe from the impressions of nature, will be strongly moved to thoughtful contemplation, not only by the spectacle of the brilliant phænomenon of meteoric showers at the August or November periods, but also whenever they behold a solitary falling star shoot across the sky. The profound repose of night is suddenly interrupted, and life and motion momentarily break the tranquil splendour of the firmament. The spectator sees in the glimmering light which marks the track of the falling star the visible delineation of a portion of its orbit ; and the burning asteroid brings to his mind the existence of matter pervading universal space. When we compare the volume of the innermost of Saturn's satellites, or of Ceres, with the enormous volume of the Sun, the relations of great and small disappear to our imagination. The sudden blazing up and subsequent extinction of stars in Cassiopea, Cygnus, and Ophiucus, have already led to the admission of the possible existence of non-luminous cosmical bodies. Condensed in smaller masses, the asteroids revolve around the sun, intersect like comets the paths of the great luminous planets, and become ignited when they enter or approach the outermost strata of our atmosphere.

Our intercourse with all other cosmical bodies—with all nature beyond the limits of our own atmosphere—is, exclusively, either through the medium of light, and of radiant heat intimately united with light [90], or through the mysterious force of attraction exerted by remote bodies, according to the measure of their distance and their mass, on our globe, its ocean, and its atmosphere. But if in shooting stars and meteoric stones we recognise planetary asteroids, we are enabled by their fall to enter into a wholly different, and more properly material, relationship with cosmical objects. Here we no longer consider bodies acting upon us exclusively from a distance, by exciting undulatory vibrations of light or heat, or by causing, or themselves undergoing, motion by the influence of gravitation; but we have actually present the material particles themselves, which have come to us from the regions of space, have descended through our atmosphere, and remain upon the earth. A meteoric stone affords us the only possible contact with a substance foreign to our planet. Accustomed to know non-telluric bodies solely by measurement, by calculation, and by the inferences of our reason, it is with a kind of astonishment that we touch, weigh, and analyse a substance appertaining to the world without: the imagination is stimulated, and the intellect aroused and animated, by a spectacle, in which the uncultivated mind sees only a train of fading sparks in the clear sky, and apprehends in the black stone which falls from the thundering cloud only the rude product of some wild force of nature.

If the asteroids, on the description of which I have lingered with pleasure, may seem in some degree to resemble comets by the smallness of their mass and the variety of their paths, they still differ essentially from those bodies, being visible to us only at the instant of their destruction, when, arrested by the earth, they become luminous by ignition.

To complete our view of all that belongs to the solar system, which now, since the discovery of the small planets, of the comets of short period, and of the meteoric asteroids, appears so complex and so rich in forms, we have yet to consider the *Zodiacal Light*, to which allusion has already been made. Those who have dwelt long in the zone of Palms, must retain a pleasing remembrance of the mild radiance of this phænomenon, which, rising pyramidally, illumines a portion of the unvarying length of the tropical nights. I have seen it occasionally shine with a brightness greater than that of the Milky Way near the constellation of Sagittarius; and this not only in the dry and highly rarefied atmosphere of the summits of the Andes, at elevations of thirteen to fifteen thousand feet, but also in the boundless grassy plains or *llanos* of Venezuela, and on the sea-coast under the ever clear sky of Cumana. The phænomenon is one of peculiar beauty when a small fleecy cloud is projected against the zodiacal light, and detaches itself picturesquely from the illuminated back-ground. A passage in my journal during a voyage from Lima to the West Coast of Mexico, notices such a picture. " For the last three or four nights (between 10° and 14° of North latitude), the Zodiacal light has appeared with a magnificence which I have never before seen. Judging also from the brightness of the stars and nebulæ, the transparency of the atmosphere in this part of the Pacific must be extremely great. From the 14th to the 19th of March, during a very regular interval of three-quarters of an hour after the disk of the sun had sunk below the horizon, no trace of the zodiacal light could be seen, although the night was perfectly dark; but an hour after sunset it became suddenly visible, extending in great brightness and beauty between Aldebaran and the Pleiades, and, on the 18th of March, attaining an altitude of 39° 5'. Long narrow clouds, scattered over the lovely azure of the sky, appeared low down in the horizon, as if in front of a golden curtain, while bright varied tints played from time to time on the higher clouds : it seemed a second sunset. Towards that side of the heavens the light diffused appeared almost to equal that of the moon in her first quarter. Towards ten o'clock, in this part of the Pacific, the Zodiacal light usually becomes very faint, and at midnight I could see only a trace of it remaining. On the 16th of March, when its brightness was greatest, a mild reflected glow was visible in the east." In the obscurer sky and thicker atmosphere of our so-called temperate zone, the Zodiacal light is only distinctly visible in the beginning of spring, when it may be seen after evening twilight above the western horizon, and at the end of autumn, before the commencement of morning twilight, above the eastern horizon.

It is difficult to understand how so striking a natural phænomenon could have failed to attract the attention of astronomers and physical philosophers before the middle of the seventeenth century, or how it should have escaped the observant Arabs in ancient Bactria, on the Euphrates, and in Southern Spain. We are almost equally surprised at the late period at which the nebulæ in Andromeda and Orion, first described by Simon Marius and Huyghens, were observed. The earliest distinct description of the Zodiacal light is contained in Childrey's Britannia Baconica ([91]), of the year 1661 ; its first observation may have been two or three years earlier. Dominic Cassini has, however, incontestably the merit of having been the first (in 1683) who investigated its relations in space. The luminous appearance which was seen by him in 1668 at Bologna, and at the same time in Persia by the celebrated traveller Chardin (and which the Court astrologers of Ispahan, who had never seen it before, named "*nyzek*," or "small lance"), was not, as has often been said, tho Zodiacal light, but the enormous tail of a comet ([92]), the head of which was concealed by the vapours near the horizon, and which, in its position and appearance, presented .many points of resemblance to the great comet of 1843. But it may be conjectured with much probability, that the remarkable light rising pyramidally from the earth, which, in 1509, was seen in the eastern part of the sky for forty nights in succession from the high table land of Mexico (and which I found mentioned in an ancient Aztec manuscript, in the Codex

Telleriano-Remensis ([93]), in the Royal Library at Paris), was the Zodiacal light.

This phænomenon, doubtless of primeval antiquity, but which in Europe was first discovered by Childrey and Dominic Cassini, is not the luminous atmosphere of the sun itself, which, according to the laws of mechanics, cannot be more oblate than in the ratio of 2 : 3, and could not therefore extend to a greater distance than nine-twentieths of the distance of Mercury from the Sun. The same laws determine that the height of the extreme limit of the atmosphere of a rotating cosmical body above its equator, or the point at which gravity and the centrifugal force are in equilibrium, can only be that at which a satellite would complete its revolution in the same time that the central body rotates around its own axis ([94]). This restricted limit of the solar atmosphere, in its present concentrated condition, is particularly striking when we compare the central body of our system with the nucleus of other nebulous stars. Herschel discovered several in which the semidiameter of the nebula surrounding the star subtends an angle of 150″. Assuming a parallax of not quite one second, we find the outermost nebulous stratum of such a star to be 150 times farther from its centre than the distance of the Earth from the Sun. If, therefore, such a nebulous star was in the place of our Sun, its atmosphere would not only include the orbit of Uranus, but would extend eight times as far ([95]).

The solar atmosphere being thus limited in extent, we may with great probability attribute the Zodiacal light to the existence of an extremely oblate ring ([96]) of nebulous matter, revolving freely in space between the orbits of Venus and Mars. We can, indeed, at present, form no certain judgment concerning the true dimensions of the supposed ring; its possible augmentation (⁹⁷) by emanations from the tails of many millions of comets when at their perihelia; the singular variability of its extension (which seems sometimes not to exceed that of our own orbit); or concerning its not improbable intimate connection with the more condensed cosmical vapour in the vicinity of the Sun. The nebulous particles of which the ring consists, and which revolve around the Sun according to the same laws as the planets, may either be self-luminous, or may reflect the light of the Sun. The first supposition is not inadmissible; even a terrestrial fog (and it is a very remarkable fact), shewed itself in 1743, at the time of the new moon, and in the middle of the night, so phosphorescent, that objects could be distinctly recognised at a distance of above 600 feet (⁹⁸).

In the tropical regions of South America, I have sometimes observed with astonishment the variations in intensity of the Zodiacal light. Having during several months passed the nights in the open air, and under a serene sky, on the banks of the great rivers, or in the midst of the wide grassy plains or llanos, I had frequent opportunities of carefully observing the phænomenon. Sometimes in a few minutes after the Zodiacal light had been at the strongest, it would become sensibly weakened, then suddenly reappear in full brilliancy. In a few instances I thought that I perceived,—not indeed a tinge of red colour, or a dark arch beneath, or, as Mairan describes, a jet of sparks,—but an undulatory motion of the light. Are there, then, processes going on in the ring of vapour itself? or is it not more probable that,—though near the ground, and in the lower part of the atmosphere, I could detect no changes of temperature or moisture by the meteorological instruments,—and even though small stars of the fifth and sixth magnitude appeared still to shine with undiminished and equable light,—condensations were taking place in the higher regions of the atmosphere, which modified the transparency of the air, or rather its reflecting power, in some peculiar and to us unknown manner ?. The assumption of such meteorological processes near the limits of our atmosphere is favoured by observations made by the acute Olbers [99], " of sudden flashings and pulsations which, in the course of a few seconds, vibrate throughout the whole of a comet's tail, which is seen at the same time to lengthen several degrees, and again to contract. As the different portions of a comet's tail, which is millions of miles in length, are at very unequal distances from the Earth, it is not possible, according to the laws of the velocity and propagation of light, that actual alterations in a cosmical body filling so immense a space, should be seen by us to take place in such short intervals of time." These considerations by no means exclude the reality of variable emanations around the denser envelopes of the nucleus of a comet, or of sudden brightenings of the Zodiacal light from internal molecular movements, or of changes due to variations in the light reflected by the nebulous matter of which the ring is composed ; but they should make us careful to distinguish between effects which should be referred to the cosmical ether and to the regions of space, and those which are referrible to the terrestrial atmospheric strata through which the bodies existing in space are beheld by us. There are well observed facts, which shew us that we are not able to

explain completely all that takes place at the uncertain and
much contested limit of our atmosphere. The wonderful
lightness of whole nights in the year 1831, in which, in the
latitudes of Italy and Northern Germany, small print could
be read at midnight, appears in manifest contradiction to all
that we learn ([100]) from the most recent and exact researches on
the theory of twilight, and on the height of the atmosphere. The phænomena of light which astonish us in the
variability of the crepuscular limits, and in the changes in
the Zodiacal light, must be dependent on conditions which
have not yet been successfully investigated.

Thus far we have been occupied in considering the
world of forms governed by our Sun, or the solar
system; comprising planets, satellites, comets of shorter or
longer periods of revolution, meteoric asteroids moving either
sporadically, or in crowded streams in continuous rings,
and, finally, a luminous nebulous ring, revolving round the
sun near the orbit of the earth, and for which, from its
position, the name of Zodiacal light may be retained. In
the movements of all, the law of periodic return everywhere
prevails, however different may be the measure of velocity,
or the quantity of the aggregated particles : the asteroids
alone, as they enter our atmosphere from the regions of
space, have their planetary revolution checked, and are
themselves united to the larger planet. In the solar system,
of which the limits are determined by the force of attraction
of the central body, comets return in their elliptic orbits
from distances equal to forty-four elongations of Uranus ;—
nay, in those very comets in which the nucleus, from the
smallness of its mass, appears to us but as a cosmic cloud,
it yet retains by its attraction the most remote particles

of the tail, which streams from it to a distance of many millions of miles. Thus the central forces are the maintaining as well as the constituent forces of the system.

Our Sun, viewed in relation to all the bodies so various in magnitude and density which revolve around it, may be regarded as at rest, although it revolves around the common center of gravity of the whole system, which center remains within the body of the Sun itself, notwithstanding the varying positions of the planets. Altogether distinct in its nature is the movement of translation of the Sun,—the progressive motion of the center of gravity of the whole solar system in universal space,—which is supposed to take place with such prodigious velocity, that, according to Bessel ([101]), the relative motion of the Sun and the star 61 Cygni amounts, in a single day, to no less than 3336000 miles. We should be unconscious of the change of place of the solar system, were it not that, by the perfection of our astronomical instruments, and by improved methods of observation, we are enabled to note our progress by reference to distant stars,—as in a vessel we estimate its speed by the apparent motion of objects on shore. The proper motion of 61 Cygni is, nevertheless, so considerable, as to produce a displacement of a whole degree in 700 years.

We can measure the amount of changes in the relative positions of the stars to one another,—in their proper motions as these changes are called,—with far greater certainty than we can explain their cause. After deducting all that depends on the precession of the equinoxes, and the nutation of the Earth's axis consequent upon the influence of the Sun and Moon on the spheroidal form of the earth, — all that results from the propagation and aberration of light, or from the parallax produced by the opposite positions of the Earth in its orbit round the Sun,—we find a residual annual motion of the fixed stars, which includes both the translation of the whole solar system in space, and the actual proper motions of the stars themselves. The very difficult numerical separation of these two elements has been rendered possible, by a careful specification of the directions in which the movements of the different stars take place, and by the consideration that, if they were all absolutely at rest, they would appear perspectively to recede from the point towards which the Sun's course is directed. The result of the investigation, confirmed by the theory of probabilities, is, that both our solar system and the stars are changing their place in space. It appears from the admirable investigations of Argelander([102]), (who has been engaged at Abo in extending, and in carrying to much greater perfection, the work commenced by William Herschel and Prevost,) that the Sun is moving towards the constellation of Hercules, and very probably towards a point which, from a combination of the observations of 537 stars, was situated (equinox of 1792·5) in 257° 49′·7 Right Ascension, and in 28° 49′7 North Declination. In this class of investigation there is still great difficulty in separating the absolute from the relative motion, and in determining what portion belongs to the solar system only.

If we consider the proper motions of stars, as contradistinguished from their apparent or perspective motions, their directions are various and even opposite in different groups; it is not, therefore, a necessary conclusion, either that all parts of our astral system, or that all the systems which fill universal space, revolve around one great undiscovered luminous or non-luminous central body, however naturally we may be disposed to an inference which would gratify alike the imaginative faculty, and that intellectual activity which ever seeks after the last and highest generalization. Even the Stagyrite has said, " All that moves leads us back to the cause of the motion which we perceive; and it would be but an endless derivation of causes, were there not a primary unmoving mover ([103])."

Amongst the manifold changes of place of stars in groups, not occasioned by parallax depending on the place of the observer, but actual changes taking place progressively and uninterruptedly in space, we have revealed to us in the most incontestable manner by one phenomenon,—the movements of double stars, and their slower or more rapid motion in different parts of their elliptical orbits,—the dominion of the laws of gravitation extending far beyond the limits of our solar system, even to the remotest regions of creation. On this subject man's desire of knowledge need no longer rest on vague conjecture, or seek satisfaction in the boundless but uncertain field of analogy, for here also the progress of astronomical observation and calculation has at length placed us on firm ground. It is not so much the astonishing number already discovered of double or multiple stars revolving round a common center of gravity (a number which, in 1837, amounted to 2800), as it is the extension of our knowledge of the fundamental forces of the whole material world,—and the evidence thus afforded of the universal prevalence of the law of gravitation,— which excites our admiration, and constitutes one of the most brilliant discoveries of our epoch. The time of revolution of differently coloured double stars varies exceedingly in different instances, from 43 years in η Coronæ, to many thousands of years in 66 Ceti, 38 Geminorum, and 100 Piscium. In the triple system of ζ Cancri, the nearest companion of the principal star has already more than accomplished one entire revolution since Herschel's measurement in 1782. By means of skilful combinations of the changes of distance, and of angles of position ([104]), the elements of the orbits have been assigned, and conclusions have even been drawn respecting the absolute distance of the double stars from the earth, and their mass as compared with the mass of the Sun. But whether the attracting forces depend solely on the quantity of matter in these systems as in ours, or whether there may not coexist with gravitation other specific forces which do not act according to mass, is, as Bessel has been the first to shew, a question of which the solution is reserved for later ages ([105]).

If we desire to compare our Sun with others of the fixed stars, or self-luminous suns, within the lenticular sidereal stratum to which we belong, we find that in the case of some of them at least, there are methods by which we may arrive approximately, and within certain limits, at a knowledge of their distance, their volume, their mass, and the velocity of their motions in space. If we take the distance of Uranus from the Sun at 19 times the solar distance of the Earth, then the central body of our system is 11900 such spaces, or solar distances of Uranus, from α Centauri, 31300 from 61 Cygni, and 41600 from α Lyræ. The comparison of the volume of the Sun with the volume of stars of the first magnitude is dependent on an optical element which is subject to extreme uncertainty, viz. the apparent diameter of the fixed stars. If, with Herschel, we assume the apparent diameter of Arcturus even at only the

tenth part of a second, it will result that the true diameter
of the star is eleven times greater than that of the Sun (106).
The distance of the double star, 61 Cygni, determined by
Bessel, has led to an approximate knowledge of the quantity of
matter contained in it.   Although the portion of the apparent
path passed through by the smaller star since Bradley's observations is not yet sufficiently large to enable us to infer the
true path, and its major semi-axis; yet the great Königsberg
astronomer (107) considers it probable that "the mass of this
double star is neither much more nor much less than half
the mass of our Sun."   This result is from actual measurement.   Analogies derived from the greater mass of those
planets of our solar system which are attended by satellites,
and from the fact that Struve has observed the proportion
of double stars to be six times greater among the brighter
than among the telescopic stars, have led other astronomers to
conjecture (108), that the average mass of the greater number
of double stars exceeds the mass of the Sun.   On this subject, however, general results are far from being yet attainable.
In respect to proper motion in space, our Sun belongs,
according to Argelander, to the class of rapidly moving stars.

The aspect of the sidereal heavens, the relative position of
stars and nebulæ, the distribution of the masses of light
formed by them, the picturesque beauty, if I may use the
expression, of the whole firmament, depend, in the course
of thousands of years, conjointly on the actual proper motion in space of stars and nebulæ, on the movement of
translation of our solar system, on the appearance of new
stars, and the extinction or diminution in intensity of the
light of others; and lastly and especially, on the changes
which the Earth's axis undergoes from the attraction of the

Sun and Moon. The beautiful stars of the Centaur and of
the Southern Cross, will at some future day be visible in
our northern latitudes, whilst other stars (Sirius and the
stars forming the belt of Orion) will no longer appear
above the horizon. The place of the North Pole will be
successively marked by β and α Cephei, and δ Cygni, until
after the lapse of 12000 years, when α Lyra will become
the brightest of all possible pole stars. These statements
serve in some degree to realise in the mind the magnitude
of the movements, which proceed uninterruptedly in infinitely small divisions of time in the great Chronometer of
the Universe. If, for a moment, we imagine the acuteness
of our senses preternaturally heightened to the extreme limits
of telescopic vision, and bring together events separated by
wide intervals of time, the apparent repose which reigns in
space will suddenly vanish ; countless stars will be seen
moving in groups in various directions ; nebulæ wandering, condensing, and dissolving like cosmical clouds ;
the milky way breaking up in parts, and its veil rent
asunder. In every point of the celestial vault we
should recognise the dominion of progressive movement,
as on the surface of the earth where vegetation is constantly putting forth its leaves and buds, and unfolding
its blossoms. The celebrated Spanish botanist, Cavanilles,
first conceived the possibility of " seeing grass grow," by
placing the horizontal micrometer wire of a telescope with a
high magnifying power at one time on the point of a
bamboo shoot, and at another on the rapidly unfolding
flowering stem of an American aloe ; precisely as the astronomer places the cross of wires on a culminating star.
Throughout the whole life of physical nature—in the organic

as in the sidereal world—existence, preservation, production, and development, are alike associated with motion as
their essential condition.

The breaking up of the Milky Way, to which I have
alluded, requires a more particular notice.　William
Herschel, our safe and admirable guide in the regions
of space, has found by his star-gaugings, that the telescopic breadth of the Milky Way is six or seven degrees
wider than is laid down in our celestial maps, or than it
appears to the naked eye ([109]). The two bright nodes in which
the two branches of the zone unite, near the constellations
of Cepheus and Cassiopea, and those of Scorpio and Sagittarius, appear to exercise a powerful attraction on the
neighbouring stars; but in the brightest portion, between
β and γ Cygni, 330000 stars are found in a breadth of $5°$,
of which half appear to be attracted towards one side, and
half towards the other. It is here that Herschel surmises
that a disruption may take place ([110]).

The number of distinguishable telescopic stars in the
Milky Way, apart from nebulæ, is estimated at eighteen
millions. In order, I will not say to realise the magnitude
of this number, but to compare it with something analogous,
I would recal to the reader, that the whole number of stars
in the firmament from the first to the sixth magnitude,
visible to the naked eye, is only about 8000. In the unpro-
ductive astonishment which is excited by the relation of
mere numerical values, unconnected with applications
affecting the higher powers of the intellect, the imagination,
or the feelings, the extremes in point of dimension meet;—
namely, the cosmical bodies of the vast regions of space,
and the smallest forms of animal existence. A cubic inch of

the polishing slate of Bilin contains, according to Ehrenberg, 40000 millions of the siliceous shells of Galionellæ.

Nearly at right angles to the Milky Way formed of stars,
in which, as Argelander remarks, brilliant stars are more
numerous than in any other part of the heavens, there is
another milky way consisting of nebulæ. The first of these,
or the galaxy of stars, according to Sir John Herschel's
views, forms around our sidereal system, and at some distance from it, a detached ring or zone, similar to the ring
of Saturn. The situation of our planetary system is
eccentric, nearer to the region of the Cross than to the
opposite region, that of Cassiopea ([111]). In a nebula
discovered by Messier, but which has been only imperfectly seen, we seem to discover the image of our own
sidereal system, and the divided ring of our Milky Way
reflected, as it were, with wonderful similarity ([112]). The
galaxy of nebulæ does not belong to our sidereal zone, but
surrounds it at a vast distance, and without any physical
connection, passing almost in a great circle through the
nebulæ in Virgo (which are particularly numerous in the
northern wing), through the Coma Berenicis, the Ursa
Major, the girdle of Andromeda, and the Northern Fish.
It probably intersects the galaxy of stars in Cassiopea, and
connects the poles, which are situated where the thickness
of the stratum is least, and which are poor in stars, owing
possibly to the action of those forces which have formed the
stars into groups ([113]).

It follows from these considerations, that our sidereal
cluster—which, in its projecting branches, shews traces of
great progressive changes of form—is surrounded by two
rings, one of which, the Nebulous Milky Way, is very

remote; while the other, composed of stars alone, is less distant. The latter ring, or that to which we usually apply the term "Milky Way," consists of stars averaging from the 10th to the 11th degree, but appearing, when viewed singly, very various in point of magnitude; whereas detached clusters, or groups of stars, almost always shew throughout great similarity in magnitude and brilliancy ([114]).

In whatever quarter the celestial vault has been examined with powerful space-penetrating telescopes, either stars, though it may be only telescopic ones from the twentieth to the twenty-fourth degree of magnitude, or nebulæ, are seen. It is probable that, with still more powerful optical instruments, many of the nebulæ would be found resolvable into stars. The sensation of light impressed on the retina by single isolated points, is less, as Arago has recently shewn, than when the rays proceed from several points extremely near to each other ([115]). It is probable that the production of heat by the condensation of the cosmical nebulous matter, whether existing in definite forms, or simply in its general state of distribution, may modify the equable intensity of light, which, according to Halley and Olbers, should arise from every point in the heavens being occupied by an infinite series of stars ([116]). Observation, however, contradicts the hypothesis of uniform distribution, shewing us instead, extensive regions wholly devoid of stars —"openings in the heavens," as William Herschel calls them—one four degrees in width in Scorpio, and another in Ophiucus. We find near the margin of both these openings resolvable nebulæ, of which the one on the western edge of the opening in Scorpio is amongst the richest and

most crowded groups of small stars with which the heavens
are adorned.   Herschel, indeed, ascribes to the attractive
force of these marginal groups ([117]), the starless openings themselves; of which he says, in his finely animated style, " they are parts of our sidereal stratum
which have already suffered great devastation from time."
If we consider the telescopic stars, situated one behind
another, as forming a canopy of stars covering the
whole apparent celestial vault, we may, I think, regard the
starless portions of Scorpio and Ophiucus as tubes through
which we look into the remote regions of space.   The strata
which form the canopy are there interrupted ; other still
remoter stars may indeed lie beyond, but our instruments
cannot reach them.   The ancients had also been led, by the
apparition of igneous meteors, to the idea of rents or chasms
in the canopy of the skies; but the chasms were supposed
to be only transitory, and, instead of being dark, to be
bright and fiery, from affording a glimpse of the burning
ether beyond ([118]).   Derham, and even Huygens, appeared not
indisposed to explain, ˙in a somewhat similar manner, the
tranquil light of the nebulæ ([119]).

When we compare the stars of the first magnitude,
which, on an average, are certainly the nearest to
us, with the non-nebulous telescopic stars,—and the nebulous stars with unresolvable nebulæ (for example, with
the nebula in Andromeda, or even with the so-called
planetary nebulæ),—and when we thus enter on the consideration of distances so diverse in the boundless regions
of space,—there presses itself on our notice a fact, which
governs the relation of the phænomena as perceived by us,
to the realities which are their actual basis, viz. the *successive propagation of light*.   The velocity of this propagation,

according to Struve's most recent investigations, is 166072 geographical miles in a second—a velocity almost a million times greater than that of sound. From all that we learn from the measurements of Maclear, Bessel, and Struve, of the parallaxes and distances of three fixed stars of very different magnitudes, α Centauri, 61 Cygni, and α. Lyræ, a ray of light from each of these three bodies requires respectively 3, 9¼, and 12 years, to reach the Earth. In the short but memorable period between 1572 and 1604, from the time of Cornelius Gemma and Tycho Brahe to that of Kepler, three new stars suddenly appeared in the constellations of Cassiopea, Cygnus, and in the foot of Ophiucus. A similar phænomenon shewed itself in the constellation of Vulpis, in 1670, but in this case the light of the new star was intermitting. In very modern times, Sir John Herschel, at the Cape of Good Hope, saw the star η Argus increase in brightness from the second to the first magnitude ([120]). But such events or occurrences in the vast regions of cosmical space, belong, in their historic reality, to other epochs than those at which the phænomena of light first reveal them to the inhabitants of the Earth ; they reach us as voices of the past. It has been justly said, that with our large telescopes we penetrate at once into space and time. We measure space by time ; the ray of light requires one hour to travel 592 millions of miles. Whilst in the Hesiodic Theogony, the dimensions of the universe were expressed by the fall of bodies, and the iron anvil was only nine days and nine nights in falling from heaven to earth, it was thought by the elder Herschel ([121]), that the light of the most distant nebulæ discoverved by his forty-foot refractor requires two millions of years to reach our eyes. Thus, much may have disappeared even before it became visible to our eyes, and in much the arrangement and order may have varied. The spectacle of the starry heavens presents to our view objects not contemporaneous ; and however much we may diminish both the supposed distance whence the faint light of the nebulæ, or the barely discernible glimmer of the remotest cluster of stars, reaches us,—and the thousands of years which serve as the measure of that distance,—it will still remain true that, according to the knowledge which we possess of the velocity of light, it is more than probable that the light of the most distant cosmical bodies offers us the oldest sensible evidence of the existence of matter. Thus, resting on simple premises, the reflecting mind rises to graver and loftier views of nature's forms, in those boundless fields which light traverses, and where " myriads of worlds spring like grass in the night." ([122])

We will now descend from the region of celestial forms to the more restricted sphere of terrestrial forces ; from the children of Uranus to those of Gea. A mysterious bond unites the two classes of phænomena. In the ancient symbolical meaning of the Titanic Mythus ([123]), the forces of the universe, and the systematic order of nature, depend on the union of the heavens and the earth. If our terrestrial spheroid, as well as each of the other planets, belongs originally to the Sun, as having been formed from detached nebulous rings of the solar atmosphere, a connection is still maintained, by means of light and radiant heat, both with the Sun of our own system, and with all those remoter suns which glitter in the firmament. The very different measure of these effects must not prevent the

physical philosopher, engaged in tracing a general picture of
nature, from noticing the connection and co-extensive dominion of similar forces.   A minute fraction of the Earth's
heat belongs to the part of space through which our planetary system is moving, the temperature of which is
supposed to be nearly equal to the mean temperature of the
poles of the earth, and is regarded by Fourier as the product
of calorific radiation from all the bodies of the universe.
Far more powerful undoubtedly are the effects of the Sun's
rays on the atmosphere, and on the upper strata of our
globe, in the electric and magnetic currents occasioned by
his heat-producing powers, and in the magical and beneficent
influence which awakes and nourishes the germs of life in
the organic forms on the surface of the earth : these will
form the subject of our subsequent consideration.

In now turning our attention exclusively to the telluric
sphere of nature, we will first consider the relative extent of
liquid and solid surface of the Earth ; its figure ; its mean
density, and the partial distribution of this density in the interior of the planet ; its temperature, and electro-magnetic
tension.   These relations and forces will lead us to consider
the reaction of the interior of our globe on its exterior ; and,
through the special agency of subterranean heat, the phænomena of earthquakes, occurring in districts of varying extent ;
the breaking forth of hot springs ; and the more powerful
action of volcanic forces.   Movements in the crust of the earth,
sometimes sudden and in shocks, sometimes continuous and
almost imperceptible, alter in the course of centuries the
relative elevation of the land and sea, and the configuration
of the land beneath the ocean ; while, at the same time,
communications are formed between the interior of the

earth and the atmosphere, either through temporary clefts or more permanent openings. Molten masses, issuing from unknown depths, flow in narrow streams down the declivities of mountains, sometimes with an impetuous, and sometimes with a slow and gentle motion, until the fiery subterranean fount is dry, and the lava solidifies under a crust which it has itself formed. We thus see new rocks produced under our eyes; whilst those of earlier formation are altered by the influence of heat, rarely in immediate contact, more often in proximity. Even when no disruption takes place, the crystalline particles in superincumbent rocks are displaced, and re-arranged in a denser texture. The waters present formations of an entirely different nature; concretions of the remains of plants and animals; deposits of earthy, calcareous, and aluminous matter; aggregations of finely pulverized rocks, covered with beds of siliceous-shelled infusoriæ, and with transported soil containing the bones of animals belonging to an earlier state of our globe. These processes of formation and stratification going on before our eyes, in modes so different,—and the disruption, flexure, and elevation of rocks and strata, by mutual pressure and by the agency of volcanic forces,—lead the thoughtful observer, by simple analogies, to compare the present with the past, to combine actual phænomena, to generalize, and to amplify in thought the extent and intensity of the forces now in operation. Thus we arrive at the domain of that geological science, long desired and obscurely anticipated, but which, in the last half century, has been placed on the firm basis of legitimate induction.

It has been acutely remarked, " that much as we have gazed on the planets through large telescopes, we know less of their exterior than of their interior." They have been weighed and measured; thanks to the progress of astronomical observation and calculation, their volumes and their densities are known with constantly increasing numerical exactness; but (with perhaps an exception in some degree in the case of the Moon), a profound obscurity still veils from us their physical properties. It is only on our own globe that immediate proximity places us in relation with all the elements both of the organic and the inorganic creation. The rich diversity of materials, their admixtures and transformations, and the ever changing play of the forces elicited, offer to the spirit of investigation appropriate and welcome food; and the immeasurable field of observation in which the intellectual activity of the human mind can here expatiate, lends to it a portion of its own elevation and grandeur. The world of sensible phænomena reflects itself into the depths of the world of ideas, and the rich variety of nature gradually becomes subject to our intellectual domain.

I here touch again upon an advantage to which I have already repeatedly alluded, possessed by that portion of our knowledge which is especially connected with our terrestrial habitation. Uranography, or the description of the heavens, from the remotely gleaming nebulous stars to the central body of our own system, is limited to general conceptions of volume and mass; no vital activity is there revealed to our senses; it is only by means of resemblances, and often fanciful combinations, that even conjectures have been hazarded respecting the specific nature of the material elements, and their presence or absence in this or that cosmical body. The heterogeneity of matter, its chemical diversity, the regular forms into which its particles arrange

themselves, whether crystalline or granular; its relations to
the deflected, or decomposed waves of light by which it is
penetrated; to radiating, and to transmitted, or polarised
heat; to the brilliant, or the not less energetic because
invisible, phænomena of electro-magnetism,—all this in-
calculable treasure of physical knowledge by which our
contemplation of the universe is enriched and exalted, we
owe to investigations concerning the surface of the planet
which we inhabit, and more to its solid than to its liquid
portion. I have already noticed how greatly this extensive
knowledge of natural objects and forces, and the measure-
less variety of objective perceptions, stimulates the cultivation and promotes the activity of the human intellect; it is
as needless, therefore, to dwell farther on this topic, as on
that of its connection with the causes of the superiority in
material power, which particular nations derive from their
command of a portion of the elements. If, on the one
hand, I have been desirous of calling attention to the difference between the nature of our telluric knowledge, and of
that which we possess concerning the regions of space, I
wish, on the other hand, to indicate the limited extent of
the field from whence our whole knowledge of the heterogeneous properties of matter is derived. It is from that
which has been rather inappropriately termed the " crust"
of the earth, or the thickness of so much of the strata
nearest to the surface of our planet, as is opened to our
view either by deep natural valleys, or by the labours of
man in boring or in mining operations. These operations ([124]) attain a perpendicular depth below the level of
the sea of little more than two thousand feet, about one-third
of a geographical mile, or $\frac{1}{9800}$ of the Earth's radius. The

crystalline masses erupted from active volcanoes, and mostly
resembling the rocks at the surface of the earth, come
from absolute depths, which, though they cannot be accurately determined, are assuredly sixty times greater than
any which have been reached by our artificial works. In
situations where strata of coal dip beneath the surface,
and rise again at distances determined by careful measurement, we are enabled to assign numerically the depth of
the basin formed by them; and we thus learn that such
coal measures, together with the ancient organic remains
which they contain, often reach (as in Belgium, for example) depths exceeding five and six thousand feet([125])
below the present level of the sea; and that the mountain
limestone, and the strata of the Devonian basin, attain a
depth fully twice as great.   If we now combine these depths
beneath the surface with those mountain summits which
have hitherto been regarded as the highest portions of the
crust of the earth, we obtain nearly 40000 English feet,
or a measure equalling about $\frac{1}{524}$ of the Earth's radius.
This, therefore, would be the whole range in a vertical
direction of our geological researches, or of our knowledge
of superimposed rocks,—even if the general elevation of the
surface of the earth equalled the height of the Dhawalagiri
in the Himalaya, or of the Sorata in Bolivia.   All that is
situated at a greater depth beneath the level of the sea than
the deepest wells or mines, or the basins I have referred
to,—or than the bed of the sea where it has been reached
by soundings (James Ross sounded with 4600 fathoms, or
27600 feet of line, without finding bottom),—is as unknown
to us as the interior of the other planets of our solar system.
In the case of the Earth, as in that of the other planets, we

know the mass and the mean density, and we are able to compare the latter with the density of the materials constituting the upper terrestrial strata, which alone are accessible to us. Where the knowledge of the chemical and mineralogical properties of substances in the interior of the Earth fails us, we find ourselves again limited to the field of mere conjecture, as in the case of the remotest planetary bodies. We can determine nothing with certainty respecting the depth at which the materials of which our rocks are composed exist, either in a softened though still tenacious state, or in complete fusion,—respecting cavities filled with elastic vapours,—the condition of fluids heated under enormous pressure,—or the law of the increase of density from the surface to the center of the earth.

The notice of the increase of heat with increasing depth in the interior of our planet, and of the reaction of the interior on the surface, leads us to the consideration of the long series of volcanic phænomena : these manifest themselves to us as earthquakes, emissions of gas, thermal springs, mud volcanoes, and streams of lava flowing from craters of eruption. The reaction of the elastic internal forces shews itself also in alterations of the configuration and of the level of the surface of the globe. Vast plains and deeply indented continents are elevated or depressed, and thus the reciprocal limits of land and sea, of solid and liquid surface, are frequently and variously modified. Plains have undergone an oscillatory motion, being alternately elevated and depressed. Subsequently to the elevation of continents above the sea, mountain chains have risen from long clefts, and these are mostly parallel, in which case the elevations were probably cotemporaneous. Salt lakes and

great inland seas, long inhabited by the same species of animals, have been violently separated, their original connection
being still evidenced by the fossil remains of shells and
zoophytes. Thus in following phænomena in their mutual
dependence, we are conducted from the consideration of
forces operating in the interior of our globe, to movements
and disruptions of its surface, and to the pouring forth of
molten streams forced up by the expansive energy of elastic
vapours. The same forces which elevated the lofty chains
of the Andes and the Himalaya to the regions of perpetual
snow, have occasioned new compositions and textures in the
mineral masses, and have altered strata which had been
previously deposited from fluids containing many organic
substances. We thus perceive the dependence of the series
of formations, divided and superposed according to their
ages, on changes of configuration of the surface, on dynamic
relations of the upheaving forces, and on the chemical
action of the vapours which issue from the fissures.

The form and distribution of the dry land, or of that
portion of the earth's crust which is suited to the luxuriant
development of vegetable life, are connected by intimate
relations, and by reciprocal action, with the surrounding sea,
in which organic life seems almost limited to the animal world.
The liquid element is again covered by the atmosphere—an
aerial ocean into which the mountain chains and plateaus of
the dry land rise like shoals, and occasion a variety of currents and changes of temperature. Collecting moisture from
the region of clouds, these loftier tracts contribute also to
the spread of life and motion, by the beneficial influence of
the streams of water which flow down their declivities.

Whilst the geography of plants and animals depends on

with their times of rotation, shews to exist also in those two large planets. Thus the knowledge of the external form of planetary bodies affords a basis for conclusions respecting their internal constitution.

The northern and southern terrestrial hemispheres appear to present nearly the same curvature under equal latitudes; (135) but, as has been already remarked, pendulum experiments and measurements of degrees give such different results for different parts of the Earth's surface, that it is impossible to assign any regular figure which shall satisfy all the results obtained by these methods. The actual figure of the Earth is probably to a regular figure, " as the uneven surface of agitated water is to the even surface of water in repose."

Having thus measured the Earth, it had to be weighed; and here also the vibrations of the pendulum, and the employment of the plumb line, have served to determine its mean density. This has been attempted in three ways: 1st, by combining astronomical and geodesical operations, for the purpose of ascertaining the deflection of the plumb line caused by the vicinity of a mountain; 2d, by comparing the length of the pendulum vibrating seconds in a plain, and on the summit of a mountain; 3d, by employing a balance of torsion, which may be regarded as a horizontal pendulum, as a measure of the relative density of neighbouring strata. Of these three methods (136) the last is the most certain, because it is independent of the very difficult determination of the density of the mineral masses of which the mountain consists, near which the observations of the other two methods must be made. The most recent experiments with the balance of torsion are those of

Reich, and give for their result 5·44 to 1 as the ratio of the mean density of the Earth to that of distilled water. Now, we know from the general nature of the rocks and strata which form the dry or continental portion of the Earth's surface, that their density can hardly amount to 2·7, or that of the land and sea surfaces taken together to 1·6 ; it follows, therefore, that, either by pressure, or from the heterogeneity of the substances, the elliptical strata in the interior must undergo a great increase of density towards the center of the Earth. Here, again, the horizontal (as before the vertical) pendulum, shews itself to be justly entitled to the name of a geological instrument.

The results thus obtained have led physical philosophers of celebrity to form to themselves, according to the different hypotheses from which they proceeded, wholly opposite views respecting the nature of the interior of the globe. It has been computed at what depths liquid and even gaseous substances, from the pressure of their own superimposed strata, would attain a density exceeding that of platinum, or of iridium ; and in order to bring the actual degree of ellipticity, which was known within very narrow limits, into harmony with the hypothesis of the infinite compressibility of matter, Leslie conceived the interior of the Earth to be a hollow sphere, filled with " an imponderable fluid of enormous expansive force." Such rash and arbitrary conjectures have given rise, in wholly unscientific circles, to still more fantastic notions. The hollow sphere has been peopled with plants and animals, on which two small subterranean revolving planets, Pluto and Proserpine, were supposed to shed a mild light. A constantly uniform temperature is supposed to prevail in these inner regions, and the air being rendered self-luminous by compression, might well render the planets of this lower world unnecessary. Near the north pole, in 82° of latitude, an enormous opening is imagined, from which the polar light visible in Auroras streams forth, and by which a descent into the hollow sphere may be made. Sir Humphry Davy and myself were repeatedly and publicly invited by Captain Symmes to undertake this subterranean expedition: so powerful is the morbid inclination of men to fill unseen spaces with shapes of wonder, regardless of the counter-evidence of well-established facts, or universally recognised natural laws. Even the celebrated Halley, at the end of the 17th century, hollowed out the Earth in his magnetic speculations. A freely rotating subterranean nucleus was supposed to occasion, by its varying positions, the diurnal and annual changes of the magnetic declination. It has been attempted in our own day, in tedious earnest, to invest with a scientific garb that which, in the pages of the ingenious Holberg, was an amusing fiction.

The figure of the Earth, its degree of solidification, and its density, are intimately connected with forces which act in its interior; in so far, at least, as those forces are not excited or caused by external influence, or by the position of the planet relatively to the luminous central body. We have considered the compression or ellipticity as a consequence of the centrifugal force acting on a rotating mass, and as evidencing an earlier condition of fluidity in our planet : in the course of the solidification of this fluid, (which some have been inclined to assume to have been gaseous, and originally heated to a very high degree of temperature,) an enormous quantity of latent heat would have been disengaged ; and supposing, with Fourier, the process of consolidation to have commenced by radiation into space from the cooling surface, the particles nearer to the center of the Earth would have continued fluid and incandescent. After long transmission of heat from the center towards the surface, a stable condition of the temperature would have been established, when the heat would increase uninterruptedly with increasing depth. The high temperature of water which rises in very deep borings in Artesian wells; direct observation of the temperature of rocks in mines; and, above all, the volcanic activity of the Earth, ejecting molten masses from opened clefts or fissures, bear unquestionable evidence to this increase for very considerable depths in the upper terrestrial strata. Inferences, which are indeed founded only on analogy, render the extension of this increase farther towards the center more than probable. That which has been learnt respecting the propagation of heat in homogeneous metallic spheroids, by means of an ingenious analytical calculus, perfected expressly for this class of investigations, ([137]) can only be applied with great caution to the actual constitution of our planet, considering our ignorance of the substances of which the Earth may be composed,—of the different capacity for heat and conducting power of the superimposed masses,—and of the chemical changes which solid and fluid matter may undergo from enormous pressure. That which is most difficult for us to conceive and to represent to ourselves, is the boundary line between the fluid interior mass and the solidified rocks which form the outer crust; or the gradual change from the solid strata to the condition of semi-fluidity; a condition to which the known laws of hydraulics can only

be applicable under considerable modifications.   It seems
highly probable that the action of the Sun and Moon, which
produces the ebb and flow of the ocean, is also felt in these
subterranean depths.   We may suppose periodic heavings
and subsidings of the molten mass, and consequent variations in the pressure against the vaulted covering formed by
the solidification of the upper rocks.   The amount and effects
of such oscillations must, however, be small; and though the
relative position of the heavenly bodies may here also occasion " spring tides," yet it is certainly not to these, but to
more powerful internal forces, that we must attribute the
movements which shake the surface of the Earth.   There are
groups of phænomena, to the existence of which it may be
useful to refer, for the purpose of illustrating the universality
of the attraction of the Sun and Moon on the external and
internal condition of our globe, however little we may be
able to assign numerically the amount of such influence.

Tolerably accordant experience has shewn that in Artesian wells, the average increase of temperature in the strata
pierced through, is 1° of the Centigrade thermometer for
92 Parisian feet of vertical depth (54·5 English feet for 1°
of Fahrenheit); or if we suppose this increase to continue
in an arithmetical ratio, a stratum of granite would, as I have
already remarked ([138]), be in a state of fusion at a depth of
nearly 21 geographical miles; or, between four and five times
the elevation of the highest summit of the Himalaya.

In the globe of the Earth three varieties in the mode of the
propagation of heat are to be distinguished.   The first is
periodical, and causes the temperature of the strata to vary,
as, according to the position of the Sun and the season of
the year, the warmth penetrates from above downwards,

or, inversely, escapes by the same path. The second is also an effect of the Sun; its action is extremely slow: part of the heat which has penetrated into the Earth in the equatorial regions travels along the interior of the Earth's crust to the vicinity of the poles, where it escapes into the atmosphere, and thence into space. The third is the slowest of all; it consists in the secular cooling of the terrestrial globe; in the escape of the very small quantity of the primitive heat of the planet which is now given out from its surface. The loss of central heat is supposed to have been very great at the time of the early terrestrial revolutions, but within historic periods it has hardly been appreciable by our instruments. The temperature of the surface of the earth is intermediate between the glowing temperature of the inferior strata, and that of space which is probably below the freezing point of mercury.

The periodic variations of the temperature, produced at the surface by the position of the Sun and by meteorological processes, propagate themselves towards the interior of the Earth, but only to a very inconsiderable depth. The slow conducting power of the soil diminishes the loss of heat in winter, and is favourable to trees having deep roots. At points placed at different depths on the same vertical line, the maximum and minimum of the imparted temperatures are attained at very different seasons; and the greater the distance from the surface, the less is the difference between the extremes of temperature. In our temperate latitudes (48°—52°) the stratum of invariable temperature is found at a depth of from 55 to 60 French feet (59 to 64 English feet nearly); and at half that depth the oscillations of temperature from the influence of season are already

diminished to less than 1° of Fahrenheit.   In tropical climates the invariable stratum is only one foot below the surface; and Boussingault has ingeniously availed himself of this
fact to obtain a very convenient, and, as he thinks, certain
mode of determining the mean temperature of the air at a
station ([139]).   This mean temperature of the air, either at a
fixed point, or at a group of points not far removed from each
other on the surface of the Earth, is, to a certain degree, the
fundamental element of the relations which determine the
climate, and the appropriate cultivation of a district; but
the mean temperature of the whole surface of the Earth is
very different from that of the Earth itself.   The often
repeated questions, whether the superficial temperature has
undergone any considerable change in the course of centuries,—whether the climate of a country has deteriorated,—
whether the winter may not have become milder, and the
summer at the same time cooler,—are all inquiries which
can only be decided by means of the thermometer, an instrument only invented about two centuries and a half ago,
and of which the intelligent scientific employment scarcely
dates back to 120 years.   The nature and the novelty of the
means, therefore, restrict within very narrow limits our
inquiries concerning the temperature of the air ; but it is
quite otherwise with the solution of the larger problem
regarding the internal temperature of the whole globe.   As
from the unaltered time of vibration of a pendulum we are
able to conclude that the equality of its temperature has
been maintained, so the unchanged velocity of the Earth's
rotation furnishes a measure of the stability of its mean
temperature.   This insight into the relation between the
length of the day and the *heat of the globe,* leads to a

most brilliant application of the long knowledge we possess of the movements of the heavens to the thermic condition of our planet. The velocity of the Earth's rotation depends on her volume; and since, therefore, by the gradual cooling of the mass from the effects of radiation, the axis of rotation would become shorter, such decrease of temperature would be accompanied by increased velocity of rotation, and diminished length of day. Now the comparison of the secular inequalities in the Moon's motion with eclipses observed by the ancients, shews that since the time of Hipparchus, or during an interval of two thousand years, the length of the day has certainly not been diminished by one-hundredth part of a second : we know, therefore, that the mean temperature of the Earth has not altered, during that period, so much as the $\frac{1}{170}$ part of a Centigrade degree, or the $\frac{1}{300}$ of a degree of Fahrenheit ([140]).

This invariability of form presupposes also great invariability in the distribution of density in the interior of the globe. Such transference of matter as is effected by the action of our present volcanoes, the eruption of ferruginous lavas, and the filling up of previously empty fissures and cavities with dense mineral masses, are therefore to be regarded merely as inconsiderable superficial phænomena, wholly insignificant when considered in relation to the dimensions of the Earth.

I have described the internal heat of our planet, both in respect to its cause and distribution, almost exclusively from the results of Fourier's admirable investigations. Poisson doubted the uninterrupted increase of the Earth's temperature from the surface to the center ; he believed that its heat had penetrated from without, and that the temperature of the globe was dependent on the high or low temperature of the part of space through which the solar system has moved. This hypothesis, imagined by one of the profoundest mathematicians of our time, has been satisfactory to few, if indeed to any one except himself, and has certainly not been received by physicists and geologists.

But whatever may be the cause of the internal heat of our planet, and its limited or unlimited increase at increasing depths, it conducts us, in this general contemplation of nature, through the intimate relation of all the primary phæ-nomena of matter, and through the common bond which unites the molecular forces, into the obscure domain of Magnetism. Changes of temperature elicit magnetic and electric currents. Terrestrial magnetism, of which, in its threefold manifestation, incessant periodical variation is the leading characteristic, is ascribed either to inequalities in the temperature of the globe ([141]), or to those galvanic currents which we regard as electricity moving in a circuit ([142]). The mysterious march of the magnetic needle is dependent both on time and space,—on the course of the Sun, and on its own change of place on the surface of the Earth. The hour of the day may be known between the tropics by the direction of the needle, as well as by the height of the mercury in the barometer. It is affected instantly, though only transitorily, by the distant Aurora—by the rich streams of coloured light which shoot in bright flashes across the polar sky. When the ordinary horary movement of the needle is interrupted by a magnetic storm, the perturbation manifests itself, often simultaneously in the strictest sense of the word, over land and sea, over hundreds and thousands of miles ; or propagates itself gradually, in short intervals of time, in every direction over the surface of the Earth ([143]). In the first case, the simultaneity of the phænomena may serve, like occultations of Jupiter's satellites, or like fire signals and shooting stars, to determine within certain limits geographical differences of longitude. We recognise with wonder and admiration, that the movements of two small magnetic needles, even if suspended at depths beneath the surface of the Earth, should measure the distance which divides them from each other; that they should tell us how far Kasan is situated east of Göttingen, or of the banks of the Seine. There are parts of the Earth where the mariner, who has been enveloped for many days in fog, seeing neither Sun nor stars, and having no means of determining time, may know with certainty, by an observation of the magnetic Inclination, whether he is to the north or south of the port which he desires to enter ([144]).

When the sudden interruption or disturbance of the horary movement of the needle announces the presence of a magnetic storm, we are unhappily still unable to determine the seat of the perturbing cause, whether it be in the crust of the earth, or in the upper regions of the atmosphere. If we regard the earth as an actual magnet, we know from the profound investigator of a general theory of terrestrial magnetism, Friedrich Gauss, that to each portion of the globe one-eighth of a cubic metre in volume, we must assign an average amount of magnetism equal to that contained in a magnetic bar of 1 lb. weight ([145]). If iron and nickel, and probably cobalt (but not chrome ([146]), as was long believed), are the only substances which become permanently magnetic, and by a certain coercive force retain polarity, the phænomena of Arago's magnetism of rotation, and of Faraday's induced currents, on the other hand, shew the probability that all terrestrial substances are capable of assuming transitory magnetic relations. According to the rotation experiments of the former of these two great physicists, water, ice (147), glass, carbon, and mercury, affect the vibrations of a needle. Almost all substances shew themselves in a certain degree magnetic when they are acting as conductors ; that is to say, when a current of electricity is passing through them.

Although a knowledge of the attracting power of the loadstone, or of naturally-magnetic iron, appears to have existed from time immemorial among the nations of the West, yet it is a well-established and very remarkable historical fact, that the knowledge of the directive power of a magnetic needle, resulting from its relation to the magnetism of the Earth, was possessed exclusively by a people occupying the eastern extremity of Asia, the Chinese. More than a thousand years before our era, at the obscurely known epoch of Codrus and the return of the Heraclides to the Peloponnesus, the Chinese already employed magnetic cars, on which the figure of a man, whose moveable outstretched arm pointed always to the south, guided them on their way across the vast grassy plains of Tartary ; and in the third century of our era, at least 700 years before the introduction of the compass in the European seas, Chinese vessels navigated the Indian Ocean (148) with needles pointing to the south. I have shewn in another work (149) what great advantages in topographical knowledge the magnetic needle gave to the Chinese geographers over their Greek and Roman contemporaries, to whom, for example, the true direction of the mountain Chains of the Apennines and Pyrenees always remained unknown.

The magnetic force of our planet is manifested at its surface by three classes of phænomena; one of these is the varying *intensity* of the force, and the other two its varying direction, shewn in the *inclination* of the magnetic needle in the vertical plane, and in its *declination* from the geographical meridian. The aggregate effect may therefore be represented graphically by three systems of lines, called isodynamic, isoclinal, and isogonic; or, of equal force, equal dip or inclination, and equal variation or declination. The distances apart, and the relative as well as absolute positions of these lines, are undergoing continual change. At particular points on the Earth's surface ([150]), for example in the western part of the Antilles, and in Spitzbergen, the mean declination of the magnetic needle has scarcely undergone any sensible change in the course of the last hundred years. Elsewhere, when the isogonic curves, in their secular movement, pass from the surface of the sea to that of a continent or island of considerable extent, they appear to be retained for a time, and the curves become thereby inflected. The gradual change in the form of the lines which accompanies their translation, and modifies the extent of the spaces which are occupied by east or west declination, makes it difficult to recognise the nature of the changes and the analogies of form in graphic representations belonging to different centuries. Each branch of a curve has its history; but in no case does that history reach farther back, among the nations of the West, than to that memorable epoch (13th Sept., 1492), when the re-discoverer of the new world found the line of no variation three degrees to the westward of the meridian of the Island of Flores, one of the group of the

Azores ([151]). At the present time the whole of Europe, with the exception of a small part of Russia, has west declination. This only began to be the case late in the 17th century, the needle having first pointed due north in London in 1657, and in Paris in 1669, an interval of twelve years, notwithstanding the small distance which divides these two capitals. In eastern Russia, to the east of the mouth of the Volga, of Saratow, Nishni-Nowgorod, and Archangel, the easterly declination of Asia is advancing towards us. In the wide extent of Northern Asia, two excellent observers, Hansteen and Adolph Erman, have traced the extraordinary double curvature of the declination lines, which are concave towards the pole between Obdorsk on the Obi and Turuchansk, and convex between Lake Baikal and the Gulf of Ochotsk. In this last-named portion of the earth, in North-eastern Asia, between the Werchoiansk Mountains, Iakoutsk, and Northern Corea, the isogonic lines form a remarkable closed system. This oval form ([152]) is repeated with still greater regularity, and on a larger scale, in the Pacific Ocean, nearly in the meridian of Pitcairn Island and the group of the Marquesas, between 20° N. and 45° S. latitude. We might be inclined to regard so singular a configuration as the effect of local peculiarity in those parts of the earth ; but if these apparently isolated systems are found to change their place progressively in the course of centuries, we must conclude that these phænomena, like all great natural facts, appertain to a general system, and have a general cause.

The horary variations of the declination are governed apparently by the sun whilst that body is above the horizon at any spot; they also decrease in angular value with the decrease of magnetic latitude: near the equator, in the Island of Rawak; for example, they barely amount to three or four minutes, while in middle Europe they attain to thirteen or fourteen minutes. Throughout the northern hemisphere the mean movement of the north end of the needle from 8½ A.M. to 1½ P.M. is from east to west; and, as at the same hours in the southern hemisphere, the same end of the needle moves in the opposite direction, or from west to east, attention has been justly called ([153]) to the circumstance, that there should be a region of the earth, probably between the terrestrial and magnetic equators, in which no horary variation of the declination is sensible. This fourth curve, which might be called the curve of no motion, or *line of no horary variation of the declination*, has not yet been found.

The name of *magnetic poles* has been applied to those points on the Earth's surface where the horizontal force disappears, and to these points more importance has been attached than properly belongs to them ([154]); in like manner the curve on which the needle has no inclination, but rests in a horizontal direction, has been called the *magnetic equator*. The position of this line, and its secular change of form, have of late years been objects of careful investigation. According to the excellent memoir of Duperrey ([155]), who crossed the magnetic equator six times between 1822 and 1825, the nodes or intersection of the two equators, or the two points at which the line without inclination crosses the geographical equator and passes from one hemisphere into the other, are unequally distributed; in 1825, the node near the Island of St. Thomas on the West Coast of Africa, was 188½° from the node in the Pacific, which is near the small islands called Gilbert Islands (nearly in the meridian of the Viti Islands). At the commencement of the present century, I determined the point where the magnetic equator crosses the chain of the Andes, in the interior of the new continent, between Quito and Lima, at an elevation of nearly 12000 English feet above the level of the sea, in 7° 1' S. lat., and 48° 40' W. long. from Paris. To the west of this point, throughout almost the whole breadth of the Pacific, the line without dip, or magnetic equator, though slowly approaching the geographical equator, continues in the southern hemisphere; in the vicinity of the Indian archipelago it passes into the northern hemisphere, just touches the southern point of Asia, and enters the continent of Africa near the strait of Bab-el-Mandeb, which is the point of its greatest distance from the geographic equator. Thence, traversing the terra incognita of the interior of Africa in a south-westerly direction, it re-enters the southern hemisphere in the Gulf of Guinea, and, maintaining a south-westerly course across the Atlantic, reaches the Brazilian coast near Os Ilheos, north of Porto Seguro, in 15° S. lat. From thence to the elevated plateau of the Cordilleras, where I observed the inclination between the silver mines of Micuipampa and the ancient seat of the Incas at Caxamarca, the line traverses a part of South America as unknown to us as the interior of Africa.

Recent observations, collected and discussed by Sabine([156]), have taught us that, in the interval between 1825 and 1837, the node near the Island of St. Thomas moved 4° from the east towards the west. It would be extremely important to learn whether the opposite node near the Gilbert Islands in the Pacific, has undergone a corresponding westerly movement, and advanced an equal amount towards the meridian of the Carolinas. In investigating the laws of terrestrial magnetism, it is no slight advantage that four-fifths of the magnetic equator are oceanic, and are thus easily accessible, and that we now possess the means of determining the declination and inclination on board ship with great exactness. The changes which alter the places of the nodes, and modify the form of the magnetic equator, are felt in the remotest regions of the earth, where also they produce changes of the inclination and of the magnetic latitude ([157]).

We have spoken of the distribution of magnetism on the surface of our planet according to the two forms of declination and inclination: it still remains to notice the third form, that of the intensity of the force, which is expressed graphically by isodynamic curves, or curves of equal intensity. The investigation and measurement of this force by means of the oscillations of a vertical or horizontal needle, has excited a general and lively interest, which, in its application to the distribution of the magnetic force on the surface of the globe, commenced with the present century. By the application of refined optical and chronometrical means, the measurement of the horizontal force, in particular, has become susceptible of a degree of accuracy exceeding that of all other magnetic determinations. Doubtless the isogonic lines are of the greatest practical importance, from their use in navigation; but in respect to the theory of terrestrial magnetism, the isodynamic lines are those from which the most fruitful results are expected([158]). The first fact in reference to these lines which direct observation made known was the increase of the intensity of the total force in proceeding from the equator towards the pole ([159]).

It is to the unwearied activity of Edward Sabine, from the

year 1819 to the present time, that we are principally in-debted for the knowledge of the variations of the magnetic intensity over the whole surface of the globe, and for their laws so far as we are yet able to infer them. After having himself vibrated the same needles in the vicinity of the North American pole, in Greenland, in Spitzbergen, on the coast of Guinea, and in Brazil, he has been constantly engaged in collecting and co-ordinating all the materials capable of elucidating the great question of the isodynamic lines. The first sketch of an isodynamic system divided into zones was given by myself, for a small portion of South America. The isodynamic lines are not parallel with the isoclinal lines; the intensity is not, as was first supposed, weakest at the magnetic equator, nor is it even equal at all parts of that line. If we compare Erman's observation (0.706), in the southern part of the Atlantic ocean, where a zone of weak intensity extends from Angola past the Island of St. Helena to the Coast of Brazil, with the most recent observations of the great navigator James Clark Ross, we find that on the surface of our planet the force augments almost in a ratio of 1 : 3. The highest intensity which has been measured is 2.071, in lat. 60° 19′ and long. 131° 20′ E. from Greenwich ([160]). These values are expressed in terms of the scale of which the unity is the intensity which I observed on the magnetic equator, in the north of Peru. In Melville Island (74° 27′ N.), in the neighbourhood of the northern magnetic pole, the force was found by Sabine only 1.624, while at New York in the United States, almost in the same latitude as Naples, he found it 1.803.

The brilliant discoveries of Oersted, Arago, and Faraday, have established intimate relations between the electric tension of the atmosphere and the magnetic charge of the Earth. According to Oersted, a conductor is rendered magnetic by the electrical current which passes along it. According to Faraday, magnetism gives rise, by induction, to electrical currents. Thus magnetism is one of the manifold forms under which electricity shews itself; and the ancient obscure presentiment of the identity of electric and magnetic attraction has been realised in our own days. Pliny ([161]), in accordance with Thales and the Ionic school, says, "The electrum (amber), when animated by friction and warmth, attracts fragments of bark and dry leaves, just as the magnetic stone does iron." We find a remark to the same effect in a speech of the Chinese philosopher Kuopho in praise of the virtues of the magnet ([162]). It was not without surprise that I noticed, on the shores of the Orinoco, children, belonging to tribes in the lowest stage of barbarism, amusing themselves by rubbing the dry, flat, shining seeds of a leguminous climbing plant (probably a Negretia), for the purpose of causing them to attract fibres of cotton or bamboo. It was a sight well fitted to leave on the mind of a thoughtful spectator a deep and serious impression. How wide is the interval which separates the simple knowledge of the excitement of electricity by friction, shewn in the sports of these naked copper-coloured children of the forest, from the invention of a metallic conductor, which draws the swift lightning from the storm-cloud,—of the voltaic pile, capable of effecting chemical decomposition,—of a magnetic apparatus evolving light,—and of the magnetic telegraph! Such intervals of separation are equivalent to thousands of years in the progress and intellectual development of the human race.

The perpetual fluctuation observed in all the magnetic phenomena, in the inclination, declination, and intensity of the force, according to the hours of the day and even of the night, the season of the year, and the lapse of years, leads to the belief in the existence of very various and complicated systems of electric currents in the crust of the Earth([163]). Are these, as in Seebeck's experiments, simple thermo-magnetic currents, the immediate effect of unequal distribution of heat, or currents induced by the calorific action of the Sun? Has the rotation of the Earth, and the velocity of its different zones according to their distance from the equator, any influence on the distribution of magnetism? Is the source of magnetic action to be sought in the atmosphere, or in the interplanetary spaces, or in a polarity of the Sun and Moon? Galileo, in his celebrated "Dialogo," ascribes the constant parallel direction of the Earth's axis to a center of magnetic attraction existing in space.

If we conceive the interior of the Earth to be molten, subject to enormous pressure, and raised to a temperature for which we possess no measure, we must renounce the idea of a magnetic nucleus. Though at a white heat all magnetism disappears ([164]), it is still sensible in iron heated to a dark-red glow; and whatever may be the modifications which, in these experiments, the molecular condition and consequently the coercitive force undergo, there must still remain a considerable thickness of terrestrial strata, in which we might seek the seat of the magnetic currents. In the old explanation of the horary variations of the declination, by the progressive warming of the Earth by the Sun in his apparent course from east to west, the action would indeed be limited to the extreme exterior surface; for thermometers sunk in the Earth, which are now accurately observed at so many places, shew how slowly the heat of the Sun penetrates even to the small depth of a few feet. Moreover, the thermic condition of the surface of the sea, which covers two-thirds of the planet, is but little favourable to an explanation assuming immediate influence, and not induced action exercised by the gaseous and aqueous strata of the atmosphere.

In the present state of our knowledge, no satisfactory reply can be given to questions respecting the ultimate physical causes of phænomena so complex. On the other hand, that part of the subject which, in the threefold manifestations of the Earth's magnetic force, presents relations admitting of measurement in regard to space and time,—and which leads us to discern, amidst constant and apparent irregular change, the regularity and dominion of laws,—has recently made the most brilliant progress in the determination of mean numerical values. From Toronto, in Canada, to the Cape of Good Hope and Van Diemen's Land, and from Paris to Pekin since 1828, the globe has been covered by *magnetic observatories,* in which every movement or manifestation, regular and irregular, of the Earth's magnetic force, is watched by uninterrupted and simultaneous observation. A variation of $\frac{1}{40000}$ of the magnetic intensity is measured. At certain epochs observations are taken at intervals of two minutes and a half, and are continued during twenty-four consecutive hours. A great English astronomer and physical philosopher, has computed[165] that the mass of observations to be discussed amount in three years to 1958000. Never before has an effort so grand, and so worthy of admiration, been made to investigate the *quantitative* in the laws of one of the great phænomena of nature. We may therefore justly hope, that these laws, when compared with those which prevail in the atmosphere and in still more distant spaces, will gradually conduct us nearer to the genetical explanation of the magnetic forces. As yet we can only boast of having opened a greater number of paths which may possibly lead to such explanation. In the physical theory of terrestrial magnetism (which must not be confounded with its purely mathematical theory), as in that of the meteorological processes of the atmosphere, a premature satisfaction can only be obtained by those who permit themselves to set aside as erroneous, phænomena inconsistent with their own views ([166]).

Telluric magnetism, and the electro-dynamic forces measured by the ingenious Ampère ([167]), are intimately connected both with the terrestrial or polar light (Aurora), and with the external and internal temperature of our planet, whose magnetic poles have been regarded by some philosophers as poles of cold ([168]). That which, 128 years ago, Halley ([169]) put forward as a bold conjecture, viz. that the Aurora is a magnetic phænomenon, has, by Faraday's brilliant discovery of the evolution of light by the action of magnetic forces, been raised from a mere conjecture to an experimental certainty. There are precursors of the Aurora; the luminous nocturnal appearance is usually foretold by antecedent irregularity in the diurnal march of the magnetic needle, indicating a disturbance in the equilibrium of the distribution of the Earth's magnetism. When the disturbance has reached a great degree of intensity, the equilibrium is restored by a discharge accompanied by an evolution of

light. The Aurora ([170]) is not, therefore, to be itself regarded as a cause of the perturbation, but as the result of
a state of telluric activity excited to the production of a
luminous phænomenon ; an activity which manifests itself,
on the one hand, by the fluctuations of the needle, and, on
the other, by the appearance of the brilliant auroral light.
The magnificent phænomenon of coloured polar light is the
act of discharge, the termination of a magnetic storm,—as
in the electric storm, an evolution of light (lightning) indicates the restoration of the equilibrium in the distribution
of the electricity. The electrical storm is usually confined
to a small space, beyond which the state of electricity in the
atmosphere remains unchanged. The magnetic storm, on
the other hand, manifests its influence on the march of the
needle, over large portions of continents, and far from the
place where the evolution of light is visible, as was first re-
discovered in our own age by Arago. It is not improbable
that, as clouds of threatening appearance and heavily
charged with electricity, do not always proceed to the point
of discharge by lightning, owing to frequent transitions in
the electrical state of the atmosphere, so magnetic storms
may produce great disturbances in the ordinary diurnal
march of the magnetic needle over a wide range, without
its necessarily following that the equilibrium of distribution
must be restored by explosion, or by luminous effusions from
the pole to the equator, or from pole to pole.

If we desire to collect into one view all the features of the
phænomenon, we may describe the commencement and
successive phases of a complete appearance of the Aurora as
follows :—Low down on the horizon, about the part where
it is intersected by the magnetic meridian, the sky, which

was previously clear, is darkened by an appearance resembling a dense bank or haze, which gradually rises and attains a height of eight or ten degrees. The colour of the dark segment passes into brown or violet, and stars are visible through it as in a part of the sky obscured by thick smoke. A broad luminous arch, first white, then yellow, bounds the dark segment; but as the bright arch does not appear until after the segment, Argelander considers that the latter cannot be attributed to the mere effect of contrast with its bright margin ([171]). The azimuth of the highest point of the luminous arch, when carefully measured ([172]), has been usually found not quite in the magnetic meridian, but from five to eighteen degrees from it, on the side towards which the magnetic declination of the place is directed. In high northern latitudes in the near vicinity of the magnetic pole, the dark segment appears less dark, and sometimes is not seen at all; and in the same localities, where the horizontal magnetic force is weakest, the middle of the luminous arch deviates most widely from the magnetic meridian. The luminous arch undergoes frequent fluctuations of form; it remains sometimes for hours before rays and streamers are seen to shoot from it and rise to the zenith. The more intense the discharges of the Aurora, the more vivid is the play of colours, from violet and bluish-white through all gradations to green and crimson. In the common electricity excited by friction, it is also found that the spark becomes coloured only when a violent explosion follows high tension. At one moment the magnetic streamers rise singly, and are even interspersed with dark rays, resembling dense smoke; at another they shoot upwards simultaneously from many and opposite points of the hori-

zon, and unite in a quivering sea of flame, the splendour of
which no description can reach, for every instant its bright
waves assume new forms.   The intensity of this light is
sometimes so great, that Lowenorn (29th January, 1786)
discerned its corruscations during bright sunshine: motion
increases the visibility of the phænomenon.   The rays
finally cluster round the point in the sky corresponding to
the direction of the dipping needle, and there form what is
called the corona—a canopy of light of milder radiance,
streaming, but no longer undulating.   It is only in rare
cases that the phænomenon proceeds so far as the complete
formation of the corona; but whenever this takes place, the
display is terminated.   The streamers now become fewer,
shorter, and less intensely coloured; the corona and the
luminous arches break up, and soon nothing is seen but
irregularly scattered, broad, pale, shining patches of an
ashy-grey colour; and even these vanish before the trace
of the original dark segment has disappeared from the
horizon.   The last trace that remains of the whole spectacle
is often merely a white delicate cloud, feathered at the edges,
or broken up into small round masses, like cirro cumuli.

   This connection of the polar light with the most delicate
cirrous clouds deserves particular attention, because it
shews us the electro-magnetic evolution of light as part of a
meteorological process.   The magnetism of the Earth is
here exhibited in its influence on the atmosphere, and on the
condensation of aqueous vapour.   The observation of
Thienemann, in Iceland, who regarded the light detached
fleecy clouds as the substratum of the Aurora, has been
confirmed in modern times by Franklin and Richardson,
in the neighbourhood of the American magnetic pole, and

by Wrangel on the Siberian coast of the polar sea. They all remark, that the Aurora shoots forth the most vivid rays when masses of cirro-strati are hovering in the upper region of the atmosphere, and when they are so thin that their presence can only be discovered by the formation of a halo round the Moon. These clouds sometimes arrange themselves, in the day-time, like the rays of the Aurora; and in such cases the movements of the needle are similarly affected by them. After a great nocturnal display of Aurora, there have been recognised early in the morning the same streaks of cloud which had before been luminous ([173]). The apparently converging "polar bands" (streaks of cloud in the direction of the magnetic meridian), which constantly engaged my attention during my journeys both on the high table lands of Mexico and in Northern Asia, belong probably to the same group of diurnal phænomena ([174]).

Southern lights have been repeatedly seen in England by Dalton, and northern lights have been seen in the southern hemisphere as far as 45° S. latitude (14th January, 1831); it not unfrequently happens, also, that the magnetic equilibrium is simultaneously disturbed in the direction of both poles. I have distinctly ascertained that the polar light has been seen within the tropics, in Mexico and Peru. It is necessary to distinguish between the sphere of simultaneous visibility of the phænomena, and the zones of the Earth in which it is seen almost nightly. Every observer certainly sees his own Aurora as well as his own rainbow; but the phænomenon of the effusion of light is generated by a large portion of the Earth at once. Many nights may be cited when it was observed simultaneously in England

and in Pennsylvania, at Rome and at Pekin. When it is
stated that Auroras decrease in frequency and brilliancy
with decreasing latitude, it must be understood of magnetic
latitude. Whilst in Italy an Aurora is a very rare occurrence, it is extremely common in the same latitude in
Philadelphia (39° 57′), owing to the vicinity of the American magnetic pole ; and in Iceland, Greenland, Newfoundland, on the shores of the Slave Lake, and at Fort
Enterprise, the " merry dancers," ([175]) as the inhabitants
of the Shetland Islands call the quivering and variously-
coloured rays of the Aurora, are seen during certain seasons
of the year almost every night.   But even in those parts
of the new continent and of Siberia which are distinguished by the frequency of the phænomenon, there may be
said to be particular districts, or zones of longitude, in
which it shews itself with peculiar splendour ([176]). Wrangel
saw its brilliancy diminish at Nishni Kolymsk, as he receded
from the coast of the Polar Sea; in this and in similar
instances local influences are not to be denied.   The experience of the various North Polar expeditions seems to shew,
that in the immediate vicinity of the magnetic pole the
evolutions of light are, to say the least, not more intense or
frequent than at somewhat greater distances.

What we know of the height of the Aurora is grounded
on measurements which, from their nature, and the incessant fluctuation of the phænomenon, and from the consequent uncertainty of the parallactic angle, cannot inspire
much confidence.   Without including older statements, the
results of these measurements give heights varying from a few
thousand feet to several miles ([177]).   The most modern observers are inclined to place the seat of the phænomenon, not

at the limits of the atmosphere, but in the region of clouds :
they even believe that the rays of the Aurora may be moved to
and fro by winds and currents of air ; and this may be the case,
if the luminous phænomenon which manifests to us the presence of an electro-magnetic current, be actually connected
with groups of vesicles of vapour in motion; or, to speak
more exactly, if it traverses the group, darting from one
vesicle to another.   Franklin saw an Aurora near Great
Bear Lake, the light of which appeared to him to illuminate
the under surface of the stratum of cloud ; while, at the distance of only eighteen miles, Kendal, who was on watch all
night, and never lost sight of the sky, observed no luminous
phænomenon whatsoever.   In respect to the statements recently made from several quarters, of rays of the Aurora being
seen to shoot down in close proximity to the Earth, between
the observer and a neighbouring hill, it must be remembered
that, as in the case of lightning and of fire balls, there is in
several ways danger of optical illusion.

Whether the magnetic storms manifested by Auroral
display (of which we have just noticed one instance remarkably restricted in respect to locality) share with electric
storms the phænomena of sound as well as of light, has become extremely doubtful since the accounts of Greenland
whalers and Siberian fox-hunters have ceased to obtain im-
plicit confidence.   The Auroras have become more silent since
observers have better understood how to observe them, and
how to listen for them.   Parry, Franklin, and Richardson,
near the north magnetic pole, Thienemann in Iceland,
Giesecke in Greenland, Lottin and Bravais near the North
Cape, Wrangel and Anjou on the Siberian coasts of the
Polar Sea, have together seen thousands of northern lights

without ever hearing a noise. Even if it be considered that this negative evidence ought not to countervail the positive testimony of two observers, Hearne at the mouth of the Coppermine River, and Henderson in Iceland, it must be remembered that Richardson and Hood heard, indeed, a sound, which one terms "a hissing noise, like that of a musket bullet passing through the air," and of which the other says, "that it resembled the noise of a wand waved smartly through the air;" but though both were inclined to regard these sounds as connected with the Aurora, each adds that they were attributed "by Dr. Wentzel to the contracting of the snow from a sudden increase of cold;" and this opinion was supported "by the same sounds being heard the following morning." (Pages 585 and 628 of the Appendix to "Franklin's First Journey to the Polar Sea.") Wrangel and Giesecke arrived at the same persuasion, that sounds heard during Auroras do not proceed from them, but are to be ascribed to contractions of the ice, and of the crust on the surface of the snow, from sudden increase of cold. The belief of a crackling noise did not originate with uncultivated persons having frequent opportunities of noticing the Aurora, but with learned travellers; the cause probably being, that as electric flashes in spaces filled only with a very rare atmosphere had been observed to resemble the northern light, the latter phænomenon was regarded as an effect of atmospheric electricity, and thus people heard what they expected they ought to hear. Recent experiments, however, with regard to atmospheric electricity, made with very sensitive electrometers, have hitherto, contrary to all expectation, given only negative results, since, during the finest Auroras, no change has been detected.

On the other hand, all the three manifestations of terrestrial magnetism, the declination, inclination, and force, are affected during the appearance of the polar light; so that in the course of the same night, and during different parts of the phænomenon, the same end of the needle is sometimes attracted, and sometimes repelled. The statement that the facts collected by Parry in Melville Island, in a high magnetic latitude, indicated rather a tranquillizing than a disturbing influence of Auroras upon the magnet, has been refuted by a more careful examination of Parry's own journal ([178]), by the valuable observations of Richardson, Hood, and Franklin, and latterly by Bravais and Lottin in Lapland. As I have before remarked, the luminous phænomenon is the act of restoration of equilibrium temporarily disturbed ; the effect on the needle varies with the intensity of the discharge ; at the winter station of Bosekop it was always sensible, except when the luminous phænomenon was very faint, and appeared only low down near the horizon. The Auroral streamers have been ingeniously compared to the light which, in the Voltaic circuit, is produced between two points of carbon placed at a considerable distance from each other, (or, according to Fizeau, between a point of carbon and one of silver) ; a light which is attracted or repelled by the magnet. This analogy renders superfluous the assumption of metallic vapours in the atmosphere, which some celebrated physicists have considered to be the substratum of the Aurora.

In applying to the luminous phænomenon which we ascribe to a galvanic current the vague term of polar light, or Aurora borealis and australis, we merely indicate thereby

the direction in which the evolution of light most frequently,
but by no means always, commences. The fact which gives to
the phænomenon its greatest importance is, that the Earth
becomes self-luminous; that besides the light which, as a
planet, it receives from the central body, it shews a
capability of sustaining a luminous process proper to itself.
The intensity of the "terrestrial light," or rather of the
degree of illumination which it diffuses at the surface of
the Earth, is, when the rays are brightest, are coloured,
and ascend to the zenith, a little greater than that given
by the Moon in her first quarter. Sometimes (as on the
7th of January, 1831) it has been possible to read print by
it without effort. This terrestrial luminous process going
on almost uninterruptedly in the polar regions, leads us by
analogy to the remarkable phenomenon presented by Venus,
when the portion of that planet not illumined by the Sun
is seen to shine with a phosphorescent light of its own. It
is not improbable that the Moon, Jupiter, and the comets,
radiate a light generated by themselves, in addition to the
reflected light which they receive from the Sun, and which
is recognised by means of the polariscope. Without speaking of the enigmatical but not uncommon kind of lightning,
which, unaccompanied by thunder, is seen flickering
throughout the whole of a low cloud for minutes together,
we have yet other examples of the production of terrestrial
light. To these belong the celebrated mists, luminous at
night, seen in the years 1783 and 1831; the steady luminous
appearance in great clouds observed by Rozier and Beccaria;
and even, as Arago ingeniously remarks, the faint diffused
light which guides our steps in densely clouded moonless
and starless autumn or winter nights, and when no snow is

on the ground ([179]). In high latitudes a flood of brilliant
and often coloured light streams through the atmosphere in
polar light or electro-magnetic storm ; and in the torrid zone
many thousands of square miles of ocean are also seen to
generate at once a light of their own ; in the latter case
the magic brightness belongs to organic nature : each
breaking wave curls in luminous foam ; the whole wide
expanse sparkles, and every spark is the vital movement of
a minute and otherwise invisible world of animal existence.
Manifold, no doubt, are the sources of the terrestrial light ;
and we may even imagine it as existing latent, and not yet
set free from combination with vapours, as a means of
explaining Moser's *pictures produced at a distance,*—
a discovery in which reality as yet presents itself to us
like the unsubstantial images of a dream.

The internal heat of our planet, which appears to be connected with the excitement of electro-magnetic currents,
and the evolution of terrestrial light accompanying a magnetic storm, is also a principal source of geological phænomena. We shall trace this connection in passing from the
purely dynamic effect manifested in earthquakes, and the
elevation of entire continents and mountain masses, to the
production and issue of gases and liquids, of hot mud, and
of glowing and molten earths which harden into crystalline
rocks. It is no small advance made by modern geology (or
the mineralogical part of terrestrial physics), to have investigated the connection of the phænomena here indicated. The
insight which we thus obtain leads away from the unpro-
fitable hypotheses by which it was formerly sought to explain
each such manifestation of force singly and independently ;

and shews the relations which subsist between the ejection
of various substances on the one hand, and earthquakes
and elevations on the other; it classes together groups of
phænomena which appear at the first glance very heterogeneous : thermal springs, exhalations of carbonic acid gas
and sulphureous vapour, harmless eruptions of mud, and
the devastating phænomena of active volcanoes.   In a general view of nature, all these phænomena are comprehended
under the one idea of the action of the interior of a planet
upon its crust and surface.   Thus in the temperature
increasing in the interior of the Earth with the distance
from the surface, we recognise the germ not only of earthquakes, but of the gradual elevation of continents, of chains
of mountains from extended fissures, of volcanic eruptions,
and of the production of very various minerals and rocks.
But it is not inorganic nature only which is influenced by
the reaction of the interior on the exterior.   It is very probable that in an earlier state of the globe, far greater emissions of carbonic acid gas mingled with the atmosphere, and
heightened the process by which plants assimilate carbon;
and thus vast forests were formed, which in subsequent
revolutions were destroyed, and inexhaustible stores of fuel
(lignites and coal) were buried in the terrestrial strata then
forming at the surface.   Nor should we overlook that the
destinies of men are in part dependent also on the form of
the outer crust of the Earth, on the direction and elevation
of mountain chains, and on the divisions and articulations
of upheaved continents.   The investigating spirit is thus
enabled to ascend from link to link in the chain of phænomena, to the supposed epoch of the solidification of the
planet, when, in its first transition from a gaseous to a

liquid or a solid form, the internal heat not due to the calorific action of the solar rays was developed.

In order to give a brief general view of the causal connection of geological phænomena, we will begin with those whose principal character is dynamic. Earthquakes are distinguished by rapidly succeeding vertical, horizontal, or circular oscillations. In the not inconsiderable number of these phænomena which I have witnessed both on the old and new continents, at sea and on land, the two first kinds of movement, the vertical and horizontal, have often appeared to me to take place together. The mine-like explosion, the vertical action from below upwards, shewed itself in the most striking manner at the overthrow of the town of Riobamba, in 1797, where many corpses of the inhabitants who perished, were hurled to a height of several hundred feet on the hill of La Cullca, beyond the small river of Lican. The shock is propagated chiefly in a linear direction, by undulations having a velocity from twenty to twenty-eight geographical miles in a minute, and occasionally in circles or ellipses of commotion, in which the shocks are propagated from the centre to the circumference, but with diminishing force. There are districts which belong to two intersecting circles. In Northern Asia, where the Father of History, Herodotus ([180]), and at a later epoch Theophylactus Simocatta ([181]), spoke of Scythia as free from earthquakes, I have found the southern and richly metalliferous part of the Altai mountains subject to the double influences of the foci of commotion of Lake Baikal and of the volcanoes of the Thian-schan (" celestial mountains") ([182]). Where the circles of disturbance intersect,—where, for example, an elevated plateau is situated between two volcanoes in a state of activity,—several systems

of waves may exist simultaneously, and produce their
effects, as in fluids, without mutual disturbance. We may
even imagine *interferences*, as in intersecting waves of
sound. The magnitude of the waves propagated in the
crust of the Earth will be increased at the surface, according to the general law in mechanics by which vibrations
transmitted in elastic bodies have a tendency to detach the
superficial strata.

The undulations in earthquakes have been examined with
tolerable accuracy, in respect to their direction and intensity,
by means of pendulums and sismometers ; but in their
characters of alternation and periodical intumescence they
have by no means attracted sufficient attention. In the
city of Quito, which is situated at the foot of a still
active volcano, the Rucu-Pichincha, and at an elevation
above the sea of 8950 (9539 English) feet, and which
possesses fine cupolas, high roofed churches, and massive
houses of several stories in height, I have been often
surprised in the night by the violence of the earthquake
shocks; but these, though extremely frequent, very rarely
injure the walls, whereas, in the Peruvian plains, even low
dwellings built of reeds suffer from apparently far slighter
oscillations. Natives of those countries, who have experienced many hundred earthquakes, believe the difference to
be less in the greater or less duration of the shocks, or
the slowness or rapidity ([183]) of the horizontal oscillation,
than in the alternation of motion in opposite directions.
The circular (or gyratory) earthquakes are the most rare,
and at the same time the most dangerous. In the great
earthquake of Riobamba, in the province of Quito (4th
February, 1797), and in that of Calabria (5th February, and
28th March, 1783), walls were changed in direction without

being overthrown, straight and parallel rows of trees were inflected, and in fields having two sorts of cultivation, one crop even took the place before occupied by the other: the latter phænomenon shewing either a movement of translation, or a mutual penetration of the different strata. When making a plan of the ruined city of Riobamba, I was shewn a place where the whole furniture of one house had been found under the remains of another; the earth had evidently moved like a fluid in streams or currents, of which we must assume that the direction was first downward, then horizontal, and lastly again upward. Disputes concerning the ownership of objects which had been thus carried to distances of many hundred yards, were decided by the Audiencia, or Court of Justice.

In countries where earthquakes are comparatively rare, for example in the south of Europe, an imperfect induction has led to the very general belief that they are always preceded by calms, oppressive heat, and a misty horizon ([184]). This popular error is however refuted, not merely by my own experience, but also by the observations of all those who have lived many years in districts such as Cumana, Quito, Peru, and Chili, where the earth is frequently and violently shaken. I have felt shocks in serene weather as well as in rain, and during a fresh east wind as well as during a storm. Even the regularity of the horary variations of the magnetic declination, and of the pressure of the atmosphere ([185]), were not disturbed on the days of earthquakes. My observations were made within the tropics; and those made by Adolph Erman in the temperate zone, during an earthquake at Irkutsk near Lake Baikal, on the 8th of March, 1829, give a similar result. At Cumana, on the

4th of November, 1799, I found no change either in
the magnetic declination or intensity from a strong shock
of earthquake ; but, on that occasion, I observed with
astonishment that the inclination was diminished 48′ [186].
I had no reason to suspect any error, although, during
a great number of other shocks and earthquakes experienced by me in the highlands of Quito and Lima,
the inclination, as well as the other elements of terrestrial magnetism, always remained unaltered. If, however,
these deep-seated terrestrial movements are not generally
announced by any peculiar state of the atmosphere, or
appearance of the sky, it is, on the other hand, as we shall
soon see, not improbable, that in some very violent earthquakes the aerial strata have participated, and that the
phænomena are not, therefore, always purely dynamic.
During the long-continued trembling of the ground in the
Piedmontese valleys of Pelis and Clusson, great variations in
the electric tension of the atmosphere were remarked, quite
independently of any storm, and when the sky was perfectly
clear.

The hollow noise which most frequently accompanies
earthquakes by no means increases in proportion to the
violence of the oscillations. I have distinctly ascertained
that the great shock of the earthquake of Riobamba (4th
February, 1797), one of the most terrible phænomena in the
physical history of our globe, was unaccompanied by any
noise; the great subterranean detonation (el gran ruido),
which was heard at the cities of Quito and Ibarra, (but not
at Tacunga and Hambato which were nearer the center of
the movement,) occurred eighteen or twenty minutes *after*
the catastrophe. In the celebrated earthquake of Lima and
Callao, October 28th, 1746, a noise, resembling a sub-

terranean thunder-clap, was heard a quarter of an hour *later* at Truxillo, and was unaccompanied by any trembling of the ground. In like manner, it was not till some time after the great earthquake of New Granada, November 16, 1827, described by Boussingault, that subterranean detonations, unaccompanied by any movement, were heard with great regularity at intervals of thirty seconds, throughout the whole Cauca Valley. The nature of the noise also differs greatly; sometimes it is rolling, and occasionally like the clanking of chains; in the city of Quito it has sometimes been abrupt, like thunder close at hand, and sometimes clear and ringing, as if obsidian or other vitrified masses clashed, or were shattered in subterranean cavities. As solid bodies are excellent conductors of sound, which is propagated, for example, in burnt clay with a velocity ten or twelve times greater than in air, the subterranean noise may be heard at great distances from the place where it has originated. In the Caraccas, in the grassy plains of Calaboso, and on the banks of the Rio Apure which falls into the Orinoco, there was heard over a district of 2300 square (German) miles, a loud noise resembling thunder, unaccompanied by any shaking of the ground; whilst, at a distance of 632 miles to the north east, the crater of the volcano of St. Vincent, one of the small West India Islands, was pouring forth a prodigious stream of lava. In point of distance, this was as if an eruption of Vesuvius should be heard in the north of France. In 1744, at the great eruption of Cotopaxi, subterranean noises, as of cannon, were heard at Honda, near the Magdalena River. Not only is the crater of Cotopaxi about 18100 English feet higher than Honda, but these two points are separated from each other by a distance of 436 miles, and by the colossal mountain masses of Quito, Pasto, and Popayan, as well as by countless valleys and ravines. The sound was clearly not propagated through the air, but through the earth, and at a great depth. During the violent earthquake in New Granada, in February 1835, subterranean thunder was heard at Popayan, Bogota, Santa Martha, and Caraccas (where it lasted seven hours without any movement of the ground), and also in Hayti, in Jamaica, and near the Lake of Nicaragua.

These phænomena of sound, even when unaccompanied by sensible shocks, produce a peculiarly deep impression, even on those who have long dwelt on ground subject to frequent trembling. One awaits with anxiety that which is to follow the subterranean thunder. The most striking instance of uninterrupted subterranean noise, unaccompanied by any trace of earthquake, is the phænomenon which is known in the Mexican territory by the name of " the subterranean roaring and thundering, (bramidos y truenos subterraneos) of Guanaxuato" [187]. This rich and celebrated mountain city is situated at a distance from any active volcano. The noise began on the 9th of January, 1784, at midnight, and lasted above a month. I have been enabled to give a circumstantial description of the phænomenon from the report of many witnesses, and from the documents of the municipality, which I was permitted to make use of. From the 13th to the 16th of January, it was as if there were heavy storm clouds under the feet of the inhabitants, in which slow rolling thunder alternated with short thunder-claps. The noise ceased gradually, as it had commenced ; it was confined to a small space, for it was not heard in a basaltic district at the distance of only a few miles. Almost all the inhabitants were terrified and quitted the city, in which large masses of silver were stored; but the most courageous, when they had become somewhat accustomed to the subterranean thunder, returned and fought with the bands of robbers who had taken possession of the treasure. Neither at the surface, nor in mines 1598 English feet in depth, could the slightest trembling of the ground be perceived. In no part of the whole mountainous country of Mexico had any thing similar been ever known before, nor has this awful phænomenon been since repeated. Thus, as chasms in the interior of the Earth close or open, the propagation of the waves of sound is either arrested in its progress, or continued until it reaches the ear.

The activity of a burning mountain, however awfully picturesque the spectacle which it presents to us, is always limited to a very small space; whereas earthquakes, whose movements are scarcely perceptible to the eye, propagate their waves sometimes to distances of many thousand miles. The great earthquake which, on November 1st, 1755, destroyed Lisbon, and the effects of which have been well traced out by the great philosopher Kant, was felt in the Alps, on the coasts of Sweden, in the West India Islands (Antigua, Barbadoes, and Martinique), on the great lakes of Canada, in Thuringia, in the flat country of northern Germany, and in small inland lakes on the shores of the Baltic. Remote fountains were interrupted in their flow, a phænomenon of earthquakes which had even been noticed among the ancients by Demetrius of Calatia. The thermal springs at Töplitz dried up, and again returned, inundating every thing with water discoloured by ochre.   At Cadiz the sea rose above sixty feet ; and in the West India Islands above mentioned, where the tide usually rises only from twenty-six to twenty-eight French inches, it suddenly rose above twenty feet, the water being discoloured and of an inky blackness. It has been computed that, on that day (1st November, 1755), a portion of the Earth's surface, four times greater than the extent of Europe, was simultaneously shaken.   There is no manifestation of force yet known to us (including the murderous inventions of our own race), by which a greater number of human beings have been killed in the short space of a few seconds or minutes, than in the case of earth-quakes : sixty thousand were destroyed in Sicily in 1693 ; thirty to forty thousand at Riobamba, in 1797 ; and perhaps five times as many in Asia Minor and Syria under Tiberius and the elder Justinian, in the years 19 and 526.

Examples have occurred in the Andes of South America, in which the earth has been shaken uninterruptedly for several successive days ; but of tremblings felt almost every hour for months together, I am at present only aware of instances at a distance from any volcano ; as on the eastern declivity of the Mont Cenis portion of the chain of the Alps at Fenestrelles and Pignerol from April 1808 ; in the United States of America, between New Madrid and Little Prairie (north of Cincinnati) in December 1811, as well as in the whole winter of 1812 ([188]) ; and in the Pachalic of Aleppo in August and September 1822.   From the popular dispo-sition to ascribe great phænomena to local causes, rather than to rise to general views, wherever the shaking of the earth is long continued, fears of the breaking out of a new volcano are entertained.   In particular and rare cases, these fears have been realised by the sudden appearance of volcanic islands; and a remarkable instance occurred in the elevation of the volcano of Jorullo, 1682 English feet above the ancient level of its site and of the plain in which it now stands, which took place the 29th of September, 1759, after eighty days of earthquakes and subterranean thunder.

If we could obtain daily intelligence of the condition of the whole surface of the earth, we should very probably arrive at the conviction that this surface is almost always shaking at some one point; and that it is incessantly affected by the reaction of the interior against the exterior. The frequency and universality of a phænomenon which probably owes its origin to the high temperature of the interior and deep-seated molten strata, explain its independence of the nature of the rocks in which it manifests itself. Earthquake shocks have been felt even in the loose alluvial soil of Holland, Middelburg, and Flushing (23d Feb. 1828). Granite and mica slate are shaken, as well as limestone and sandstone, trachyte and amygdaloid. It is not the chemical nature of the constituent particles, but the mechanical structure of the rocks, which modifies the propagation of the shock or of the wave which occasions it. Where such a wave proceeds in a regular course along a coast, or at the foot of and parallel to the direction of a mountain chain, interruptions at certain points have sometimes been remarked, and continue for centuries; the undulation passes onward in the depths below, but it is never felt at those points of the surface. The Peruvians say of these upper strata which are never shaken, that they form a bridge [189]. As the mountain chains themselves appear to have been elevated over fissures, it may be that the walls of these

cavities favour the propagation of the undulations moving
in their own direction; sometimes, however, the waves
intersect several chains almost at right angles; an example
of which occurs in South America, where they cross both the
littoral chain of Venezuela and the Sierra Parime.   In Asia
shocks of earthquakes have been propagated from Lahore and
the foot of the Himalaya (22d Jan. 1832), across the chain of
the Hindoo Coosh, as far as Badakschan, or the upper Oxus,
and even to Bokhara ([190]).   The range of the undulations is
sometimes permanently extended, and this may be a consequence of a single earthquake of unusual violence.   Since
the destruction of Cumana on the 14th Dec. 1797, and only
since that epoch, every shock on the southern coast extends
to the mica slate rocks of the peninsula of Maniquarez,
situated opposite the chalk hills of the main land.   In the
great alluvial vallies of the Mississipi, the Arkansas, and the
Ohio, the progressive advance from south to north of the
almost uninterrupted undulations of the ground between 1811
to 1813, was very striking.   It would seem as if subterranean
obstacles were gradually overcome; and that the way being
once opened, the undulatory movement is propagated through
it on each occasion.

If earthquakes appear at first sight to produce solely dy-
namical effects, we learn on the other hand, from well-
established evidence, that not only are whole districts of
country elevated by them above their former level (such as
the Ulla-Bund, east of the Delta of the Indus, after the
earthquake of Cutch in June 1819,—and the coast of Chili
in November 1822) but also that during their occurrence various substances are ejected from the earth; such as hot-water
at Catania in 1818; hot steam in the Valley of the Mississipi at

New Madrid in 1812; noxious gases which injured the herds of cattle grazing on the chain of the Andes; mud, black smoke, and even flames, at Messina in 1783, and at Cumana on the 14th Nov. 1797. During the great earthquake of Lisbon (1st Nov. 1755), flames and a column of smoke were seen to issue from a newly-formed fissure in the rock of Alvidras, and the smoke was more dense as the subterranean noise became louder ([191]). At the destruction of Riobamba (1797), where the shocks were not accompanied by any eruption of the closely adjacent volcano, a singular mass (called by the natives Moya), in which carbon, crystals of augite, and siliceous shells of infusoria were intermingled, was pushed up in numerous small conical eminences. During the earthquake of New Granada (16th Nov. 1827), carbonic acid gas issuing from fissures in the valley of the Magdalena River suffocated many snakes, rats, and other animals which live in holes. Great earthquakes have sometimes been followed in Quito and Peru by sudden changes in the weather, and by a premature commencement of the tropical rainy season. Do gaseous fluids issue from the interior of the earth and mingle with the atmosphere? or are these meteorological processes the effects of a disturbance of the electricity of the atmosphere by the earthquake? In intertropical parts of America, where sometimes, for ten months together, not a drop of rain falls, repeated earthquake shocks which do no injury to the low reed-huts of the natives, are regarded by them as the welcome harbingers of abundant rain and a fruitful season.

The common origin of the different phænomena which have been thus described, is still wrapped in obscurity. Elastic fluids subjected to enormous pressure in the interior of the

K 2

globe no doubt occasion the slight and perfectly harmless tremblings of the crust of the earth, lasting several days (such as those which were experienced in 1816 at Scaccia in Sicily, before the volcanic elevation of the new island of Julia), as well as those terrible explosions which are accompanied by loud noises.   But the focus of the action, the seat of the moving force, is placed deep below the crust of the earth, and we can as little judge of the depth as we can of the chemical nature of the fluids so powerfully compressed.   At the edge of the crater of Vesuvius, and on the towering cliff which rises above the great abyss of the crater of the Pichincha near Quito, I have felt periodical and very regular shocks, oc-curring from twenty to thirty seconds before the eruption of the incandescent scoriæ or gases.   The shocks were greatest when the explosions were at long intervals, and when, there-fore, the gases were longer in accumulation.   This simple experience, confirmed by many travellers, contains the general solution of the phænomenon.   Active volcanoes may be regarded as safety-valves for the country in their im-mediate vicinity.   The danger increases when the openings of the volcanoes are stopped, and the free communication with the atmosphere impeded; but the destruction of Lisbon, of Caraccas, of Lima, of Cashmeer in 1554 ([192]), and of so many towns of Calabria, Syria, and Asia Minor, shews that on the whole the most violent shocks do not usually take place in the vicinity of still active volcanoes.

As the impeded activity of volcanoes influences the force of earthquakes, so do the latter react on volcanic phænomena. The opening of fissures favours the elevation of cones or craters of eruption, and the chemical processes which take place in the cones by free contact with the atmosphere.

A column of smoke which was seen for some months to rise from the volcano of Pasto in South America, suddenly disappeared, when, on the 4th of February, 1797, the province of Quito, one hundred and ninety-two miles to the southward, was visited by the great earthquake of Riobamba. Tremblings of the ground, which had long been felt over the whole of Syria, in the Cyclades, and in the island of Euboea, suddenly ceased when a stream of lava issued forth in the plains near Chalcis (¹⁹³).   The celebrated geographer of Amasea, from whom we have received this account, adds, " Since the craters of Etna have been opened, through which fire issues, and since glowing masses and water have been ejected from them, the lands near the sea shore have not been so often shaken as in the time when, previous to the separation of Sicily from Lower Italy, all the issues were closed."   We thus see that the force which manifests itself in earthquakes, acts also in the phænomena of volcanoes; but though as universally diffused as the internal heat of the planet, and making its presence everywhere known, it is only rarely, and at insulated points, that its accumulated energy produces the phænomenon of eruption. The formation of veins, *i. e.* the filling up of fissures with crystalline masses issuing from the interior (basalt, melaphyre, and greenstone), gradually impedes the free escape of the elastic fluids.   They then accumulate, their tension increases, and their reaction against the crust of the earth shews itself in three different ways—in earthquakes,—in sudden elevations,—or in slow and continuous elevations, which alter progressively the relative levels of the land and sea.   The last mode of action produces effects which are only sensible at intervals of long period, and was observed for the first time over a considerable portion of Sweden.

Before we quit this important class of phænomena, which I have considered not so much in its individual as in its general, physical, and geological relations, I would advert to the cause of the deep and peculiar impression produced on the mind by the first earthquake which we experience, even if it is unaccompanied by subterranean noise. I do not think that this impression is produced by the recollection at the moment of the dreadful images of destruction, which historic relations of past catastrophes have presented to our imaginations : it is rather occasioned by the circumstance that our innate confidence in the immobility of the ground beneath us is at once shaken ; from our earliest childhood we are accustomed to contrast the mobility of water with the immobility of the earth : all the evidences of our senses have confirmed this belief; and when suddenly the ground itself shakes beneath us, a natural force of which we have had no previous experience presents itself as a strange and mysterious agency. A single instant annihilates the illusion of our whole previous life ; we feel the imagined repose of nature vanish, and that we are ourselves transported into the realm of unknown destructive forces. Every sound affects us—our attention is strained to catch even the faintest movement of the air—we no longer trust the ground beneath our feet. Even in animals similar inquietude and distress are produced ; dogs and swine are particularly affected, and the crocodiles of the Orinoco, which at all other times are as dumb as our little lizards, leave the agitated bed of the river and run with loud cries into the forest.

To man the earthquake conveys a sense of danger of which he knows not the extent or limit. The eruption of a volcano, the flowing stream of lava threatening his habitation, can be fled from ; but in the earthquake, turn where he will, danger and destruction are around him and beneath his feet. Though such emotions are deeply seated, they are not of long duration. The inhabitants of countries where long series of weak shocks succeed each other, lose almost every trace of fear. On the coasts of Peru, where rain scarcely ever falls, and where hail, lightning, and thunder, are unknown, these atmospheric explosions are replaced by the subterranean thunder which accompanies the trembling of the earth. From long habit, and a prevalent opinion that dangerous shocks are only to be apprehended two or three times in a century, slight oscillations of the ground scarcely excite so much attention in Lima as a hail-storm does in the temperate zone.

Having thus taken a general view of the active internal terrestrial forces;—of the earth's heat, its electro-magnetic currents, its auroral light, and the irregular action of those forces at the surface of the earth,—we will now proceed to the production of material substances, and to the chemical changes of which the crust of the earth and the constitution of the atmosphere is the theatre. We see steam and carbonic acid gas issue from the ground almost always free from any admixture of nitrogen[194],—carburetted hydrogen gas (which has been used for more than a thousand years in the Chinese province of Sse-tchuan[195], and recently in the village of Fredonia in the North American State of New York, both for culinary purposes and for illumination),—sulphuretted hydrogen and sulphurous vapours,—and more rarely sulphurous acid and hydrochloric acid gas[196]. The fissures of the earth from whence the vapours and gases issue are not peculiar to districts of active or of long extinct volcanoes, but occur also in countries where neither trachyte nor other volcanic rocks are present at the surface. In the Cordillera of Quindiu, at an elevation of 6830 English feet above the sea, I have seen sulphur deposited in mica slate from hot sulphureous vapours ([197]) ; and to the south of Quito, in the Cerro Cuello, near Tiscan, the same rock, which was formerly regarded as primitive, contains an immense deposit of sulphur imbedded in pure quartz.

Of gaseous emissions, those of carbonic acid are, as far as we yet know, the most numerous and the most abundant. In Germany, in the deep ravines of the Eifel, in the vicinity of the Laacher-See, in the crater-like valley of Wehr, and in Western Bohemia, exhalations of carbonic acid gas appear as a last effort of volcanic activity, in and near its ancient foci in an earlier state of the globe. With the high terrestrial temperatures of that period, and the numerous fissures which were not then filled up, the processes which we have here described, and in which carbonic acid gas and hot steam mingled in considerable quantities with the atmosphere, must have acted far more powerfully ; and then, as Adolphe Brongniart has shewn, the vegetable world must have attained everywhere, almost independently of geographical position, the most luxuriant development and abundance ([198]). In this constantly warm and moist atmosphere loaded with carbonic acid gas, plants must have found both the stimulus and the superabundant nourishment, which prepared them for becoming the materials of those nearly inexhaustible stores of coal, on which the physical power and prosperity of nations are based. Beds of this fuel are accumulated in basins in particular parts of Europe, in the British Islands, in Belgium, in France, on the Lower Rhine, and in Upper Silesia. At the same early period of generally distributed volcanic activity, there also issued from the earth the enormous quantity of carbonic acid, which, in combination with lime, has formed the limestone rocks, and of which the carbon alone, in a solid form, constitutes about the eighth part of their absolute bulk ([199]). The portion of carbonic acid which was not absorbed by the alkaline earths, but still remained in the atmosphere, was gradually consumed by the luxuriant vegetation ; and the atmosphere being thus purified by the vital action of plants, retained only that extremely minute portion which we now find, and which is not injurious to the present condition of animal life. More abundant exhalations of the vapours of sulphuric acid, in the inland waters of the ancient world, appear to have occasioned the destruction of the numerous species of fish and mollusca which inhabited them, and the formation of the contorted beds of gypsum which have doubtless been subjected to the frequent action of earthquakes.

Gases, liquids, mud, and melted lavas (the last emitted through volcanic cones, and to be regarded as a kind of "intermittent springs," ([200]), all issue from the earth at the present day under similar relations. All these substances owe their temperature and their chemical nature to the place of their origin. The mean temperature of springs is less than that of the air at the points where they issue, when their waters descend from greater elevations ; and the temperature increases according to the depth of the stratum with which they are in contact at their origin : the numerical law of this increase has been already stated. But I have found, from my own observations and those of my companions in Northern Asia, that the mixture of the waters from the various sources from which springs originate— mountains, hills, or deep subterranean strata—makes it very difficult to determine the position of the "Isogeothermal lines ([201])" (lines of equal internal terrestrial heat), from the temperature of water as it issues from the earth. The temperature of springs which, for the last half century, has been so much an object of physical research, depends, indeed, like the limit of perpetual snow, on the concurrent influence of many and very complicated causes. It is a function of the temperature of the stratum in which they take their rise, of the specific heat of the soil, and also of the quantity and temperature of the water which falls, in rain, snow, or hail ([202]), and which, from the conditions of its origin, has a different temperature from that of the air in the lower portion of the atmosphere ([203]). In order that cold springs may shew the true mean temperature of the place where they issue from the ground, they must be unmixed with waters coming either from great depths, or from mountain elevations; and they must have passed through a long subterranean course, at a depth below the surface of about forty to sixty feet in our latitudes, and, according to Boussingault, of one French foot within the tropics ([204]) : these depths being those at which the temperature is supposed to be constant, or unaffected by the horary, diurnal, or monthly variations of the temperature of the atmosphere.

Hot springs issue from rocks of every kind ; the hottest permanent springs yet known are those found by myself, at a distance from any volcano,—the "Aquas calientes de las Trincheras," in South America, between Porto Cabello and New Valencia, and the "Aquas de Comangillas," in the Mexican territory near Guanaxuato. The first of these had a temperature of 90°.3 Cent. (194°.5 Fahr.), and issued in granite; the latter in basalt, with a temperature of 96°.4 Cent. (205°.5 Fahr.). According to our present knowledge of the increase of heat at increasing depths, the strata, by contact with which these temperatures were acquired, are probably situated at a depth of about 7800 English feet, or above two geographical miles. If the internal terrestrial heat be the general cause of thermal springs as well as active volcanoes, the rocks which the waters traverse can influence the temperature only by their different capacity for heat and their conducting powers. The hottest permanent springs (between 95° and 97° Cent., or 203° and 209° Fahr.) are also the purest, containing the smallest portion of mineral substances in solution, but their temperature appears to be less constant than that of springs between 50° and 74° Cent. (122° and 165° Fahr.), which, in Europe at least, have been found remarkably uniform, both in temperature and mineral contents; having undergone no change for the last fifty or sixty years, or since the application of exact thermometric measurement and accurate chemical analysis. The thermal springs of las Trincheras, on the other hand, have increased about 7° Cent. (or 12° Fahr.) in twenty-three years; their temperature having been observed by myself, in 1800, to be 90°.3 Cent. (194°.5 Fahr.), while in 1823, according to Boussingault ([205]), it reached 97° Cent. (or 206°.6 Fahr.). This gently flowing source is therefore, at the present time, almost 7° Cent. (12°.6 Fahr.) hotter than the intermitting fountains of the Geyser and the Strokr, of which the temperatures have been recently determined with great care by Krug of Nidda. The elevation of the new volcano of Jorullo, unknown before my American journey, offers a remarkable example of ordinary rain water sinking to a great depth, where it acquires heat, and afterwards re-appears at the surface of the earth as a thermal spring. When, in September 1759, Jorullo was suddenly elevated to a height of 1682 English feet above the surrounding plain, the two small streams called Rio de Cutimba and Rio de San Pedro disappeared, and some time afterwards broke forth afresh from the ground during severe earthquake shocks, forming springs, whose temperature, in 1803, I found to be 65°.8 Cent. (or 186°.4 Fahr.)

The springs in Greece still flow at the same places as in the Hellenic times : the spring of Erasinos, on the slope of the Chaon, two hours' journey to the south of Argos, was mentioned by Herodotus ; the Cassotis at Delphi, now the well of St. Nicholas, still rises on the south of the Lesche, and its waters pass under the temple of Apollo ; the Castalian fount still flows at the foot of Parnassus, and the Pirenian near Acro-Corinth ; the thermal waters of Ædepsos in Euboea, in which Scylla bathed during the war of Mithri-dates, still exist (²⁰⁶). I take pleasure in citing these details, which shew that, in a country subject to frequent and violent earthquakes, the relative condition of the strata, and even of those narrow fissures through which these waters find a passage, has continued unaltered during at least two thou-sand years. The "Fontaine jaillissante" of Lillers, in the Département du Pas de Calais, bored in 1126, still reaches the same height, and gives the same quantity of water, as at first. I may add that the excellent geographer of the Cara-manian coast, Captain Beaufort, saw, in the district of the ancient Phaselis, the same flames, fed by emissions of the same inflammable gas, which Pliny has described as the flame of the Lycian Chimera (²⁰⁷).

The remark made by Arago, in 1821, that the deepest
Artesian wells are the warmest, threw new and important
light on the origin of thermal springs, and on the investigation of the law of increase of terrestrial heat at increasing depths ([208]).　It is a striking circumstance, which has
been only recently noticed, that, at the end of the third century, Saint Patricius ([209]), who was probably bishop of Pertusa,
was led, by a consideration of the hot springs which issue
from the ground near Carthage, to form very correct views
regarding these phænomena.　To inquiries as to what might
be the cause of boiling water thus issuing from the earth,
he replied, " Fire is nourished in the interior of the earth
as well as in the clouds, as you may learn both from Mount
Etna and another mountain near Naples.　Waters rise from
beneath the ground, as in siphons ; those at a distance from
the subterranean fire are colder, but those which have their
source near the fire are heated by it, and bring with them to
the surface which we inhabit an insupportable degree of heat."

As earthquakes are often accompanied by emissions of
water and elastic fluids, we may recognise in the *Salses*, or
small "mud volcanoes," a transitional phænomenon between
issues of gaseous fluids and of thermal springs, and the grand
and awful phænomenon of streams of lava issuing from
burning mountains.　On the one hand, we may consider
the mountains as springs or fountains sending forth, instead of water, molten earths forming volcanic rocks ;
and, on the other hand, we should remember that thermal
springs, impregnated with carbonic acid and sulphurous
gases, are continually depositing successive horizontal beds
of travertin, or forming conical hills as in Algeria in Northern Africa, and in the Bânos of Caxamarca on the

western declivity of the Peruvian chain of the Andes. The travertin of Van Diemen Island (near Hobarton), contains, as we learn from Charles Darwin, the remains of vegetation belonging to the earlier ages of the world. It may be noticed, that lava and travertin, which are rocks still formed beneath our eyes, present to us the two extremes in geological relations.

The phænomena of mud volcanoes are deserving of more attention than geologists have hitherto given to them; their grandeur has been overlooked, because, of the two phases presented by them, it is only the second, or calmer state, lasting for centuries, which has usually been described: but their origin is accompanied by earthquakes, subterranean thunder, the elevation of great districts of country, and lofty jets of flame of short duration. When the mud volcano of Jokmali, on the peninsula of Abscheron, east of Baku, on the Caspian Sea, was first formed, on the 27th of November, 1827, flames blazed up to an extraordinary height for a space of three hours, and during the following twenty hours they rose about three feet above the crater from which mud was ejected. Near the village of Baklichli, west of Baku, the column of flame rose so high that it could be seen at a distance of twenty-four miles. Enormous fragments of rock, torn doubtless from depths, were hurled to a great distance round. Similar fragments are seen around the now tranquil mud volcano of Monte Zibio, near Sassuolo in Northern Italy. For fifteen centuries the Sicilian salse, near Girgenti (Macalubi), described by the ancients, has continued in the secondary stage of activity; it consists of several conical mounds, from eight or ten to thirty feet high, subject to variation both in form and height. Streams of argillaceous mud, accompanied by periodical disengage-

ments of gas, flow from very small basins containing water
at the summits of the cones. In these cases the mud is
usually cold, but sometimes it has a high temperature, as
at Damak in the province of Samarang in Java. The gaseous
eruptions, which are accompanied by noise, vary in their
nature, consisting sometimes of hydrogen gas mixed with
naphtha, sometimes of carbonic acid, and even occasionally
of almost pure nitrogen, as Parrot and myself have shewn
in the peninsula of Taman, and in the South American
volcancitos of Turbaco ([210]).

After the violent explosions and flames which accompany the first appearance of mud volcanoes, and which
may not perhaps be common to all in an equal degree,
they present to the observer an image of the constant
but feeble activity of the interior of the globe. It would
seem as if, soon after their first formation, the channels
of communication with the very deep strata having a
high temperature, became obstructed, and the coldness
of the mud emitted appears to indicate that, during the
more permanent condition of the phænomenon, the seat of
activity is situated not very far below the surface. The
reaction of the interior of the earth upon its crust manifests
itself far more powerfully in *volcanoes* properly so called,
viz. at points where there exists a communication, either
permanent or reopened from time to time, with deep-seated
volcanic foci. We must, however, carefully distinguish between volcanic phænomena of greater or less intensity,—
between earthquakes, thermal springs, and jets of steam;
mud volcanoes; the elevation of bell-shaped or dome-shaped
trachytic hills or mountains, without openings; the formation of an opening at the summits of such mountains, or of

craters of elevation in basaltic districts ; and lastly, the appearance of a permanent volcano within the crater of elevation itself, or among the debris of its earlier formation. At different epochs, and in different stages of activity and force, permanent volcanoes emit aqueous or acid vapours and ignited scoriæ, or, when the resistance is overcome, glowing streams of molten earth.

Sometimes great but local manifestations of force in the interior of our planet, acting by means of elastic vapours, upheave portions of the Earth's crust in dome-shaped unbroken masses of feldspathic trachyte, and of dolorite (Puy de Dome and Chimborazo) ; or the upheaved strata are broken through, so as to present a slope on the exterior side, and a steep precipice towards the interior which forms the inclosure or bounding wall of a crater of elevation.   When it is a part of the bottom of the sea which has been thus elevated (but this is by no means always the case), the form and character of the upheaved island are determined thereby.   In this manner have originated the circular form of Palma, so well described by Leopold von Buch, and also that of Nisyros in the Ægean Sea (211).   Sometimes a part of the annular circumference has been destroyed, and in the bay where the sea has entered, families of coral animals have built up their cellular habitations.   Even on continents, craters of elevation are often filled with water, and the lakes thus formed impart to the landscape a picturesque beauty of a very peculiar kind.   The formation of these " craters of elevation" is independent of the nature of the rock ; they are found in basalt, trachyte, leucitic porphyry (Somma), and in combinations of augite and labradorite : hence the varieties of their form and aspect.   " No phænomena of eruption, however, proceed from these craters ; nor is

there any permanent channel of communication open with
the interior; it is only rarely that any traces of modern
volcanic activity are found either within them or in their
vicinity.   The force capable of producing such considerable
effects must have been long accumulating in the interior;
before it acquires sufficient strength to overcome the resistance of the superincumbent mass, and is enabled, for example,
to raise new islands above the surface of the ocean, by
breaking through granular rocks and conglomerates (strata of
tufa containing marine plants).   The strongly compressed
vapours escape through the crater of elevation, but the great
upheaved mass again falls back, and recloses the opening
thus momentarily produced by a vast effort.   No volcano
could in such case be formed." (212)

A volcano, properly so called, exists only where a permanent communication is established between the interior of
the earth and the atmosphere : the reaction of the interior
upon the surface is in such case continued during long
periods of time, and although interrupted for centuries, as
in the case of Vesuvius (213), it may afterwards be renewed with fresh energy.   In the time of Nero there was
a disposition to class Mount Etna amongst the burning
mountains which were gradually becoming extinct (214) ;
and at a still later epoch Ælian even affirmed that the
summit of the mountain was subsiding, that mariners could
no longer discern it at so great a distance from the shore as
formerly (215).   Where traces of the first eruption exist,—or
where, if I may so express myself, the primitive scaffolding
is still preserved entire,—the volcano rises from the middle
of the crater of elevation, and the isolated cone is surrounded
by an amphitheatre of lofty precipices, composed of greatly

inclined strata : but frequently no trace of this circular rampart can be perceived ; and the volcano, which is not always
of a conical form, rises immediately from the table land like
the ridge-shaped volcano of Pichincha, at the foot of which
the town of Quito is built.

As the nature of rocks, or the mixture or association
of simple minerals which unite to form granite, gneiss, and
mica slate, trachyte, basalt, and dolerite, is wholly independent of our present climates, and is the same in all latitudes
and all regions of the earth ; so also we see that everywhere in inorganic nature the same laws regulate the superposition of the strata composing the crust of the globe, their
mutual penetrations, and their elevation by the agency of
elastic forces.   In volcanoes especially, the identity of form
and structure is peculiarly striking.   The navigator amongst
islands of remote seas, where new stars replace those on
which he has been accustomed to gaze, and where he finds
himself surrounded by palms and other unfamiliar forms of
an exotic flora, yet recognizes in the features of inorganic
nature which characterise the landscape, the forms of
Vesuvius, of the dome-shaped summits of Auvergne, of
the craters of elevation of the Canaries and the Azores, and
of the fissures of eruption of Iceland.   The analogies thus
noticed receive a still wider generalization when we view the
attendant satellite of our planet.   The maps of the moon,
which have been traced by the aid of powerful telescopes,
exhibit to us a surface devoid of air and water, abounding in
vast craters of elevation surrounding or supporting conical
eminences ; thus clearly evidencing the effects of the reaction
of the interior of the moon upon its exterior ; a reaction favoured by the feebler influence of gravitation at the surface.

Although volcanoes are justly termed in many languages "fire-emitting mountains," they are not formed by the gradual accumulation of erupted streams of lava; they appear, on the contrary, to originate generally in a sudden elevation of masses of trachytic or augitic rock in a softened state. The degree of intensity of the upheaving force is shewn by the height of the volcano, which varies from that of a mere hill such as the volcano of Cosima, one of the Japanese Kurile islands, to that of a cone of above 18000 feet of elevation. It has appeared to me that the height of volcanoes exercises a great influence on the frequency of eruptions, which are far more numerous in the lower than in loftier volcanoes. As instances I may place in a series,—Stromboli (2175 French, or 2318 English feet) : Guacamayo, in the province of Quiros, whence detonations are heard almost daily; (I have often heard them myself at Chillo, near Quito, at a distance of 88 miles :) Vesuvius (3637 French, or 3876 English feet) : Etna (10200 French, or 10870 English feet) : the Peak of Teneriffe (11424 French, or 12175 English feet) : and Cotopaxi (17892 French, or 19070 English feet).   If we suppose the seat of action to be at an equal depth below the general surface of the earth in the case of all these volcanoes, it must require a greater force to raise the molten masses in the case of the higher mountains.   It is not therefore surprising that the one whose elevation is least considerable, Stromboli (Strongyle), should have been in a state of constant activity from the Homeric times, and should still serve as a flaming beacon to the mariners who navigate the Tyrrhenian Sea, whilst the loftier volcanoes are characterised by longer intervals of repose.   Thus also we see the eruptions of most of the colossal summits which crown the chain of the Andes separated by intervals of almost a century. To this law I long since called attention; and where exceptions occur, they may perhaps be explained by the channels of communication between the volcanic seat of action and the crater of eruption not being in all cases alike permanently free; since this channel may be temporarily obstructed in some volcanoes of moderate elevations, so that eruptions may become more rare, without any immediate prospect of their absolute extinction.

These considerations, respecting the relation of the height of volcanoes to the frequency of their eruptions, naturally conduct us to an examination of the causes which determine the place at which the lava issues from the mountain. In many volcanoes eruptions from the crater are extremely rare; they more often take place (as was remarked in the case of Etna in the sixteenth century by the celebrated historian Bembo ([216]), when a youth), from lateral openings formed in those parts of the walls of the upheaved mountain, which, from their nature and shape, may offer the least resistance. Sometimes "cones of eruption" rise over these lateral fissures; and in this case the larger cones, erroneously denominated "new volcanoes," are ranged in rows, indicating the line of fissure, which is speedily reclosed; while the smaller cones, which are shaped like bells or bee-hives, form numerous crowded groups, which cover large spaces of ground. To the latter class belong the "hornitos de Jorullo" ([217]) and the cones of eruption of Vesuvius in October 1822, those of the volcano of Awatscha according to Postels, and of the lava field described by Erman, near the Baidar mountains, in the peninsula of Kamtschatka.

Where volcanoes are not isolated in the midst of plains, but are surrounded, as in the double chain of the Andes of Quito, by a table land from nine to twelve thousand feet high, this circumstance may very probably account for the non-production of streams of lava, during the most dreadful eruptions of ignited scoriæ, which are sometimes accompanied by detonations heard at distances of more than four hundred miles (218). Such are the volcanoes of Popayan, of the table land of los Pastos, and of the Andes of Quito; the volcano of Antisana may possibly form an exception.

The height of the cone of cinders, and the magnitude and form of the crater, which are the principal elements of the individual character of volcanoes, are independent of the dimensions of the mountain itself. In Vesuvius, which is only a third of the height of the Peak of Teneriffe, the cone of ashes rises to a third of the height of the whole mountain, while the cone of the Peak amounts to only 1-22d part of its altitude; in the case of the Rucu-Pichincha, a volcano much loftier than Teneriffe, the proportions much more nearly resemble those of Vesuvius. Among all the volcanoes which I have had an opportunity of seeing in both hemispheres, the conical form of Cotopaxi is at once the most regular and the most picturesque. A sudden melting of the snow on its cone of cinders announces the near approach of an eruption; even before smoke is seen to ascend through the rarefied atmosphere which surrounds the summit and the crater, the walls of the cone of cinders sometimes become glowing, and the mass of the mountain itself then assumes an aspect of awful and portentous blackness.

The crater which, except in very rare instances, always

occupies the summit of the volcano, forms a deep circular
and often accessible caldron-like valley, the bottom of which
is subject to constant change.   In many volcanoes the
greater or less depth of the crater is a sign of the greater or
less time elapsed since the last eruption.   Long narrow
fissures from which vapours escape, or small circular hollows
filled with substances in a state of fusion, alternately open
and close within the crater.   The ground intumesces and
subsides, and mounds of scoriæ and cones of eruption rise
sometimes high above the surrounding wall of the crater,
giving the volcano a peculiar character, which may last for
years, until, during a new eruption, the mounds and cones
sink or otherwise disappear.   The openings of such cones
of eruption, rising from the bottom of the crater, ought not to
be confounded, as they sometimes have been, with the crater
itself which incloses them.   When the latter is inaccessible from its great depth and precipitous descent, as is the
case of Rucu-Pichincha, (14946 French feet in height,) the
traveller may look down from the edge on the summits
which rise from the depth below, through the sulphureous
vapours by which the valley of the crater is partially filled.
This spectacle is a magnificent one.   I have never seen
nature under an aspect more grand and wonderful than in
the view from the edge of the crater of Pichincha.   In the
interval between two eruptions, a crater may either offer to
the eye no phænomenon whatever of incandescence, but
merely open fissures from which steam issues ; or the geologist who is able to approach the cones of scoriæ without
danger over a soil only slightly heated, may enjoy the view
of the eruption of burning fragments which fall back on the
flanks of the mounds from whence they have issued.   Each

such explosion is regularly announced by small and purely local earthquake shocks. Sometimes lava is poured out from the open fissures or hollows, but without making its way beyond the sides of the crater; and when it does break through them, the new stream of molten rock usually finds a course which does not prevent the great crater-valley itself from being accessible even during such minor eruptions. The margins of craters appear to undergo far less variation than might have been expected; for example, in the case of Vesuvius, Saussure's measurements compared with mine, shew that no change beyond the limits of observation error took place in the height of the north-western edge of the volcano, the Rocca del Palo, in the interval of forty-nine years, from 1773 to 1822 ([219]). I have been solicitous to give an accurate idea of the form and normal structure of volcanoes, without which it is impossible to attain a right understanding of phænomena, which have been long much disfigured by fanciful descriptions, and by the equivocal and ill-defined use of the terms, crater, cone of eruption, and volcano.

Volcanoes which, like those of the Andes, rise high above the region of perpetual snow, present peculiar phænomena; the masses of snow, by their sudden melting during eruptions, produce terrible inundations and torrents of water, by which smoking scoriæ are hurried along with blocks of ice; they also exert a continued action during the periods when the volcano is in a state of entire repose, by infiltration into the fissures of the trachytic rocks. Cavities in the declivity or at the foot of the volcano are thus gradually converted into subterranean reservoirs of water, with which the alpine torrents and rivulets of the highlands

of Quito communicate by numerous narrow channels. The
fish of these rivulets multiply by preference in the obscurity
of the caverns ; and when the whole mass of the volcano is
powerfully shaken by the earthquake shocks, which, in
the Andes, always precede eruptions, these subterranean
caves are suddenly opened, and water, fishes, and tufaceous
mud, are all ejected together. It was by this singular phænomenon that the inhabitants of the plains of Quito were
made acquainted with the little fish which they call Preña-
dilla, *Pimelodes cyclopum* ([220]). When, in the night of
the 19th of June, 1698, the summit of the Carguairazo
(18000 French feet in height) fell in, leaving two immense
peaks of rock as the sole remains of the wall of the crater,
masses of liquid tufa, and of argillaceous mud (*lodazales*),
containing dead fish, spread themselves over and rendered
sterile a space of nearly two square German miles. The
putrid fevers which seven years before prevailed in
the mountain town of Ibarra, north of Quito, were attributed to the quantity of dead fish ejected in like manner
from the volcano of Imbaburu.

Water and mud, which, in the Andes, do not issue from
the crater, but from caverns in the trachitic mass of the
mountain, cannot be strictly classed among volcanic phænomena, in the restricted sense of the expression. Their connection with the volcanic activity of the mountain is only
indirect, as is that of the singular meteorological phænomenon to which, in my earlier writings, I have given the
name of " volcanic storm." The hot steam which, during
the eruption, issues from the crater and mingles with the
atmosphere, condenses as it cools, and forms a cloud surrounding the column of fire and ashes, which rises to a

height of many thousand feet. The electric tension is in-creased by the suddenness of the condensation, and also, as Gay Lussac has shewn, by the formation of such an enor-mous surface of cloud. Forked lightnings dart from the column of ashes, and (as at the close of the eruption of Vesuvius, near the end of the month of October 1822) the rolling thunder of the volcanic storm is heard, and clearly distinguished from the sounds which issue from the interior of the volcano. We learn from Olafsen's relation, that in Iceland (at the volcano of Katlagia, 17th October, 1755), eleven horses and two men were killed by lightning from the cloud of volcanic steam.

Having thus pourtrayed, as part of the general view of nature, the structure and dynamic activity of volcanoes, we have next to glance briefly at the diversity of their material products. The subterranean forces dissolve old combinations, and form new ones; but they operate also by displacing the otherwise unchanged substances whilst in a state of liquefaction by heat. The greater or less pressure under which the solidification either of liquid or of merely softened substances takes place, appears to be the principal cause of the difference between "plutonic" and "volcanic" rocks. The molten rock which has issued in a distinct current from a volcanic opening is called lava; and where such currents meet, and are impeded in their course by opposing obstacles, they spread out in breadth, and cover large areas, in which they solidify in superposed strata. These few sentences contain all that can be affirmed generally respect-ing the products of volcanic activity.

Fragments of the rocks, which have been broken through by volcanic disturbance, are sometimes inclosed in

the igneous products.   Thus I have found angular fragments of feldspathic syenite imbedded in the black augitic
lava of the volcano of Jorullo, in Mexico.   But the masses
of dolomite and granular limestone found in the neighbourhood of Vesuvius, containing magnificent groups
of crystallized minerals (vesuvian and garnets, covered
with mionite, nepheline, and sodalite), are not substances
which have been erupted from that volcano ; " they belong
rather to a very generally distributed formation—to beds
of tufa, which are older than the elevation of the Somma
and of Vesuvius, and were probably produced by a submarine and deeply-seated volcanic action" ([221]).   We find
among the products of existing volcanoes five metals, iron,
copper, lead, arsenic, and selenium, the latter of which was
discovered by Stromeyer in the crater of Volcano.   The
vapours from the small cones contain chlorides of iron,
copper, lead, and ammonia ; specular iron ([222]), and common
salt (the latter often in large quantities) are found in
cavities of recent lava currents, and in fissures in the
margin of the crater.

The mineral composition of lava differs according to the
nature of the crystalline rock of which the volcano consists,
—according to the height of the point at which the eruption
takes place (whether at the foot of the mountain or near the
crater),—and according to the degree of heat of the interior.
Vitreous volcanic rocks, obsidian, pearl stone, and pumice,
are entirely wanting in some volcanoes ; in others they
proceed from the crater itself, or at least from inconsiderable depths beneath it.   These important and complicated
relations can only be investigated by very exact crystallo-
graphical and chemical examination.   My Siberian tra-

velling companion, Gustav Rose, and subsequently Hermann Abich, have commenced, with much ingenuity and success, the investigation of the structure of volcanic rocks.

The greater part of the vapour which rises from volcanoes is pure aqueous vapour, which condenses and forms springs, as the spring in the Island of Pantellaria, to which the goat-herds resort for a supply of water. The current which, on the morning of the 26th of October, 1822, was seen to pour from the crater of Vesuvius through a lateral opening, and was long supposed to have been boiling water, appears from the careful examination of Monticelli, to have consisted of dry ashes, or of lava pulverised by friction. The phænomenon of volcanic ashes, which darken the air for hours and even for days, and by their fall cause great damage to vineyards and olive trees by adhering to the leaves, mark by their columnar ascent, upborne by vapours, the termination of every great eruption. This was the magnificent phænomenon which, in the case of Vesuvius, the younger Pliny, in his celebrated letter to Cornelius Tacitus, compared to a lofty pine spreading out at its summit into wide shadowing branches. The appearance of flame, which has been described as accompanying the eruptions of scoriæ, and the red glow of the clouds which hover over the crater, are not certainly true flames, or to be attributed to the combustion of hydrogen ; they are rather due to reflections from the incandescent substances projected high in air, and also to the ascending vapours illuminated by the fiery sea within the crater itself. In regard to the flames seen occasionally, as in the time of Strabo, to issue from the deep sea during the activity of coast volcanoes, or a short time before the elevation of a volcanic island, we can give no explanation.

When it is asked, what it is that *burns* in volcanoes— what excites the heat, fuses the earths and metals, and im- parts to lava currents of great thickness ([223]) a heat which lasts for many years,—the question assumes by implication, that the presence of materials capable of supporting combus- tion is indispensable in volcanoes, like the beds of coal in subterranean fires. According to the different phases which chemical science has passed through, bitumen, pyrites, or a humid mixture of pulverised sulphur and iron, pyrophoric substances, and the metals of the earths and alkalies, have been successively assigned as the cause of active volcanic phænomena. Sir Humphry Davy, the great chemist to whom we owe the knowledge of these latter most inflam- mable metals, has himself renounced his bold chemical hypothesis in his last work, " Consolation in travel, and last days of a Philosopher," which cannot be read without a sen- timent of melancholy. The high mean density of the earth (5.44), compared with the much inferior specific gravity of potassium (0.865), and of sodium (0.972), or of the metals of the earths (1.2), the absence of hydrogen in gaseous emanations from the fissures of craters, and from lava cur- rents which have not yet cooled ; and lastly, many chemical considerations, oppose themselves to the earlier conjectures of Davy and of Ampère ([224]). If hydrogen were disengaged by the eruption of lava, what prodigious quantities of that gas must have been set free in the memorable eruption at the foot of the Skaptar-Jokul, in Iceland, described by Macken- zie and Soemund Magnussen, which lasted from the 11th of June to the 3d of August, and covered very many square miles of country with lava, which, where it met with obstacles to its course, accumulated to a thickness of several hundred feet. The small quantity of nitrogen emitted opposes similar difficulties to the hypothesis of the entrance of atmospheric air into the crater, or, as it has been metaphorically expressed, to the breathing or inhaling of air by the earth. An activity so general as that of volcanoes, so deeply seated, and extending itself so widely in the interior, cannot well have its source in chemical affinities, and in the contact of certain substances found only in particular localities. Modern geology prefers to seek its cause in the internal terrestrial heat manifested in every latitude by the increase of temperature with increasing depth, and which has been ascribed to the supposed condition of the earth as a body only partially cooled. If we consider volcanoes as irregular intermitting springs, supplying in tranquil flow a fluid mixture of oxidized metals, alkalies, and earths, which, upheaved by the powerful expansive force of vapours, have found a permanent outlet,—we are involuntarily reminded how nearly the rich imagination of Plato approached to the same view, when he attributed thermal springs and all volcanic phænomena to a single cause every where present in the interior of the earth, the *Pyriphlegethon,* or subterranean fire ([225]).

The geographical distribution of volcanoes is wholly independent of climatic relations; they have been arranged characteristically in two classes: " central volcanoes," and " volcanic chains;" the former term being applied to volcanoes forming the centers of numerous orifices of eruption distributed with some regularity in every direction; and the latter to those which, placed at moderate distances apart, form lines running in one direction, like chimneys or vents from a long extended subterranean fissure. The latter class of volcanoes, or those which form lines, is again subdivided into those which rise as single conical islands from the bottom of the sea, (in which case they are usually parallel to, and at the foot of a chain of primitive mountains), or they are elevated upon the highest ridge of the primitive chain, of which they then form the summits ([226]). The Peak of Teneriffe, for example, is a " central volcano ;" it is the center of the group to which we refer the volcanic islands of Palma and Lancerote. The grandest example of a continental volcanic " chain" is offered by the great rampart of the Andes, extending from the southern part of Chili to the north-west coast of America, sometimes forming a single range, sometimes divided into two or three parallel branches, which are occasionally connected by narrow cross or transversal ridges. In this chain the proximity of active volcanoes is always announced by the appearance of certain kinds of rocks (dolerite, melaphyre, trachyte, andesite, and dioritic porphyry), breaking through and dividing the primitive rocks, the transition slates and sandstones, and the more recently formed strata. The constant recurrence of this phænomenon led me long since to the belief that these sporadic rocks were the seat of volcanic phænomena, and determining conditions of volcanic eruptions. It was at the foot of the majestic Tunguragua, near Penipe (on the banks of the Rio Puela), that I first distinctly observed mica schist (resting on granite) traversed by a volcanic rock. In parts of the volcanic range of the new continent, where the single volcanoes are nearest to each other, they shew a certain mutual dependence and connection ; it even appears that the volcanic activity has progressively advanced for centuries in certain directions, as in the province of

Quito from north to south. The seat of volcanic action extends under the whole of that elevated province; (²²⁷) its several channels of communication with the atmosphere being the volcanic mountains of Pichincha, Cotopaxi, and Tunguragua, which, by their grouping, as well as by their lofty elevation and grand outlines, present the most sublime and picturesque aspect which is anywhere concentrated within so small a space in a volcanic landscape. The extremities of volcanic chains are connected with each other by subterranean communications; and this fact, which experience has made known to us in numerous instances, reminds us of the old and just statement of Seneca, that "the crater is only the issue of the deeper seated volcanic forces." (²²⁸) The Mexican volcanoes of Orizaba, Popocatepetl, Jorullo, and Colima, also appear to be connected with each other, and are situated over a transverse fissure running from sea to sea. As was first shewn by myself, (²²⁹) these mountains are all situated between 18° 59′ and 19° 12′ of North latitude; and in the exact line of the direction of these volcanoes, and over the same transverse fissure, Jorullo was suddenly elevated on the 29th of September, 1759; this last mountain has only once sent forth streams of lava, resembling in this respect Mount Epomeo, in the island of Ischia, of which likewise only a single eruption (in 1302) is recorded.

But although Jorullo, situated eighty miles from any active volcano, is in the strictest sense a "new mountain," yet its appearance must not be confounded with that of the Monte Nuovo, near Pozzuolo (19th September, 1538), which is rather to be classed with the "craters of elevation." It appears, indeed, to me, to agree better with the results of observation, if we compare the sudden appearance of the Mexican volcano with the volcanic elevation of the Hill of Methone, now Methana, in the peninsula of Trœzena, a phænomenon of which the description by Strabo and Pausanias led one of the Roman poets endowed with the richest fancy to develop views strikingly accordant with those of modern geology. " Near Trœzena is a tumulus, steep and treeless, once a plain, now a mount. The vapours, pent up in dark caverns, in vain sought an outlet; thus constrained, their powerful force caused the inflated ground to swell upwards, like a bladder or goat-skin filled with air. The ground thus raised still remains, but has been changed by time into a hard rock" (²³⁰). Thus picturesquely, and, as analogous phænomena justify us in believing, thus truly, does Ovid describe the great natural phænomenon which took place between Trœzena and Epidaurus, 282 years before our era; 45 years therefore before the volcanic separation of Thera (the island of Santorin) from Therasia. In the same spot Russegger found intersecting veins of trachyte.

Of all " islands of eruption," belonging to volcanic chains, Santorin is the most important as an object of study : " it is a complete type of islands of elevation : for more than 2000 years, or as far back as history and tradition enable us to trace, efforts of nature to form a volcano in the middle of the crater of elevation seem to have been perpetually going on." (²³¹) Near the island of St. Michael, in the Azores, similar insular elevations manifest themselves at almost regularly recurring intervals of eighty or ninety years (²³²); but in this case the bottom of the sea is not always raised at precisely the same points. The island to which Captain Tillard gave the name of Sabrina, appeared at a time (30th January, 1811) when unfortunately political events

did not permit scientific institutions to give to this great
phænomenon the attention which, at a later epoch (2d July,
1831) was devoted to the ephemeral apparition of the
igneous island of Ferdinandea, between the limestone coast
of Sciacca and the purely volcanic island of Pantellaria ([233]).

The geographical distribution of the volcanoes which
have been in a state of activity within historic times,—the
great number situated on islands or on coasts,—and the recurring phænomena of eruptions from the bed of the sea,—
early led to a belief that volcanic activity is connected with
the vicinity of the sea, and dependent on it for its continu-
ance.   "Etna and the Æolian islands have been burning for
centuries," says Justin ([234]), (or rather Trogus Pompeius,
whom Justin follows;) "and how could they have lasted so
long if the neighbourhood of the sea did not feed the fire?"
Even in recent times it has been attempted to explain the
supposed necessity of the vicinity of the sea, by the hypothesis of sea water penetrating to the foci of volcanic activity, or to very deep-seated strata.   After comprehending
in one view all that my own observation has furnished, and
all that I can gather from facts diligently collected elsewhere, it appears to me that the conclusion in this intricate
investigation must depend upon the solution of certain
questions; we must, for instance, first determine whether
the great mass of aqueous vapour unquestionably exhaled
by volcanoes, even when in a state of repose, be derived
from sea water impregnated with salt, or from fresh water
obtained from meteoric sources.   In the next place we
must decide whether the expansive force of aqueous
vapour (which, at a depth of 88000 French feet, is equivalent to 2800 atmospheres), would be sufficient, at the different depths of the foci of volcanic action, to counterba-

lance the hydrostatic pressure of the waters of the sea, and to allow them, under certain conditions, free access to the foci ([235]). We must learn whether the presence of the metallic chlorides, or even of marine salt in the fissures and crevices in the sides of craters, and the frequent admixture of hydrochloric acid in the aqueous vapours, *necessarily* imply such access of the sea water. Finally, we must inquire if the inactivity of volcanoes (whether temporary, or final and complete), is dependent on the interruption of the channels by which either the sea or fresh water previously penetrated; or if the absence of flames and of hydrogen gas (not of sulphuretted hydrogen which belongs rather to solfataras than to active volcanoes), is directly opposed to the hypothesis which attributes their activity to the decomposition of great masses of water. The discussion of these important physical questions does not belong to a work like the present: but whilst on the subject of the geographical distribution of volcanoes, it is proper to notice that all active volcanoes are *not* in close proximity to the sea; since on the new continent, Jorullo, Popocatepetl, and the Volcano de la Fragua, are respectively 80, 132, and 156 geographical miles from it; and even in central Asia, a fact to which Abel Rémusat ([236]) first called the attention of geologists, we find a great volcanic chain, the Thianschan (celestial mountains),—to which belong the Pe-schan from whence lava issues, the solfatara of Urum-tsi, and the still active "fire mountain" (Ho-tscheu), of Turfan,—almost equidistant from the shores of the Polar Sea and of the Indian Ocean (1400 and 1528 miles). Pe-schan is also fully 1360 miles from the Caspian Sea, and 172 and 208 miles respectively from the great lakes of Issikoul and Balkasch ([237]). It is deserving of notice that, of the four great parallel chains of mountains which traverse the Asiatic continent from east to west, the Altai, the Thian-schan, the Kuen-lun, and the Himalaya, it is not the Himalaya, which is nearest to the sea,—but the two interior chains (the Thian-schan and Kuen-lun), at distances of 1600 and 720 miles from the sea,—which have eruptive volcanoes like Etna and Vesuvius, and issues of ammoniacal gas like the volcanoes of Guatimala. It is impossible not to recognise currents of lava in the descriptions given by Chinese writers of smoke and flame bursting from the Pe-schan, accompanied by " burning masses of stone flowing as freely as melted fat," and devastating the surrounding district, in the first and seventh centuries of our era. The facts thus brought together, and which have not perhaps been hitherto sufficiently considered, render it at least highly probable that the vicinity of the sea and access of sea water to the focus of volcanic activity, are not essential conditions of the breaking forth of the subterranean fire; and that, if littoral situations favour such eruptions, it is only because they are on the margin of the deep sea basin, of which the bed, covered only by superincumbent water, and situated many thousand feet lower than the elevated terra firma of the interior of continents, offers less resistance to the subterranean forces.

The present active volcanoes, of which the craters establish a communication between the interior of the earth and the atmosphere, have been opened at so late an epoch, that the upper chalk strata, and all the tertiary formations, were previously existing : this is evidenced by the trachyte and the basalt which often form the sides of the craters of elevation. Melaphyres extend to the middle tertiary strata, having apparently been poured out also beneath the oolite, since they traverse the bunter or variegated sandstone ([238]).   We must not confound the present active craters with earlier outpourings of granite, of quartzose porphyry, and of euphotide, from open temporary fissures in the old transition rocks.

The cessation of volcanic activity may be only *partial*, the subterranean fire finding for itself another outlet in the same chain of mountains; or it may be *total*, as in Auvergne.   More recent examples of the latter class are known to have occurred within historic periods; the volcano of Mosychlos ([239]), on the island consecrated to Vulcan, of which the " high whirling flames" were still known to Sophocles; and the volcano of Medina, which, according to Burckhardt, sent forth a stream of lava as late as the 2d of November, 1276.   Each stage of volcanic activity, from its first excitement to its extinction, is characterised by a particular class of products : first, by ignited scoriæ, by currents of lava, consisting of trachyte, pyroxene, and obsidian, and by rapilli and tufaceous ashes, accompanied by an abundant disengagement of steam, usually quite pure : at a later period the volcano becomes a solfatara, where aqueous vapours are emitted mixed with sulphuretted hydrogen and carbonic acid gases : and lastly, the crater becomes entirely cooled, and carbonic acid exhalations only proceed from it. There is an extraordinary class of volcanoes (such as the Galungung, in the island of Java), which do not emit lavas, but only devastating streams of boiling water, accompanied by sulphur in combustion, and rocks reduced to the state of dust ; ([240]) but whether these present a normal condition, or are only a certain transitory modification of volcanic processes, must remain undecided until they are visited by geologists skilled in the doctrines of modern chemistry.

This general description of volcanoes,—one of the most
important manifestations of the internal activity of our
planet,—has been based in part on my own observations, but
still more, and in its more comprehensive outlines, on the
labours of my friend, Leopold von Buch, the greatest geologist of our age, who was the first to recognise the intimate
connection of the several volcanic phænomena, and their mutual dependence.   Volcanic action, or the reaction of the interior of a planet on its external crust and surface, was long
regarded as an isolated and purely local phænomenon, and
was considered solely in respect to its destructive agency;
it is only in modern times that, greatly to the advantage of
geological science founded on physical analogies, volcanic
forces have been contemplated as *formative of new
rocks*, and *transformative of those which were pre-
existing* We here arrive at the point I previously indicated, at which a well-grounded study of volcanic activity in its various manifestations, branches into, and connects itself with the mineralogical portion of geology
(the science of the structure and succession of terrestrial
strata), and with the configuration of continents and islands
which have been elevated above the level of the sea.   The
enlarged view presented by this connection of phænomena is
a result of the philosophical direction which the more
earnest and serious study of geology has now so generally
assumed.   The prosecution and improvement of the sciences
has the same tendency as political and social improvements,
to bring together and unite that which had long been
divided.

If instead of arranging rocks, according to their differences of form and superposition, into stratified and unstratified, schistose and compact, normal and abnormal, we trace out and study the phænomena of formation and transformation which are still going on before our eyes, they may be distributed into the four following classes, according to their mode of origin :—

1. *Erupted rocks,* which have issued from the interior of the earth, either by volcanic action in a state of fusion, or by plutonic action in a more or less softened state.

2. *Sedimentary rocks,* precipitated or deposited from liquids in which their particles were held in solution or suspended ; these form the greater part of the secondary and tertiary groups.

3. *Transformed* or *metamorphic rocks,* in which the texture and mode of stratification have been altered, either by the contact or proximity of an erupted plutonic or volcanic rock (*endogenous rocks*) [241], or, as is more frequently the case, by the action of vapours and sublimations ([242]), which accompany the issue of certain masses in a state of igneous liquefaction.

4. *Conglomerates,* coarse or fine-grained sandstones or breccias, consisting of mechanically divided fragments of the three preceding classes.

The production of these four kinds of rocks, as still going on before our eyes—by the pouring forth of volcanic masses in streams of lava,—by the influence of these masses on rocks previously hardened,—by mechanical separation, or chemical precipitation from liquids charged with carbonic acid,—and by the cementation of the detritus of rocks of every kind ;—may be regarded as presenting only a faintly reflected image of that which took place in the early chaotic period of more energetic activity, under very different conditions of pressure, and a far higher temperature of the more extended and vapour-loaded atmosphere, as well as of the crust of the earth.   The vast fissures which were then open in the solid portions of the crust have been since closed by the elevation of mountain chains protruded through them, or filled up by veins of granite, porphyry, basalt, and melaphyre.   At the present period of the globe there remain only, on an extent of the size of Europe, four volcanoes, or openings through which ignited masses may issue ; whereas formerly, channels of communication between the molten interior and the atmosphere existed at almost every part of the thinner and much fissured crust of the globe. Gaseous exhalations, rising from very unequal depths, and bringing with them different chemical substances, gave great activity to the processes of plutonic formation and of metamorphic action.   In like manner, in the case of sedimentary formations, the beds of travertin which are now in daily course of deposition from cold and warm springs and river water, near Rome, and near Hobarton in Van Diemen Island, afford but a feeble representation of the formation of the earlier mineral strata.   On the coasts of Sicily and of the island of Ascension, and in King George's Sound in Australia, small banks of limestone, of which some parts are scarcely inferior in hardness to Carrara marble ([243]), are in course of gradual formation by our present seas, under the influence of processes which have not yet been sufficiently investigated, by means of precipitation, accumulation by drift, and cementation.   On the coasts of the West India Islands, these formations of the present ocean contain pot-

tery, instruments of human art and industry; and, at Gua-
daloupe, even human skeletons of the Carib race.   The
negroes of the French colonies call these banks "Maconne-
bon-Dieu" ([244]).   In Lancerote, one of the Canary Islands,
a small bank of oolite, which, notwithstanding its recent
formation, resembles the Jura limestone, has been recognised as a product of the sea and of tempests ([245]).

The composite rocks are determinate associations of certain
simple minerals, such as feldspar, mica, silex, augite, and
nepheline.   Rocks very similar to those of the earlier periods,
and composed of the same elements, but differently grouped,
are now produced under our eyes by volcanic processes.
We have already remarked ([246]), that the mineralogical characters of rocks are wholly independent of their geographical
distribution; and the geologist recognises with surprise, in
opposite hemispheres, and in very dissimilar climates, the
familiar aspect, and the repetition, even in the most minute
details, of the successive members of the silurian series, and
the precisely similar effects of contact with erupted augitic
masses.

We will now take a nearer view of the four fundamental
classes of rocks, which correspond to the four phases or
modes of formation presented by the stratified and unstratified portions of the earth's crust: and, first, in the endogenous or erupted rocks, designated by some modern geologists by the terms massive and abnormal rocks, we distinguish as immediate products of the active subterranean
forces, the following principal groups :—

1. *Granite* and *syenite*, of very different ages.   The
granite is often the more recent rock, traversing the

syenite ([247]). Where granite is found in large insulated masses, having a slightly vaulted ellipsoidal form, whether it be in the Hartz district, in Mysore, or in Lower Peru, it is surmounted by a kind of crust divided into blocks. These " seas of rocks," as they are sometimes called, are probably occasioned by a contraction of the distended surface of the granite when first upheaved ([248]). In Northern Asia, on the romantic shores of Lake Kolivan on the north-western slope of the Altai ([249]), as well as at las Trincheras ([250]) on the declivity of the maritime chain of Caraccas, I have seen divisions in the granite, which were probably caused by similar contractions, but which appeared to penetrate deep below the surface. Farther to the south of Lake Kolivan, towards the boundary of the Chinese province of Ili (between Buchtarminsk and the river Narym), the appearance of the erupted rocks, in which there is no trace of gneiss, is more remarkable than I had ever before seen in any part of the globe. The granite, which always scales at the surface, and is characterised by tabular divisions, rises, on the steppe, in small hemispherical hillocks of six or eight feet in height, and sometimes, like basalt, in small mounds with narrow streams on opposite sides of their base ([251]). At the cataracts of the Orinoco, as in the Fichtelgebirge in Bavaria, —and in Gallicia, as on the Pappagallo between the high lands of Mexico and the Pacific,—I have seen large flattened globes of granite, which could be separated into concentric layers like certain basalts. In the valley of the Irtysch, between Buchtarminsk and Ustkamenogorsk, granite covers transition slate for a space of four miles, and penetrates it from above downwards in narrow branching veins, having wedge-shaped terminations ([252]). These details have been introduced for the purpose of indicating by examples the character of erupted rocks, as shewn in a rock most generally distributed throughout all parts of the earth. As granite covers argillaceous schists in Siberia, and in the Département de Finisterre (Ile de Mihau), so does it cover oolitic limestone in the mountains of Oisons (Fermonts), and syenite and chalk, in Saxony near Weinböhla ([253]). At Mursinsk, in the Oural, the granite is porous, and as in the later volcanic rocks, the cavities are filled with magnificent crystals, particularly beryls and topazes.

2. *Quartzose porphyry*, frequently imbedded as veins in other rocks. The matrix is usually a fine-grained mixture of the same elements as those which form the larger disseminated crystals. In granitic porphyry, which is very poor in quartz, the feldspathic base is almost granular, and laminated ([254]).

3. *Greenstone, Diorite*, granular mixtures of white albite and dark-green hornblende, forming *dioritic porphyry* when the crystals of albite are disseminated in a compact paste. The greenstones, either pure, or, as in the Fichtelgebirge, containing laminæ of diallage, and passing into serpentine, have sometimes penetrated, in the form of beds, between ancient strata of green argillaceous schist; but they more often traverse the rock in the form of veins, or appear as domes of greenstone, analogous to domes of basalt and of porphyry ([255]).

Hypersthene rock is a granular mixture of labrador feldspar, and hypersthene.

Euphotide and serpentine, containing sometimes, instead of diallage, crystals of augite and uralite, and thus becoming nearly allied to a more abundant, and, I might almost say, a more active eruptive rock, viz. augitic porphyry ([256]).

Melaphyre, and the porphyries containing crystals of augite, uralite, and oligoklas; to which latter species the celebrated verd-antique belongs.

Basalt, containing olivine and its elements (which, treated with acids, give gelatinous precipitates), phonolite (argillaceous porphyry), trachyte, and dolerite : the first of these rocks is partially, and the second always, divided into tabular laminæ, which gives to them an appearance of stratification, even when covering a large extent.   Mesotype and nepheline form, according to Girard, an important part in the composition and internal texture of basalts.   The nepheline in basalt reminds the geologist of the miascite of the Ilmen mountains in the Ural ([257]), a mineral which has been confounded with granite, and which sometimes contains zircon; it also reminds him of the pyroxenic nepheline discovered by Gumprecht near Lobau and Chemnitz.

To the second or sedimentary class of rocks belongs the greater part of the formations comprehended under the old systematic but incorrect denominations of floetz rocks, or of transition, secondary, and tertiary formations.   If earth-quakes and erupted rocks had not exerted an upheaving and disturbing influence on the sedimentary formations, the surface of our planet would have consisted of horizontal strata, regularly superimposed one over the other. Deprived of our mountain chains,—the declivities of which may be said to reflect, in the picturesque gradation of the different vegetable forms with which they are clothed, the scale of diminishing atmospheric temperature from their base to their summits,—the only features of variety in the disposition of the ground would have been the occasional presence of ravines hollowed out by the feeble erosive force of currents of fresh water, and slight eminences of transported detritus due to the same cause. From pole to pole, under every region, continents would have presented to the eye the dreary uniformity of the llanos of South America, or the steppes of Northern Asia; the vault of heaven would have everywhere appeared to rest on the unbroken plain, and the stars to rise and set as on the horizon of the ocean. Such a state of things, however, cannot have had a long duration even in the primitive world, for at all periods subterranean forces have exerted their modifying influence.

Sedimentary strata have been either precipitated or deposited from liquids, according as the materials which constitute them were chemically dissolved or mechanically suspended. But in the case of earths dissolved in fluids impregnated with carbonic acid, their descent after precipitation, and their accumulation in strata, require to be regarded as a true mechanical process; a view of some importance in respect to the envelopment of organic bodies in calcareous beds. The oldest sedimentary strata of the transition and secondary series were probably formed from water of high temperature, at a time when the heat of the upper surface of the globe was still very considerable. In this point of view a plutonic influence may be said to have acted in a certain degree even on the sedimentary strata, especially on the older ones ; but these strata appear to have hardened, and to have acquired their schistose structure, under great pressure, whereas the rocks which issued from the interior (granite, porphyry, and basalt), solidified by cooling. As the heat of the waters gradually diminished, they could absorb a larger portion of the carbonic acid gas existing in the atmosphere, and were thus fitted for holding a larger quantity of lime in solution.

The sedimentary rocks, excluding all other exogenous purely mechanical deposits of sand or detritus, are as follow :—

Argillaceous schist of the lower and upper transition series, comprehending the silurian and devonian formations, from the lower silurian strata, once termed cambrian, to the upper strata of the old red sandstone or devonian period, immediately below the mountain limestone.

Carboniferous deposits.

Limestones, included in the transition and carboniferous formations, the zechstein, the muschelkalk, the Jura or oolitic limestone, the chalk, and various beds of the tertiary period which cannot be classed amongst the sandstones or conglomerates.

Travertin, including fresh water lime-stone, and siliceous concretions from hot springs, formations which have not been produced under the pressure of a great body of sea water, but almost at the surface, in shallow marshes and streams.

Infusorial masses, a geological phænomenon of great importance, as it has revealed to us the influence which organic life has exercised on the formation of a part of the solid crust of the earth; the existence of such masses is a very recent discovery, for which science is indebted to my distinguished friend and travelling companion Ehrenberg.

If in this short but general review of the mineralogical constituents of the crust of the earth, I do not place immediately after the simple sedimentary rocks, those conglomerates and sandstones, which are also partially sedimentary deposits, and which alternate with argillaceous schists and chalks in the secondary and older formations,—it is only because these conglomerates and sandstones are not composed solely of the debris of eruptive and sedimentary rocks, but contain also the detritus of gneiss, mica slate, and other metamorphic masses. The obscure process of metamorphism, and the influence it exerts, should therefore form the third class of fundamental rocks.

I have already had occasion to remark that the endogenous or erupted rocks (granite, porphyry, and melaphyre), not only act dynamically, shaking, elevating, inclining, and laterally displacing the superincumbent strata, but they also modify the chemical combinations of their elements, and the nature of their internal structure : thus forming new kinds of rocks, of which the gneiss, mica slate, and granular or saccharoidal limestone (Carrara and Parian marble) may be cited as examples. The schists of the silurian or devonian periods, the belemnitic limestone of the Tarantaise, the dull grey calcareous sandstone, containing fucoids of the northern Apennines (*macigno*), often assume in their altered state a new and brilliant appearance, which renders their recognition difficult. The metamorphic theory has been established by following step by step the successive phases of transformation, and by bringing to the aid of inductive conclusions, direct chemical experiments on the effects of different degrees of fusion and pressure, and different rates of cooling. When the study of chemical combinations is pursued under the guidance of leading ideas ([258]), a bright light may be thrown on the wide field of geology, and on the operations of the great laboratory of nature, in which subterranean forces have formed and modified the terrestrial strata. But to avoid being misled by apparent analogies to entertain too narrow a view of the processes of nature, the philosophical inquirer must ever keep in view the complicated conditions, and the unknown intensity, of the forces which, in the primitive world, modified the reciprocal action of the several substances. It cannot, however, be doubted that the elementary substances always obeyed the same laws of affinity; and I am fully persuaded, that where apparent contradictions are met with, the chemist will generally succeed in explaining them, by ascending in thought to the primary conditions of nature, which cannot be identically reproduced in his experimental researches.

Observations made with great care, and over considerable tracts of country, shew that erupted rocks have acted in a regular and systematic manner. In parts of the globe most distant from each other ([259]), granite, basalt, and diorite, are seen to have exerted, even in the minutest details, a perfectly similar metamorphic action on the argillaceous schists, the compact limestone, and the grains of quartz in sandstone. But whilst the same kind of erupted rock exercises almost every where the same kind of action, the different rocks belonging to this class, present, in this respect, very different characters. The effects of intense heat are indeed apparent in all the phænomena; but the degree of fluidity has varied greatly in all of them from the granite to the basalt : and at different geological epochs, eruptions of granite, basalt, greenstone, porphyry, and serpentine, have been accompanied by the issue of different substances in a state of vapour. According to the views of modern geology, the metamorphism of rocks is not confined to the effects of simple contact, or of the juxta-position of two kinds of rock; but it comprehends all the phænomena that have accompanied the issuing forth of a particular erupted mass; and even where there has been no immediate contact, the mere proximity of such a mass has frequently sufficed to produce modifications in the cohesion of the particles and texture of the rock, in the proportions of the silicious ingredients, and in the forms of crystallization of the pre-existing rocks.

All eruptive rocks penetrate as veins into sedimentary strata, or into other previously existing endogenous masses; but there is an essential difference in this respect between plutonic rocks,— granites, porphyries, and serpentines,— and those called volcanic in the most restricted sense,—trachytes, basalts, and lavas. The rocks produced by the still existing volcanic activity present themselves in narrow streams, and do not form beds of any considerable breadth except where several meet together and unite in the same basin. Where it has been possible to trace basaltic eruptions to great depths, they have always been found to terminate in slender threads, examples of which may be seen in three places in Germany, —near Marksuhl, eight miles from Eisenach,—near Eschwege, on the banks of the Werra, —and at the Druidical stone on the Hollert road (Siegen). In these cases the basalt, injected through narrow orifices, has traversed the bunter sandstone and greywacke slate, and has spread itself out, in the form of a cup; sometimes forming groups of columns, and sometimes divided into thin laminæ. This, however, is not the case with granite, syenite, porphyritic quartz, serpentine, and the whole series of unstratified rocks, to which, by a predilection for mythological nomenclature, the term plutonic has been applied. With the exception of occasional veins, all these rocks have been forced up in a semi-fluid or pasty condition, through large fissures and wide gorges, instead of gushing in a liquid stream from small orifices; and they are never found in narrow streams like lava, but in extensive masses (²⁶⁰). Some groups of dolerites and trachytes shew traces of a degree of fluidity resembling that of basalt; others, forming vast craterless domes, appear to have been elevated in a simply softened state; others again, like the trachytes of the Andes, in which I have often remarked a striking analogy to the greenstone and syenitic porphyries (argentiferous without quartz), are found in beds like granite and quartzose porphyry.

Direct experiments (²⁶¹) on the alterations which the texture and chemical constitution of rocks undergo, from the action of heat, have shewn that volcanic masses (diorite, augitic porphyry, basalt, and the lava of Etna) give different products according to the pressures under which they are melted, and the rate at which they are cooled; if the cooling has been rapid, they form a black glass, homogeneous in the fracture; if slow, a stony mass, of granular or crystalline structure; and in this latter case crystals are formed in cavities, and even in the body, of the mass in which they are imbedded. The same materials also yield products very dissimilar in appearance; a fact of the highest importance in the study of eruptive rocks, and the transformations which they occasion; since, for example, carbonate of lime, melted under high pressure, does not part with its carbonic acid, but becomes when cooled granular limestone or saccharoidal marble, when the operation is performed by the dry method, while in the humid process, calcareous spar is produced with a less, and arragonite with a greater, degree of heat [262]. The mode of aggregation of the particles which unite in the act of crystallization, and consequently the form of the crystal itself, are also modified by differences of temperature [263]; and even where the body has not been in a state of fluidity, the particles, under particular circumstances, may undergo a new arrangement, manifested by different optical properties [264]. The phænomena presented by devitrification,—by the production of steel by casting or cementation,—by the passage from the fibrous to the granular texture of iron, occasioned by increased temperature [265], and possibly by the influence of the long-continued repetition of slight concussions,—may e ucidate the geological study of metamorphism. Heat sometimes elicits opposite effects in crystalline bodies; for Mitscherlich's beautiful experiments have established the fact, that without altering its condition of aggregation, calcareous spar, under certain conditions of temperature, expands in one of its axial directions while it contracts in the other [266].

Passing from these general considerations to particular examples, we may mention the case of schist converted by the vicinity of plutonic rocks into roofing slate of a dark blue colour and glistening appearance; the planes of stratification are intersected by other divisional planes, often almost at right angles with those of stratification, indicating an action posterior to the alteration of the schist (²⁶⁷). The silicic acid which has penetrated into the mass causes it to be traversed by veins of quartz, and transforms it in part into whetstone and siliceous schist; the latter sometimes containing carbon, and then perhaps capable of producing galvanic phænomena. The most highly silicified rocks of this kind are known as ribbon jasper (²⁶⁸), a material valuable in the arts, produced in the Oural mountains by the eruption and contact of augitic porphyry (as at Orsk), of dioritic porphyry (as at Aufschkul), or of a rounded mass of hypersthene rock (as at Bogoslowsk). In the island of Elba (at Monte Serrato), according to Friedrich Hoffman, and in Tuscany, according to Alexandre Brongniart, the ribbon jasper is formed by contact with euphotide and serpentine.

Sometimes (as observed by Gustav Rose and myself in the Altai, within the fortress of Buchtarminsk) (²⁶⁹), the contact and plutonic action of granite have rendered argillaceous schists granular, and transformed the rock into a mass resembling granite itself, consisting of a mixture of feldspar and mica, in which larger laminæ of mica are found imbedded (²⁷⁰). We are told by Leopold von Buch, " that all the gneiss between the Icy Sea and the Gulf of Finland has been produced by the metamorphic action of granite upon the silurian strata. In the Alps near the St. Gothard, calcareous marl has been similarly changed by the influence of granite, first into mica slate, and subsequently into gneiss (²⁷¹)." Similar phænomena of gneiss and mica slate, formed under the influence of granite, present themselves in the oolitic group of the Tarantaise (²⁷²), in which belemnites are found in rocks which have already in great measure assumed the character of mica slate,—in the schistose group of the western part of the island of Elba, not far from Cape Calamita,—and in the Fichtelgebirge near Baïreuth, between Lomitz and Markleiten (²⁷³).

I have already alluded to the jasper employed in the arts, and which the ancients could not obtain in large masses (²⁷⁴), and have described it as produced by the volcanic action of augitic porphyry; there is also another material of which ancient art made the noblest and most extensive use, *i. e.* granular or saccharoidal marble, which is to be regarded as a sedimentary rock, altered by terrestrial heat and the vicinity of erupted rocks. This assertion is justified by a careful observation of the phænomena which result from the contact of igneous rocks, and by the remarkable experiments made by Sir James Hall on the fusion of mineral substances. These experiments, made more than half a century ago, together with the attentive study of the phænomena of granitic veins, have contributed in a very high degree to the recent progress of geological science. Sometimes the metamorphic action of the erupted rock extends only to a very small distance from the surface of contact, and produces a partial transformation, or a sort of penumbra, as in the chalk of Belfast in Ireland traversed by veins of basalt, and as in the compact calcareous beds, partially inflected by the contact of syenitic granite, near the bridge of Boscampo, and at the cascade of Canzocoli, in the Tyrol, brought into notice by Count Marsari Pencati (²⁷⁵). Another mode of transformation is, when the whole of the beds of compact limestone become granular by the action of granite, syenite, or dioritic porphyry (⁷⁶).

Let me here make a special mention of the Parian
and Carrara marbles, to which the noblest works of
sculpture have given such celebrity, and which were so long
regarded in our geological collections as the types of primitive limestone. The action of granite has been exerted
sometimes by immediate contact, as in the Pyrenees ([277]);
sometimes through intermediate beds of gneiss or of mica
slate, as in Greece, and the islands of the Ægean
Sea. In both cases the transformation of the calcareous
rock has been cotemporaneous with the granite, but the
process has been different. It has been remarked that in
Attica, in the island of Eubœa, and in the Peloponnesus,
"the limestone superposed on mica slate is more beautiful and
more crystalline, as the mica slate is most pure, or least argillaceous," and it is known that mica slate and beds of gneiss
shew themselves at many points beneath the surface in Paros
and Antiparos ([278]). Xenophanes of Colophon (who supposed
the whole surface of the earth to have been originally covered
by the sea), remarked, in a notice preserved by Origen ([279]),
that marine fossils had been found in the quarries of Syra-
cuse, and the impression of a small fish (a sardine) at the
bottom of that of Paros; supposing the latter statement to
have been correct, we might infer the presence of a fossiliferous
bed but partially metamorphosed. The Carrara (Luna) marble,
which, from the Augustan era, and even from an earlier period,
has afforded the principal supply of statuary marble, and will
probably continue to do so unless the quarries of Paros are
reopened, is a bed, altered by plutonic action, of the same
calcareous sandstone (macigno), which, in the insulated Alp
of Apuana, shews itself between micaceous and talcose
schists ([280]). A very different origin has indeed been
assigned for marble in some other localities; and whether in

some cases granular limestone may not have been formed in the interior, and raised to the surface by gneiss and syenite ([281]), where it occupies fissures, as at Auerbach or in the Bergstrasse, is a question on which I do not venture to express an opinion, because I have not personally visited the localities.

The most remarkable instance of metamorphism produced by erupted rocks on compact calcareous strata, is that which Leopold von Buch has pointed out in masses of dolomite, especially on the southern Tyrol, and on the Italian declivity of the Alps. The alteration of the limestone appears to have been effected by means of fissures traversing it in every direction; the cavities are everywhere covered with rhomboidal crystals of magnesia, and the whole formation consists of a granular agglomeration of crystals of dolomite, without any trace of the original stratification, or of the fossils which were previously contained in it. Laminæ of talc are partially disseminated in the new rock, and it is interspersed with masses of serpentine. In the valley of the Fassa the dolomite rises perpendicularly in smooth walls of dazzling whiteness to a height of several thousand feet. It forms groups of numerous sharply-pointed conical mountains, clustered but separate. These features recal the lovely mountain landscape with which the imagination of Leonardo da Vinci has adorned the background of the portrait of Mona Lisa.

The geological phænomena which we are here describing, and which interest both the imagination and the intellect, result from the action of augitic porphyry, which has elevated, shattered, and transformed the beds which are above it ([282]). The illustrious geologist who first brought into notice the conversion of limestone into dolomite, does not attribute it to the introduction of a certain portion of talc derived from the black porphyry, but considers it a modification of the limestone, contemporaneous with the projection of the erupted rock through wide fissures filled with vapours. But in certain localities, beds of dolomite are found interposed between limestone strata, and it is yet to be explained how the transformation can have taken place without the presence of an erupted rock; and where are we to look for the concealed channels of the plutonic action? We ought not to resort, however, even in this case, to the old Roman adage, "that much that is alike in nature has been formed in different ways:" since, over widely extended parts of the earth, we have seen two phænomena associated,—the protrusion of a certain igneous rock,— and the transformation of compact limestone into a crystalline mass, possessing new chemical properties,—we may well suppose that, in the few cases in which the latter phænomenon alone is visible, future observation will remove the difficulty, and will shew that the apparent anomalies are due to the conditions under which the general causes of metamorphism have acted in the particular cases. We cannot doubt the volcanic nature and igneous fluidity of basalt, because some rare instances occur of basaltic dykes traversing a bed of coal, without reducing it to charcoal,— of sandstone without producing the usual effect of heat,— or of chalk without converting it into granular marble. If we have as yet obtained only an imperfect light to guide us in the obscure domain of mineral formations, it would surely be unwise to abandon it, because there are some points in the history of the transformation of rocks, and some difficulties connected with the intcrposition of beds of altered rock between unaltered strata, which we cannot wholly explain.

Having described the transformation of compact carbonate of lime into granular limestone and dolomite, we have to notice a third mode of alteration in the same rock, occasioned by the emission at some ancient epoch of the vapours of sulphuric acid. The gypsum thus produced offers analogies with beds of rock salt and sulphur (the latter deposited from aqueous vapour charged with that mineral). In the lofty Cordilleras of Quindiu, far from any volcano, I have observed deposits of sulphur in fissures in gneiss, while in Sicily (at Cattolica, near Girgenti), sulphur, gypsum, and rock salt, are found in the most recent secondary formations (283). At the edge of the crater of Vesuvius also I have seen fissures filled with rock salt, sometimes in masses sufficiently considerable to occasion a contraband trade; while on the northern and southern declivities of the Pyrenees, one cannot doubt the connection of dioritic (pyroxenic?) rocks with the occurrence of dolomite, gypsum, and rock salt (284). In these phænomena everything indicates the action of subterranean forces on the sedimentary strata deposited by the ancient sea.

There is much difficulty in assigning the origin of those vast masses of pure quartz which are characteristic of the Andes of South America (285). In descending towards the Pacific, from Caxamarca to Guangamarca, I have found beds of quartz from seven to eight thousand feet in thickness, resting sometimes on porphyry devoid of quartz, and sometimes on diorite. Can these be a metamorphosed sandstone, such as Elie de Beaumont conjectures to be the origin of the beds of quartz of the Col de la Poissonière, to the east of Briançon? ([286]).   In the diamond districts of Minas Geraes and St. Paul, in Brazil, which have been recently studied with great care by Clausen, the plutonic action of dioritic veins has produced ordinary mica and specular iron in quartzose itacolumite.   The diamonds of Grammagoa are contained in silicious beds, and are sometimes enveloped in laminæ of mica, like the garnets found in mica slate. The most northern diamonds yet known, which have been discovered since 1829, in 58° lat., on the European declivity of the Oural, are geologically related to the black carboniferous dolomite of Adolphskoi ([287]), and to augitic porphyry; but these relations have not yet been sufficiently elucidated by exact observation.

Among the most remarkable phænomena of contact, I may also allude to the formation of garnets in argillaceous schist in contact with basalt and dolerite in Northumberland and the Island of Anglesea, and the production of a large quantity and variety of beautiful crystals of garnet, vesuvian, augite and ceylanite, at the surface of contact of erupted and sedimentary rocks; namely, at the junction of the syenite of Monzon with dolomite and compact limestone ([288]). In the island of Elba, masses of serpentine, which perhaps nowhere present such clear evidences of their eruptive character, have produced sublimations of specular iron, and of red oxide of iron, in fissures of calcareous sandstone ([289]). We still see specular iron formed daily from sublimation on the sides of open fissures in the craters, and in recent currents of lava, of the volcanoes of Stromboli, Vesuvius, and Etna ([290]). The veins thus formed under our eyes by volcanic forces, in rocks which have attained a certain degree of solidity, teach us how, in earlier terrestrial or geological epochs, metalliferous or mineral veins may have been produced, wherever the crust of our planet, then thinner, and frequently rent by earthquakes, was fractured and fissured in every direction by change of volume in cooling, presenting communications with the interior, and a means of escape for ascending vapours and sublimations of the metals and the earths. The arrangement of the particles in layers parallel with the bounding surfaces of veins,—the regular repetition of layers of the same materials on opposite parts,—for example, on the walls of veins,—and the elongated drusy cavities occupying the middle space, often furnish direct evidence of the plutonic act of sublimation in metalliferous veins. As the veins or dykes which traverse rocks are more recent than the rocks which are traversed by them, the relative positions of the porphyry and of the argentiferous ores in the mines of Saxony, which are the richest and most important in Germany, teach us that they are at least more recent than the remains of vegetation in the coal measures and in the lower portions of the new red sandstone (Rothliegendes) [291].

Our geological hypotheses of the formation of the crust of the earth, and of the metamorphism of rocks, have derived unexpected elucidation, from the comparison of minerals elaborated by nature with the products of our smelting furnaces, and from the endeavours to reproduce the former artificially from their elements [292]. In all these operations, the same affinities which determine chemical combinations are in action both in our laboratories and in the bosom of the earth. Amongst the minerals formed artificially are the most important simple minerals which characterise very widely distributed eruptive plutonic and volcanic rocks, as well as metamorphic rocks altered by them; and these have been produced in a crystalline state, and with complete identity. We must distinguish, however, between minerals formed accidentally in the scoriæ, and those produced by chemical operations purposely devised: to the first class belong feldspar, mica, augite, olivine, blende, crystallized oxide of iron (specular iron), octahedral magnetic oxide of iron, and metallic titanium ([293]); to the second, garnet, idocrase, ruby (as hard as the oriental ruby), olivine, and augite ([294]). These minerals form the principal constituents of granite, gneiss, and mica slate, of basalt, dolerite, and many porphyries. The artificial production of feldspar and mica is of singular geological importance, in reference to the theory of the metamorphic conversion of argillaceous schist into quartz. We have in the schist all the elements of granite, without even excepting potash ([295]). It would not be very surprising, therefore, as our ingenious geologist von Dechen has justly remarked, if on some occasion we were to find a fragment of gneiss formed on the inside wall of a furnace built of argillaceous schist and greywacke.

Having passed in review the three great classes of erupted, sedimentary, and metamorphic rocks, we have still to notice the fourth class, comprising conglomerates, or rocks formed of detritus. The terms which we employ to designate this class recal the revolutions which the crust of the earth has undergone, as well as the cementing action which has consolidated, by means of the oxide of iron, or argillaceous or calcareous pastes, the sometimes rounded, and sometimes sharply angular masses of fragments. Conglomerates and breccias, in their widest acceptation, present the characters of a double origin. Their mechanical constituents have not been accumulated solely by the action of the sea, or by streams of fresh water, and there are some of these rocks to the formation of which the action of water has not contributed. " When basaltic islands or trachytic mountains have been elevated through large fissures, the friction of the ascending masses against the sides of the fissures has occasioned the basalt or the trachyte to be surrounded by conglomerates, formed from fragments of their own substance. In the sandstones of many formations, the grains of which they are composed have been separated, rather by the friction of erupted plutonic or volcanic rocks, than by the erosive action of a neighbouring sea. The existence of this species of conglomerate (which is found in immense masses in both continents), testifies the intensity of the force with which the eruptive masses were impelled from the interior towards the surface. The pulverized materials must have been subsequently conveyed away by the waters, and disseminated in the beds where they are now found." (296) Formations of sandstone are found every where interposed between other strata, from the lower silurian series to the tertiary formations above the chalk. On the margins of the vast plains of the new continent, both within and beyond the tropics, we find these beds of sandstone extending in long ramparts or walls, as if indicating the ancient shore against which the billows of the sea once broke.

When we glance at the geographical distribution of rocks, and at the extent which each occupies of the portion of the crust of the earth accessible to our researches, we recognise that the most generally prevailing chemical substance is silica, usually opaque, and variously coloured: the substance next in abundance is carbonate of lime; then combinations of silicic acid with alumina, potash, and soda, with lime, magnesia, and oxide of iron.   The substances to which we give the generic name of rocks are definite associations of a small number of minerals, to which some other minerals attach themselves as it were parasitically, but always under definite laws.   These elements are not confined to particular rocks: thus quartz (silicic acid), feldspar, and mica, are the substances which, in their association, essentially constitute granite; but they are also found in many other rock formations, either singly or two of them combined.   A single example will suffice to shew how the proportions of these elements may vary in different rocks, and how quantitative relations distinguish a feldspathic from a micaceous rock.   Mitscherlich has shewn, that if we add to feldspar three times the quantity of alumina, and one-third of the proportion of silex which previously belonged to it, we obtain the composition of mica.   Both these minerals contain potassium; a substance of which the existence in many kinds of rock was no doubt anterior to vegetation.

The succession and the relative age of different formations are traced, partly by the order of superposition of sedimentary strata, of metamorphic beds, and of conglomerates, and partly by the nature of the formations which the erupted rocks have reached or traversed, but most securely by the presence of organic remains, and their diversities of structure. The application of botanical and zoological evidence in determining the age of rocks, and in fixing points in the chronology of the crust of the globe, which the genius and

sagacity of Hooke led him to anticipate, marks one of the most
brilliant eras in the progress of modern geology, into which
palæontological studies have, as it were, breathed new life,
investing it with fresh charms and richly varied interests.

In the fossiliferous strata are inhumed the remains of the
floras and faunas of past ages.   As we descend from stratum
to stratum to study the relations of superposition, we ascend
in the order of time, and new worlds of animal and vegetable existence present themselves to the view.   Widely
extended changes of the surface of the globe, elevations of
the great mountain chains of which we are able to deter
mine the relative age, have been accompanied by the destruction of existing species, and by the appearance of new
forms of organic life; a few only of the older remaining
for a time amongst the more recent species.   In our ignorance of the laws under which new organic forms appear from
time to time upon the surface of the globe, we employ the
expression of " new creations," when we desire to refer to
the historical phænomena of the variations which have taken
place at intervals, in the animals and plants which have
inhabited the basins of the primitive seas and the uplifted
continents.   It has sometimes happened that extinct species
have been preserved entire, even to the minutest detail of
their tissues and articulations. In the lower beds of the secondary period, the lias of Lyme Regis, a sepia has been found so
wonderfully preserved, that a part of the black fluid with which
the animal was provided myriads of years ago to conceal itself
from its enemies, has actually served, at the present time, to draw
its picture ([297]).   In other cases such traces alone remain, as
the impression which the feet of animals have left on wet sand
or mud over which they may have passed when alive, or the

remains of their undigested food (coprolites). Some strata furnish only the impression of a shell; but if it be one of a characteristic kind ([298]), we are able, on its production, to recognise the formation in which it was found, and to state other organic remains which were buried with it. Thus the shell brought home by the distant traveller acquaints us with the geological character of the countries which he has visited.

The analytical study of the animal and vegetable kingdoms of the primitive world has given rise to two distinct branches of science; one purely morphological, which occupies itself in natural and physiological descriptions, and in the endeavour to fill up from extinct forms the chasms which present themselves in the series of existing species; the other branch, more especially geological, considers the relations of the fossil remains to the superposition and relative age of the sedimentary beds in which they are found. The first long predominated; and the superficial manner which then prevailed of comparing fossil and existing species, led to errors of which traces still remain in the strange denominations which were given to certain natural objects. Writers attempted to identify all extinct forms with living species; as, in the sixteenth century, the animals of the New World were confounded by false analogies with those of the Old Continent. Camper, Sömmering, and Blumenbach, were the first to enter on a more rational course, and have the merit of having first applied the resources of comparative anatomy, in a strictly scientific manner, to that part of palæontology (the archæology of organic life), which treats of the bones of the larger vertebrated animals. But it is pre-eminently to the admirable works of George Cuvier and

Alexander Brongniart, that we owe the establishment of the science of fossil geology, by the successful combination of zoological types with the order of superposition and the relative age of strata. The oldest sedimentary strata present to us, in the organic remains which they contain, a variety of forms occupying very different gradations in the scale of progressive development. Of plants, we find a few Fuci, Lycopodiaceæ which were perhaps arborescent, Equisetaceæ, and tropical ferns : but of animals we discover a strange association of Crustacea (including Trilobites with reticulated eyes), Brachiopoda (Spirifers, Orthis), elegant Sphæronites allied to Crinoidea ([299]), Orthoceratites of the family of Cephalopoda, and numerous corals ; and, mingled with these animals of inferior order, we already find in the upper beds of the silurian system fish of singular form. The family of Cephalaspidæ, animals of which the head was defended with large bony enamelled plates, and fragments of one genus of which, (the Pterichtys), were long mistaken for trilobites, belong exclusively to the devonian formation (old red sandstone) ; this family constitutes, according to Agassiz, as distinctly marked a type in the series of fishes, as that which includes the Ichthyosaurus and Plesiosaurus among reptiles ([300]). Goniatites, belonging to the tribe of ammonites ([301]), also begin to shew themselves in the limestone, and in the greywacke of the devonian formation; and even appear in the lower silurian strata.

In respect to *invertebrate* animals, no very clear relation has yet been recognised between the age of rocks, and the physiological gradation of the species which they contain ([302]) ; but this relation manifests itself in a very systematic manner in *vertebrate* animals. In these, the most ancient forms, as we have just seen, are those of certain fishes; ascending in the scale of superposition, we find first reptiles, and then mammalia. The first reptile (a Saurian of the genus Monitor, according to Cuvier), and which had already attracted the notice of Leibnitz ([303]), is found in the copper slate (Kupferschiefer) of the zechstein in Thuringia; the Paleosaurus and the Thecodontosaurus of Bristol belong, according to Murchison, to the same period. The number of Saurians continues to augment in the muschelkalk ([304]), the keuper sandstone, and the oolite in which they reach their maximum. At the oolitic period (including under this name the lias) lived several species of Plesiosaurus, an animal having a long swan-like neck, formed in some cases of upwards of thirty vertebræ; the Megalosaurus, a gigantic reptile of forty-five feet in length, with bones of the extremities resembling those of the heavy terrestrial quadrupeds; eight species of Ichthyosauri with enormous eyes; the Geosaurus (Sömmering's Lacerta gigantea); and seven species of hideous Pterodactyles, or reptiles with membranous wings ([305]). There existed also towards the latter part of the period the colossal Iguanodon, an herbivorous animal; and in the chalk where the number of crocodilian Saurians begins to diminish, we find the Mososaurus of Conybeare, a crocodile of Maestricht. We learn from Cuvier, that animals belonging to the present race of crocodiles are found in the tertiary formations; and Scheuchzer's supposed human skeleton (homo diluvii testis), a great salamander allied to the axolotl which I brought from the lakes round the city of Mexico, belongs to the most recent fresh water formations of Œningen.

In studying the relative age of fossils by the order of superposition of the strata in which they are found, impor-

tant relations have been discovered between families and
species (the latter always few in number) which have disappeared, and those which are still living.   All observations
concur in shewing, that the fossil faunas and floras differ
from the present animal and vegetable forms the more widely,
in proportion as the sedimentary beds to which they belong
are lower or more ancient.   Thus great variations have successively taken place in the general types of organic life;
and these grand phænomena, which were first pointed out by
Cuvier ([306]), offer numerical relations, which Deshayes and
Lyell have made the object of important researches, by which
they have been conducted to decisive results, especially as regards the numerous and well observed fossils of the different
groups of the tertiary formation.   Agassiz, who has examined 1700 species of fossil fishes, and who estimates at
8000 the number of living species which have been described or which are preserved in our collections, affirms in
his great work, that, with the exception of one small fossil
fish peculiar to the argillaceous geodes of Greenland, he
has never met in the transition, secondary, or tertiary strata,
with any animal of this class specifically identical with any
living fish ; and he adds the important remark, that even
in the lower tertiary formations, a third of the fossil fishes
of the calcaire grossier, and of the London clay, belong to
extinct families.   Below the chalk we no longer find a single
genus of the present period ; and the singular family of the
Sauroid fishes (fishes with enamelled scales, almost approaching reptiles in some of their characters, and extending upwards from the carboniferous rocks, where the larger species
among them are found, to the chalk, where only a few individuals are met with), presents relations with two species
which now inhabit the Nile and some of the American rivers

(the Lepidosteus and the Polypterus), similar to those exhibited by the Mastodon and Anoplotherium of the ancient world when compared with our elephants and tapirs (307).

We learn from Ehrenberg's remarkable discoveries, that the beds of chalk, which still contain two species of these sauroid fishes, and in which are found gigantic reptiles, and a considerable number of corals and shells which no longer exist, are nevertheless composed in great part of microscopic polythalamia, many of which are still living even in the seas of our own latitudes, in the North Sea, and in the Baltic. Thus we see that the tertiary group of deposits resting immediately upon the chalk, and to which the name of Eocene group is usually given, is not strictly entitled to that designation, for the dawn of the world in which we live extends much farther back in the history of our planet than has been hitherto supposed (308).

We have seen that fishes, which are the oldest vertebrated animals, first appear in the silurian strata, and are found in all the succeeding formations up to the beds of the tertiary period. Reptiles begin in like manner in the zechstein (magnesian limestone) ; and if we now add, that the first mammalia (the Thylacotherium prevostii and Thylacotherium bucklandi, allied, according to Valenciennes, to marsupial animals) (309), are met with in the Stonesfield slate, a member of the oolite,—and that the first remains of birds have been found in the deposits of the cretaceous period (310), we shall have indicated the inferior limits, according to our present knowledge, of the four great divisions of the vertebrata.

In regard to invertebrate animals, we find corals and some shells associated, in the oldest formations, with very

highly organised cephalopodes and crustaceans, so that
widely different orders of this part of the animal kingdom
appear intermingled ; there are, nevertheless, many isolated
groups belonging to the same order, in which determinate
laws are discoverable. Whole mountains are sometimes found
to consist of a single species of fossil, goniatites, trilobites,
or nummulites. Where different genera are intermingled,
there often exists a systematic relation between the series of
organic forms and the superposition of the formations ; and
it has been remarked, that the association of certain families
and species follows a regular law in the superposed strata
of which the whole constitute one formation. Thus Leopold von Buch, after classifying the immense variety of
ammonites in well defined families, by observing the disposition of the lobes, has shewn that the ceratites belong to
the muschelkalk, the arietes to the lias, and the goniatites
to the transition limestone and older rocks [311]. Belemnites have their lower limit [312] in the keuper sandstone,
below the Jura or oolitic limestone, and their upper limit in
the chalk. It has been found that the waters in the most
distant parts of the globe were inhabited at the same epochs
by testaceous animals corresponding at least in generic
character with European fossils : Von Buch, for example,
has described species of exogyra and trigonia from the
southern hemisphere (from the volcano of Maypo in Chili),
and D'Orbigny has indicated species of ammonites and
gryphæa from the Himalaya, and the Indian plains of Cutch,
which have been supposed even specifically identical with
those of the ancient oolitic beds of France and Germany.

Strata defined by their fossil contents, or by the fragments of other rocks which they include, form a geological

horizon by which the geologist may recognise his position, and obtain safe conclusions in regard to the identity or relative antiquity of formations,—the periodical repetition of certain strata,—their parallelism,—or their entire suppression. If we would thus comprehend in its greatest simplicity the general type of the sedimentary formations, we find in proceeding successively from below upwards :—

1. The *transition group*, divided into lower and upper greywacke : this group includes the silurian and devonian systems ; the latter formerly called the old red sandstone.

2. The *lower trias* ([313]), comprising the mountain limestone, the coal measures, the lower new red sandstone (todtliegendes), and the magnesian limestone (zechstein).

3. The *upper trias*, comprising the bunter or variegated sandstone ([314]), the muschelkalk, and the keuper sandstone.

4. The *oolitic* or *jurassic series ;* including the lias.

5. The *cretaceous series*, the quadersandstein, the lower and the upper chalk. This group includes the Floetz formations of Werner.

6. The *tertiary group*, as represented in its three stages by the calcaire grossier and other beds of the Paris basin, the lignites or brown coal of Germany, and the sub-apennine group in Italy.

To these succeed transported soils (alluvium), containing the gigantic bones of ancient mammalia, such as the Mastodons, the Dinotherium, and the Megatheroid animals, among which is the Mylodon of Owen, an animal upwards of eleven feet in length, allied to the sloth. Associated with these extinct species are found the fossil remains of animals still living : elephants, rhinoceroses, oxen, horses, and deer. Near Bogota, at an elevation of 8200 French feet above the level of the sea, there is a field filled with the bones of Mastodon (Campo de Gigantes), in which I have had careful excavations made ([315]). The bones found on the table lands of Mexico belong to true elephants of extinct species. The minor ranges of the Himalaya, the Sewalik hills, which Major Cautley and Dr. Falconer have examined with so much zeal, contain, besides numerous Mastodons, the Sivatherium, and the gigantic land tortoise (Colossochelys), more than twelve feet in length and six in height, as well as remains belonging to still existing species of elephants, rhinoceroses, and giraffes. It is worthy of notice, that these fossils are found in a zone which still enjoys the same tropical climate which is supposed to have prevailed at the period of the Mastodons ([316]).

Having thus viewed the series of inorganic formations which compose the crust of the earth in combination with the animal remains interred in them, we have still to consider the vegetable kingdom of the earlier times, and to trace the epochs of the successive floras which have accompanied the increasing extent of dry land, and the progressive modifications of the atmosphere. As we have already remarked, the oldest strata contain only marine plants exhibiting cellular tissue, the devonian strata being the first in which some cryptogamic forms of vascular plants are found (calamites and lycopodiaceæ) ([317]). It had been inferred from certain theoretical views concerning the simplicity of the primitive forms of organic life, that in the ancient world, vegetable had preceded animal life, and that the former was in fact always the indispensable condition of the latter. But there do not appear to be any facts to justify this hypothesis; and the circumstance that the Esquimaux, and other tribes who

live on the shores of the Polar Sea, subsist at the present
day exclusively on fish and cetaceæ, is alone sufficient to
shew that vegetable substances are not absolutely indispensable to the support of animal life. After the devonian
strata and the mountain limestone, we come to a formation
in which botanical analysis has recently made the most
brilliant progress ([318]). The coal measures contain not
only cryptogamic plants analogous to ferns, and phænogamous monocotyledons (grasses, yucca-like liliaceæ, and palms),
but also gymnospermic dicotyledones (coniferæ and cycadeæ).
We have already distinguished nearly four hundred species in
the coal formation, but of these I will here merely enumerate
arborescent calamites and lycopodiaceæ; the scaly lepidodendron; the sigillaria sixty feet long, distinguished by a
double system of vascular fascicles, and sometimes found
upright and apparently having its roots attached; the stig-
maria approaching in some respects the cactaceæ; an immense number of fronds and sometimes stems of ferns, which,
by their abundance, indicate the insular character of the dry
land of that period ([319]); cycadeæ ([320]), and especially
palms ([321]), though fewer in number than the ferns; astero-
phyllites, with verticillate leaves, allied to naiades; and
coniferæ, resembling araucarias ([322]), having some faint
traces of annual rings. All this vegetation was luxuriantly
developed on those parts of the older rocks which had risen
above the surface of the water, and the characters which distinguish it from our present vegetation were maintained
through all subsequent epochs up to the last of the cretaceous
strata. The flora of the coal formation, comprising these
remarkable forms, presents a very striking uniformity of
distribution in genera, if not in species, over all the

parts of the surface of the earth which were then existing, as in New Holland, Canada, Greenland, and Melville
Island ([323]).

The vegetation of the ancient world presents forms
which, by their affinities with several families of living
plants, shew that in their extinction we have lost many
intermediate links in the organic series.   Thus, for example,
the lepidodendra find their place, according to Lindley, between the coniferæ and the lycopodiaceæ ([324]) ; whilst the
araucariæ and the pines present differences in the junction of
their vascular fascicles.   Even limiting our consideration to
the present vegetable world, we perceive the great importance of the discovery of cycadeæ and coniferæ, in the
flora of the coal measures, by the side of sagenariæ and
lepidodendra.   Coniferæ are allied not only to cupuliferæ
and betulinæ, with which they are associated in the lignites, but also to the lycopodiaceæ.   The family of the
cycadeæ, in their external aspect approach palms, while
in the structure of their flowers and seeds they present
material points of accordance with coniferæ ([325]).   Where
many beds of coal are placed one above another, the genera
and species are not always intermingled, but are more often
so disposed that only lycopodiaceæ and certain ferns are
found in one bed, and stigmariæ and sigillariæ in another.
To give an idea of the luxuriance of vegetation in the primitive world, and of the immense vegetable masses accumulated in particular places by streams or currents, and
transformed ([326]) into coal, I will notice the coal measures
of Saarbruck, where a hundred and twenty beds are found
one above another, exclusive of a great number which
are not more than a foot thick ;   and there are beds

exceeding thirty, and even exceeding fifty feet in thickness, as at Johnstone in Scotland, and in the Creuzot
in Burgundy. In the forests of the temperate zone at the
present period, the carbon contained in the trees which
grow upon a given surface would hardly suffice to cover it
with an average thickness of seven French lines in a century (327). It should also be remarked, that the masses of
drift-wood transported by rivers, or by marine currents, such
as those which are found at the mouth of the Mississippi, and
those which have formed the "hills of wood," described by
Wrangel, on the shores of the Polar Sea, may give us some
idea of the accumulations which must have taken place near
projecting points of land in inland waters, and along the
island shores of the ancient world, and which have produced
our present coal-beds : there can be no doubt, also, that
these beds owe a considerable portion of their substance,
not to large trunks of trees, but to grasses, to low branching shrubs, and to small cryptogamia.

The association of palms and coniferæ, which we have
noticed in the coal measures, continues through all the succeeding formations until far into the tertiary period. In
the present day it may almost be said that these families
avoid each other's presence. We have become so accustomed (although without sufficient ground), to regard
coniferæ as a northern form, that I well remember experiencing a feeling of surprise, when, in ascending from the
coast of the Pacific towards Chilpansingo and the elevated
valleys of Mexico, between the Venta de la Moxonera and
the Alto de los Caxones, at a height of above 4000 English
feet above the level of the sea, I rode for an entire day
through a thick forest of Pinus occidentalis, and saw
amongst these trees, resembling the Weymouth pine, fan

palms (Corypha dulcis) ([328]), covered with parrots of many-
coloured plumage.   South America has oaks, but not a
single species of pine; and the first of these familiar forms
of my native land which presented itself to my sight, was
thus in strange association with a fan palm.   Columbus, in
his first voyage of discovery, saw coniferæ and palms growing together on the north-eastern point of the island of
Cuba ([329]), and consequently within the tropics and nearly at
the level of the sea.   That wonderful man, whom nothing
escaped, notices this fact in his journal as remarkable ; and
his friend Anghiera, the secretary of Ferdinand the Catholic,
recounts with astonishment, that "in the newly discovered
lands palmeta and pineta are found together."   The comparison of the present distribution of plants over the surface
of the earth, with that disclosed by fossil floras, is of great
geological interest.   The southern temperate zone, according to Darwin's beautiful and animated description ([330]) of
its ocean-covered surface, its numerous islands, and its
wonderful intermixture of tropical forms of vegetation with
those of colder regions, offers us the most instructive examples for the study of the past and present geography of
plants.   The past is undoubtedly an important portion of
the history of the vegetable kingdom.

Cycadeæ, which, from the number of their fossil species,
must have occupied a far more important place in the
ancient than in the present vegetable world, accompany
their allies the coniferæ from the coal formation upwards,
but are almost entirely wanting in the variegated sandstone,
which contains the remains of a luxuriant growth of certain
coniferæ of peculiar form, Voltzia, Haidingera, and Albertia.
The cycadeæ attain a maximum in the keuper and the lias,
which contain twenty different species.   In the cretaceous

rocks, marine plants and naiades predominate. Thus, the forests of cycadeæ of the oolitic period had long disappeared, and in the oldest groups of the tertiary formation this family is very subordinate to the coniferæ and the palms [331].

The lignites, or beds of brown coal, which are found in each division of the tertiary period, contain, amongst the earliest land cryptogamia, some palms, a great number of coniferæ with well-marked annual rings, and arborescent forms (not coniferous) of a more or less tropical character. The middle tertiary period is marked by the re-establishment, in full numbers, of the families of palms and cycadeæ, and, finally, the most recent shews great similarity to our present vegetation, exhibiting suddenly and abundantly various pines and firs, cupuliferæ, maples, and poplars. The dicotyledonous stems in lignite are occasionally characterised by colossal size and great age. In a trunk found near Bonn, Nöggerath counted 792 annual rings [332]. In the turf bogs of the Somme, at Yseux, near Abbeville, a trunk of an oak tree has been found above fourteen feet in diameter, which is an extraordinary thickness for the extra-tropical parts of the old continent. It appears from Göppert's excellent investigations that " all the amber ' of the Baltic comes from a coniferous tree, which, judging from the remains of its wood and bark at different ages or stages of growth, seems to have been a peculiar species, approaching nearest to our white and red pines. The amber tree of the ancient world (Pinites succinifer) was far more resinous than any conifer of the present period, the resin being deposited not only as in our present trees within and upon the bark, but also in the wood itself, following the course of the medullary rays, which, as well as the cells,

are still distinctly recognisable under the microscope, and
large masses of white and yellow resin are sometimes found
between the concentric ligneous rings.   Among the vegetable substances inclosed in amber there are male and female blossoms of native needle-leaved trees and cupuliferæ;
but distinctly recognisable fragments of Thuia, Cupressus,
Ephedera, and Castania vesca, intermingled with those of
Junipers and Firs, indicate a vegetation different from that
now subsisting on the coasts and plains of the Baltic."

We have now passed through the whole series of formations in the geological portion of the general view of nature,
from the eruptive rocks and most ancient sedimentary
strata, to the alluvial soils on which are scattered the fragments of rock known by the name of " erratic blocks."   The
dissemination of these blocks has been the subject of much discussion; it has been attributed to glaciers, and to floating
masses of ice; but I am inclined to ascribe it rather to the
impetuous flow of waters from reservoirs in which they had
long been detained, and from which they were set free by
the elevation of mountain chains ([333]).   It is a point which
will probably long continue undecided, and to which I only
incidentally allude.   The most ancient members of the
transition series with which we are acquainted are the
schists and greywacke, containing remains of marine plants
of the silurian, or, as it was before called, the cambrian sea.
On what do these oldest formations, or,—supposing the
gneiss and mica schist beneath them to be merely metamorphic sedimentary rocks,—on what were the oldest sedimentary strata deposited?   May we venture a conjecture on
that which cannot be the subject of actual geological observation?   According to an Indian mythus, the earth is sup-

ported by an elephant, who, that he may not fall, is in his turn supported by a gigantic tortoise; but the credulous Brahmins are not permitted to ask on what the tortoise rests. We here venture on a somewhat similar problem, though aware that we cannot hope to escape criticism. In the astronomical portion of this work, reasons were given which seemed to make it probable that our planet has been formed from nebulous rings, separated from the solar atmosphere, agglomerated into spheroids, and consolidated by progressive condensation, beginning at the exterior and proceeding towards the center. We will suppose a first solid crust of the earth to have been thus formed, of which the oldest silurian strata were the upper part. The eruptive rocks, which broke through and upheaved these strata, rose from depths inaccessible to our research; they must have existed, therefore, below the silurian strata, and were composed of the same association of minerals which, when they are brought to the surface, become known to us as granite, augitic rock, and quartzose porphyry. Guided by analogy, we may assume that the substances which traverse the sedimentary strata, and fill up their wide fissures, are only ramifications of a great inferior mass. The foci of our present volcanoes are situated at enormous depths; and, judging from the fragments which, in various parts of the globe, I have found imbedded in volcanic currents of lava, I consider it as more than probable that a primitive granitic rock is the substratum and support of the whole edifice of superimposed fossiliferous strata ([334]). Basalt, containing olivine, does not shew itself before the cretaceous period, and trachyte still later; but we find that eruptions of granite certainly belong to the epoch of the oldest sedimentary strata of the transition formation, as is indeed shewn by the effects of their metamorphic action. Where knowledge cannot be obtained from direct evidence, it may well be permitted, after a careful comparison of the facts which are accessible, to take analogy as our guide, and to advance a conjecture which would restore to the old granite a part of its disputed claim to the title of *primordial rock*.

The recent progress of geology, and the extended knowledge of geological epochs, characterised and determined by the mineralogical composition of rocks, by the peculiarities and succession of the organic remains which they contain, and by the circumstances of their stratification, whether uplifted, inclined, or with its horizontality undisturbed, conduct us, pursuing the intimate causal connection of phenomena, to the distribution of the solid and liquid portions of the surface of our planet, its continents and seas. We here indicate a connecting point between the history of the revolutions which the globe has undergone, and the description of its present surface; between geology and physical geography, which are thus combined in the general consideration of the form and extent of continents. The boundaries which separate the dry land from the liquid element, and the relative areas of each, have varied greatly during the long series of geological epochs; they have been very different, for example, when the strata of the coal formation were deposited horizontally upon the inclined strata of the mountain limestone and the old red sandstone;—when the lias and the oolite were deposited on the keuper and the muschelkalk;—and when the chalk was precipitated on the slopes of the greensand and oolitic limestone. If, with

Elie de Beaumont, we give the names of *jurassic* or *oolitic sea*, and of *cretaceous sea*, to the waters from which the oolite and the chalk were respectively deposited in soft beds, the outlines of those formations will indicate, for the two corresponding geological epochs, the boundary between the dry land, and the ocean in which these rocks were then forming. Maps have been drawn representing the state of the globe in respect to the distribution of land and water at these periods. They rest on a more sure basis than the maps of the wanderings of Io, or even than those of Ulysses, which at best but represent legendary tales, whilst the geological maps are the graphical representations of positive facts.

The following are the results of the investigations which have had for their object the determination of the extent of the dry land at different epochs. In the most ancient times,—during the silurian, devonian, and carboniferous epochs, and even as lately as the triassic period,—the portion of the surface supporting land vegetation was exclusively insular. At a subsequent epoch these islands became connected with each other, forming numerous lakes and deeply indented bays. Finally, when the mountain chains of the Pyrenees, the Apennines, and the Carpathians, were elevated,—about the epoch, therefore, of the older tertiary formations,—the great continents possessed nearly their present form and extent. During the silurian epoch, when the cycadeæ were in the greatest abundance, and the gigantic saurians were living, the whole surface of dry land from pole to pole must have been less than it now is in the Pacific and Indian oceans. We shall see presently how this great preponderance of oceanic surface must have contributed, together with other causes, to equalise climates, and to maintain a high temperature. It is only necessary to add here, in reference to the progressive extension of dry land, that a short time before the cataclysms which, at longer or shorter intervals, caused the destruction of so many gigantic vertebrated animals, part of the continental masses presented the same divisions as at present. There prevails, both in South America and in Australia, a great analogy between the living animals and the extinct species of those countries. Fossil species of Kangaroo have been discovered in New Holland; and in New Zealand, the semi-fossilized bones of a gigantic struthious bird, the Dinornis of Owen, closely allied to the present Apteryx of the same islands, and remotely so to the recently extinct Dodo of the island of Rodriguez.

A considerable part of the height of the present continents above the surrounding waters may perhaps be due to the eruption of the quartzose porphyry, which overthrew with violence the first great terrestrial flora, the material of our coal-beds. The level portions of our continents, to which we give the name of plains, are the broad summits of mountains, of which the feet are at the bottom of the ocean : considered in respect to submarine depths, these plains are elevated plateaus, of which the original inequalities have been partially filled up by horizontal layers of later sedimentary deposits, and covered over with alluvium.

Amongst the leading considerations in this part of the general contemplation of nature, we must regard, first, the quantity of land raised above the water; next, the configuration of each great continental mass in horizontal extension

and vertical elevation. We then come to the two coverings
or envelopes of our planet, of which one, composed of
elastic fluids, the atmosphere, is general; and the other, the
sea, is local, or restricted to portions of the earth's surface. These two envelopes, air and sea, constitute a natural
whole, materially affecting the distribution of the various
climates of the surface of the globe, according to the relative
extent of land and sea, the form and aspect of the land, and
the height and direction of the mountain chains. It results
from this reciprocal influence of the atmosphere, the sea,
and the land, that great meteorological phænomena cannot
be well studied apart from geological considerations. Meteorology, as well as the geography of plants and of animals,
have only begun to make real progress, since the mutual
connection and dependence of the phænomena to be investigated have been recognised. It is true that the word climate has especial reference to the condition of the atmosphere; but this condition is itself subjected to the double
influence of the ocean, whose agitated waters are traversed
by currents differing greatly in temperature, and of the dry
land, which radiates heat with very different degrees of intensity, according to its varying characters of form, elevation, and colour, and whether bare or clothed with forests,
or with grasses or other low growing plants.

In the present state of our knowledge, the superficial extent of dry land compared to that of the liquid element, is
as 1 : 2·8 ; or, according to Rigaud, as 100 : 270 ([335]). The
islands form scarcely one-twenty-third part of the continental
masses, which are so unequallÿ distributed, that the northern
hemisphere contains three times as much land as the southern,
which is pre-eminently oceanic. From 40° South latitude

to the Antarctic Pole, its surface is principally covered with
water.   The liquid element predominates equally in the
space comprised between the eastern shores of the old, and
the western shores of the new continent, where it is only
interrupted by a few widely-scattered groups of islands.   The
learned hydrographer Fleurieu has very justly given to this
great basin, which, under the tropics, extends over 145
degrees of longitude, the name of " the Great Ocean," to
distinguish it from all other seas.   The southern hemisphere,
and the western (from the meridian of Teneriffe), are
therefore the most oceanic portions of the globe.

Such are the leading points in the comparison of the relative areas of land and sea,—a relation which exercises a
powerful influence on the distribution of temperature, the
variations of atmospheric pressure, the direction of the winds,
and the hygrometric state of the air which materially influences the development of vegetation.   When we consider
that nearly three-fourths of the entire surface of the globe
are covered by water ([336]), we shall be less surprised at the
imperfect state of meteorology before the commencement of
the present century; but since that epoch a considerable
mass- of exact observations on the temperature of the sea
in different latitudes and at different seasons has been
obtained, and numerically compared.

The philosophers of ancient Greece indulged in general
speculations with regard to the horizontal configuration of
the dry land; they discussed its greatest extent from east
to west, which, according to the testimony of Agathemerus,
was placed by Dicearchus in the latitude of Rhodes, and in the
direction of a line passing from the Pillars of Hercules to Thine.

This line has been termed the " parallel of the diaphragm of Dicearchus ;" and the exactness of its geographical position, which I have made the subject of discussion in another work, may well excite our astonishment ([337]). Strabo, guided no doubt by the views of Eratosthenes, appears to have been so fully persuaded that the 36th parallel of latitude, as the line of greatest extent of the then known world, had some intimate connection with the form of the earth, that he places under it, between Iberia and the coast of Thinæ, the land of which he divined the existence ([338]).

We have noticed the great inequality in the extent of dry land in the two hemispheres, whether we divide the sphere at the Equator, or at the Meridian of Teneriffe; the two great insular masses, or continents, eastern and western, old and new, present also some striking contrasts, and at the same time some analogies, deserving of notice. Their major axes are in opposite directions ; the eastern, or old, continent, extending, in its greatest dimension, from east to west, or more precisely from north-east to south-west; whilst the western continent extends from north to south, or more exactly from N.N.W. to S.S.E. On the other hand, both continents are terminated towards the north by a line coinciding nearly with the 70th parallel ; and to the south they both run into pyramidal points, having submarine prolongations, which are indicated by islands and shoals; such are, the Archipelago of Tierra del Fuego, the Lagullas bank south of the Cape of Good Hope, and Van Diemen's Island, separated from New Holland by Bass's Straits. The northern part of Asia passes beyond the above-mentioned (70th) parallel towards Cape Taimura (78° 16′ according to Krusenstern), and falls short of it from the mouth of the larger Tschukotschia

river to Behring's Straits, where the eastern extremity of
Asia (Cooke's East Cape) is, according to Beechey, in
66° 3′ N. lat. ([339]). The northern shore of the new continent
follows the 70th parallel with tolerable exactness, for the
lands to the north and south of Barrow's Straits are detached
islands.

The pyramidal form of all the southern terminations of
continents belong to those "similitudines physicæ in con-
figuratione mundi," to which Bacon called attention in the
Novum Organum, and with which Reinhold Forster, one of
the companions of Cook on his second voyage of circumnavigation, connected some ingenious considerations. Directing our attention eastwards from the meridian of Teneriffe,
we perceive that the terminations of the three continents,
i. e. the southern extremities of Africa, of Australia, and
of America, successively approach nearer to the South Pole.
New Zealand, which is fully twelve degrees of latitude in
length, seems to form a regular intermediate member between
Australia and South America; its southern termination is
likewise marked by an island, New Leinster.   We may
notice, further, as a remarkable circumstance, that the
projecting points of the old continent, both to the north and
to the south, are nearly under the same meridian ; thus the
Cape of Good Hope and the Lagullas bank are situated nearly
in the same meridian as the north cape of Europe; and the
peninsula of Malacca nearly in that of Cape Taimura ([340]).
We know not whether the two poles of the Earth are surrounded by land, or by an ice-covered sea; towards the
North Pole, the parallel of 82° 55′ has been reached, and
towards the South Pole, that of 78° 10′.

The pyramidal terminations of the great continents are
frequently repeated on a smaller scale, not only in the Indian

ocean, in the peninsulas of Arabia, Hindostan, and Malacca, but also in the Mediterranean, where Eratosthenes and Polybius had compared in this respect the Iberian, Italic, and Hellenic Peninsulas ([341]). Europe itself, having an extent of surface equalling only one-fifth part of that of Asia, may be considered as the western peninsula of the compact mass of the Asiatic continent, to which it bears, in point of climate, a relation somewhat similar to that of the peninsula of Brittany to the rest of France ([342]). The favourable influence of the articulated and varied form of a continent on the civilization and intellectual cultivation of its inhabitants was recognised by Strabo ([343]), who extolled as a special advantage the richly varied form of our little Europe. Africa ([344]) and South America, which also offer many other features of analogy in their configuration, are the two continents which have the simplest and least indented outlines, while the eastern side of Asia, as if it were rent by the force of the currents of the ocean ([345]), (fractas ex æquore terras,) presents a richly varied coast line; peninsulas and islands alternate along its shores, from the equator to 60° N. latitude.

Our Atlantic Ocean presents the characteristics of a valley. It is as if the flow of the waters had been directed first towards the north-east, then towards the north-west, and then again towards the north-east. The parallellism of the coasts north of 10° of South latitude, the projecting and re-entering angles, the convexity of Brazil opposite to the Gulf of Guinea, and the convexity of Africa to the Gulf of Mexico, all favour this view, which at first may seem too hazardous ([346]). In the Atlantic Valley, as is indeed usually the case in the form of large masses of land, coasts deeply

indented and fringed with many islands are placed opposite
to those of a contrary character.   It is long since I called
attention to the geognostical interest of the comparison of the
west coasts of Africa and of South America.   Within the
tropics, the deeply re-entering curve of the African coast
near Fernando Po (4°½ N. lat.), is repeated on the shores of
the Pacific (in 18°¼ S. lat.), where, between the Valle de
Arica and the Morro of Juan Diaz, the Peruvian coast suddenly alters the direction from south to north which it had
previously followed, and turns to the north-west.   This
change of direction in the coast of Peru is participated in
by the chain of the Andes ([347]) ; and not only by the maritime
branch of the two parallel ranges into which the mountains
are there divided, but also by the eastern Cordillera, in which
civilisation had its earliest seat in the South American
highlands, and where the small alpine lake of Titiaca lies at
the foot of the giant mountains of Sorata and Illimani.
Farther to the south, from Valdivia and Chiloe in 40° and
42° S. lat., through the Archipelago of los Chonos, to Tierra
del Fuego, we find the same peculiar "fiord" character,
which distinguishes the west coasts of Norway and
Scotland.

    Such are the most general considerations respecting the
form of continents in the horizontal direction, which an
examination of the present surface of our planet suggests.
We have brought facts together, so as to call attention to
certain analogies of form in regions remote from each other,
without, however, venturing to give to these analogies the
title of laws.   When an observer at the foot of an active
volcano, Vesuvius for example, notices the not unfrequent
phenomenon of partial elevations, in which, previously to, or

during the occurrence of an eruption, small portions of ground have their level permanently altered several feet, and are converted into shelving or flattened eminences, he perceives how greatly small accidental variations in the intensity of the subterranean force, and in the resistance which it has to overcome, must modify the form and direction taken by the upheaved particles. In like manner slight disturbances in the equilibrium of the elastic forces in the interior of our planet may have determined their action more towards the northern than the southern hemisphere, and have occasioned the elevation of the dry land in the eastern hemisphere in the form of a wide connected mass, having its major axis almost parallel to the equator,—and in the western and more oceanic hemisphere, in a comparatively narrow band, following the direction of the meridian.

But little can be ascertained by investigation respecting the causal connection of the great phænomena appertaining to the formation of our continents, and to the analogies and contrasts presented by their configuration. We know that a subterranean force was the agent; that the continents were not suddenly formed in their present shape, but that they gradually acquired it by progressive enlargement, or by the junction of smaller masses, effected by a series of successive elevations and depressions, commencing with the silurian epoch, and continuing to the tertiary. The present form has been produced by two causes, which have acted in succession, the one after the other : the first is a subterranean reaction, of which the measure and direction are unknown to us ; the second comprises all the causes acting at the surface, as volcanic eruptions, earthquakes, elevations of mountain chains, and oceanic currents. How different would have been the present state of temperature, of vegetation, of agri-

culture, and even of human society, if the major axes of the
old and new continents had been given the same direction;
if the chain of the Andes, instead of following a meridian,
had been directed from east to west; if no heat-radiating
mass of tropical land extended to the South of Europe; or if
the Mediterranean, which was once in connection both with
the Caspian and Red Sea, and which has so powerfully
favoured the social establishment of nations, were not in
existence: that is to say, if its bed had been raised to
the level of the plains of Lombardy and of the ancient
Cyrene.

The changes in the relative heights of the solid and liquid
portions of the surface, which have determined the emersion or
submersion of the lower lands, and the present outlines of
continents, must be referred to various causes acting at
different times. The most powerful among these have
no doubt been, elastic forces acting in the interior of the
earth; sudden changes of temperature affecting great
masses of rock ([348]); the unequal secular loss of heat in the
terrestrial crust and in the nucleus, causing ridges and con-
tortions in the solid crust; and local modifications of gravitation ([349]), with consequent changes of curvature in the
surface of equilibrium of certain portions of the liquid element. According to the opinion generally received among
the geologists of the present day, the elevation of continents above the sea is a real, and not merely an apparent or
relative elevation, such as would be occasioned by a depression of the general level of the sea. The merit of
this view, which has been derived from long observation
of connected facts, and from the analogy of important
volcanic phænomena, belongs to Leopold von Buch, who
advanced it for the first time in the narrative of his memo-

rable journey through Norway and Sweden, in 1806 and
1807 ([350]). While the whole coast of Sweden and Finland, from Sœlvitsborg at the limit of Northern Scania,
to Torneo, and from Torneo to Abo, is undergoing a
gradual rise amounting to four French feet in a century,
the southern part of Sweden, according to Nilsson, is being
depressed ([351]). The elevating force appears to attain its
maximum in the north of Lapland, and to diminish gradually southwards towards Calmar and Sœlvitsborg. Lines
marking the ancient level of the sea, in pre-historic times,
may be traced throughout Norway ([352]), from Cape Lindes-
næs to the North Cape, by banks of shells identical
with those of the present sea; these lines have been
recently examined and measured with great exactness by
Bravais, during his long winter sojourn at Bosekop. The
banks rise as high as 600 (640 English) feet above the
present level of the sea, and reappear, according to Keilhau
and Eugene Robert, on the coast of Spitzbergen, opposite to
the North Cape, in a N.N.W. direction. Leopold von
Buch, who was the first to call attention to the high
bank of shells at Tromsoe (lat. 69° 40'), has shewn,
that the more ancient elevations of the shores washed
by the North Sea belong to a series of phænomena wholly
different from the gentle and gradual elevation of the
Swedish coast in the Gulf of Bothnia. Nor must this latter
phænomenon, which has been established by sure historical
evidence, be confounded with changes of level caused by
earthquakes, as on the coasts of Chili and of Cutch: it has
recently led to similar observations in other countries, and
instances have been found, in which a sensible subsidence in
one part takes place, corresponding to an elevation else-

where; this has been observed in West Greenland (by Pin. gel and Graah), in Scania, and in Dalmatia.

As it appears more than probable that, in the earlier periods of our planet, the elevations and depressions of its surface were much more considerable than at present, we ought not to be surprised at finding, even in the interior of continents, portions of the surface depressed below the general level of the present sea; as the small natron lakes described by General Andreossy, the small bitter lakes of the isthmus of Suez, the Caspian Sea, the lake or sea of Tiberias, and above all, the Dead Sea [353].   The levels of the two last-named seas are respectively 625 (666 English) and 1230 (1311 English) feet below the level of the Mediterranean.   If we could remove the alluvial soil which so frequently covers the rocky strata in the level portions of the surface of the earth, we should discover how many parts of the denuded crust are below the present level of the sea.   The periodical, but irregularly recurring, rise and fall of the Caspian, of which I observed distinctly marked traces in the northern portion of the basin of that sea [354], and the observations of Darwin in the Coral Sea [355], appear to indicate that, without earthquakes properly so called, the surface of the earth still undergoes in some places slow and continued oscillations, similar to those which, in earlier times, must have been of very general occurrence in the then thinner solid crust.

The phænomena to which we are now directing our attention remind us of the instability of the present order of things, of the changes to which the outline and form of continents are probably still subject in long intervals of time.   These variations, though hardly sensible from one generation to the next, accumulate and become so in periods similar in duration to those of the remoter heavenly bodies. The eastern coast of the Scandinavian peninsula may have risen about 320 French feet in 8000 years; in 12000 years, if the present rate of movement were uniformly continued, the parts of the bed of the sea nearest to the shore, now covered with 45 fathoms of water, would begin to emerge. Such intervals of time, however, are short, when compared with the length of the geological periods disclosed to us by the series of superposed formations, and by the successive groups of extinct organic forms. We have hitherto considered only the phænomena of elevation; but the analogy of observed facts will justify us in assuming the equal possibility of the depression of large tracts of country: the mean height of the parts of France which are not mountainous is less than 480 (512 English) feet: geological changes, of moderate amount compared with those which took place in the earlier periods of the globe, would therefore be sufficient to effect the permanent submersion of large portions of north-western Europe, and would alter materially its general configuration.

All variations in the form of continents are produced either by the elevation or the depression of the land or sea, and these are complementary phænomena, since the real elevation of the one element produces the appearance of a depression in the other. In a general contemplation of nature such as is here presented, we should not overlook the *possibility* at least of a *real* depression of the general level of the ocean, or a diminution of the quantity of its waters. No one in the present day can doubt that, at the periods when the surface of the earth possessed a higher temperature,—when wide fissures, capable of receiving large masses of water, were frequent,—and when the constitution of the atmosphere differed materially from what it now is,—great changes of the oceanic level may have taken place from variations in the actual quantity of water: but in the present state of our planet, there is no direct evidence whatsoever of a progressive augmentation or diminution of the waters of the sea; nor is there any evidence of a progressive change in the mean height of the barometer at the level of the sea. According to Daussy and Antonio Nobile's experimental investigations, an increase in the mean height of the barometer would itself occasion a depression of the level of the sea. As, however, the mean pressure of the atmosphere is not the same in all latitudes, owing to meteorological causes, such as the general direction of the wind and the hygrometric state of the air, the barometer alone would not afford certain evidence of variations in the general level of the ocean. The remarkable fact, that several of the ports of the Mediterranean were repeatedly left dry for several hours about the commencement of the present century, appears to shew that, without any actual diminution of the general mass of water, or any general depression of the oceanic surface, changes in the direction and strength of currents *may* occasion a *local* retreat of the sea, and possibly a permanent emersion of a small portion of coast; but we cannot be too cautious in the interpretation of our very recently acquired knowledge of these complicated phænomena, lest we should be led to attribute to one of the old "four elements," water, that which really belongs to two others, viz. the air and the earth.

As the external form of continents, in the varied and deeply-indented outline of their coasts, exercises a beneficial influence on climate, trade, and the progress of civilisation, so also, in the interior, the variations of form in the vertical direction, by mountains, hills, valleys, and elevated plains, have consequences no less important. Whatever causes diversity of form or feature on the surface of our planet,—mountains, great lakes, grassy steppes, and even deserts surrounded by a coast-like margin of forest,—impresses some peculiar mark or character on the social state of its inhabitants. Continuous ridges of lofty mountains covered with snow impede intercourse and traffic ; but where lowlands are interspersed with discontinuous chains, and with groups of more moderate elevation ([356]), such as are happily presented by the south-west of Europe, meteorological processes and vegetable products are multiplied and varied ; and different kinds of cultivation, even under the same latitude, give rise to different wants, which stimulate both the industry and the intercourse of the inhabitants. Thus those formidable terrestrial revolutions, in which, by the reaction of subterranean forces, portions of the oxidized crust of the globe were upheaved, and lofty mountain chains were suddenly formed, have served, when repose was re-established and organic life re-awakened, to furnish a more beautiful and richer variety of individual forms, and to rescue the greater part of the dry land in both hemispheres from a dreary uniformity, which tends to impoverish both the physical and the intellectual powers of man.

The great views of Elie de Beaumont assign a relative age to each system of mountains, on the principle that their elevation must necessarily have intervened between two periods ;—between the deposition of the strata which have been upheaved and inclined, and of those which extend in undisturbed hori-

zontality to the base of the mountains, and which must therefore have been deposited subsequently to the strata which
are inclined ([357]).   The ridges of the crust of the earth,
which are of the same geological age, appear to follow a common direction.   The line of strike of the uplifted strata is not
always parallel to the axis of the chain, but sometimes intersects it; whence I am led to infer ([358]), that the phænomenon of the inclining of the strata, or of the disturbance of
their horizontality, (which sometimes extends to the adjacent
plain,) must, in such cases, be older than the elevation of the
mountains.   The general direction of the land of the European
continent is from south-west to north-east, and is at right
angles to the direction of the great fissures, which is from
north-west to south-east, extending from the mouths of the
Rhine and the Elbe through the Adriatic and Red Seas,
and the mountain system of Puschti-koh in Luristan, and
terminating in the Persian Gulf and Indian Ocean.   This
rectangular intersection of the continent, in the direction of
its principal extent, has powerfully influenced the commer-
cial relations of Europe with Asia and the north of Africa,
as well as the progress of civilization on the formerly more
flourishing shores of the Mediterranean ([359]).

The more the imagination and the intellect are impressed
by lofty and massive mountain chains, from the evidences
they afford of great terrestrial revolutions,—as the boundaries
of different climates,—as the lines of separation from whence
the waters flow to opposite regions,—and as the sites of
a peculiar vegetation,—the more necessary is a correct numerical estimation of their volume ; in order to show the smallness
of their mass when compared either to that of the continents, or

to that of the adjacent countries. Let us take, for example, the chain of the Pyrenees, in which both the mean elevation and the area covered have been measured with great exactness, and if we suppose the whole mass of these mountains to be spread equably over France, we find that its surface would be only raised thereby 108 (115 English) feet. In like manner, if the material which forms the chain of the Alps were spread equably over the whole surface of Europe, it would not raise it more than 20 (21.3 English) feet. I have found by a laborious investigation, which from its nature can only give a maximum limit, that the center of gravity of the land at present above the surface of the ocean is, in Europe 630, in North America 702, in Asia 1062, and in South America 1080 French feet (or 671, 748, 1132, and 1151 English feet) above the level of the sea [360]. These results shew that the more northern regions are comparatively of lower altitude. In Asia, the low elevation of the extensive plains, or steppes, of Siberia, is compensated by the mountain masses between the parallels of $28\frac{1}{2}°$ and $40°$, from the Himalaya to the Kuen-lun of Northern Thibet, and to the Tian-schian or celestial mountains. We may in some degree form an idea, from these calculations, in what portions of the surface of the globe the action of the subterranean plutonic forces, as exhibited in the upheaval of continental masses, has been most intense.

There is no sufficient reason why we should assume that the subterranean forces may not, in ages to come, add new systems of mountains to those which already exist, and of which Elie de Beaumont has studied the directions and relative epochs. Why should we suppose the crust of the earth to be no longer subject to the agency which has formed the ridges now perceived on its surface? Since Mont Blanc, and Monte Rosa, Sorata, Illimani, and Chimborazo, the colossal summits of the Alps and the Andes, are considered to be amongst the most recent elevations, we are by no means at liberty to assume that the upheaving forces have been subject to progressive diminution. On the contrary, all geological phenomena indicate alternate periods of activity and repose ([361]); the quiet which we now enjoy is only apparent; the tremblings which still shake the surface in every latitude, and in every species of rock,—the pro. gressive elevation of Sweden, and the appearance of new islands of eruption,—are far from giving us reason to suppose that our planet has reached a period of entire and final repose.

The two ambient coverings of the solid surface of our planet, the liquid ocean and the aerial atmosphere, present, by reason of the mobility of their particles, many analogies with each other with respect to currents and thermometric relations; whilst, at the same time, they also offer contrasts arising from the great difference in their conditions of aggregation and elasticity. The height and depth of both are unknown: soundings of 27600 English feet (or above four geographical miles), have been taken without finding any sea bottom; and if we assume with Wollaston that the atmosphere has a limit from which waves reverberate, the phenomena of twilight will indicate a height at least nine times as great. The aerial ocean rests partly on the solid earth, whose mountain chains and high table lands may be considered to represent shoals; and partly on the sea, whose surface forms a liquid base or floor in immediate contact with the lower and denser strata of humid air.

Proceeding both upwards and downwards from the common limit or plane of contact of the atmosphere and the ocean, we find the strata of air and of water subject to definite laws of decreasing temperature. The decrease is much more rapid in the ocean than in the atmosphere; the sea has, in all zones, a tendency to keep up the temperature of its superficial strata by the sinking of the cooled particles which are the heaviest. It has been found, by an extensive series of careful observations, that the surface temperature of the ocean, throughout a band extending forty-eight degrees on either side the equator, is, in its usual and mean condition, somewhat warmer than the adjacent strata of the air (362). By reason of the diminishing temperature at increasing depths, fish and other inhabitants of the sea, whose organs are fitted for deep water, may find even under the tropics the low temperature of cooler latitudes. This circumstance, which is analogous to that of the temperate and even cold Alpine climate of the elevated plains of the torrid zone, has an important influence on the migration and on the geographical distribution of many marine animals. The depths also at which fishes live, exercise, by reason of the increased pressure, a modifying influence on their cutaneous respiration, and on the quantity of oxygen and nitrogen contained in their swimming bladders.

The saline particles of sea water causing it to attain its maximum of density at a lower temperature than is the case with fresh water, we comprehend why water brought up from oceanic depths, in the voyages of Kotzebue and Dupetit

Thouars, was found at the temperature of 2°.8 and 2°.5 Cent. (37° and 36°.5 Fah.), and even lower. The almost icy temperature found at great sea depths within the tropics first led to the knowledge of a submarine polar current flowing from either pole towards the equator; as without this current, the depths of the tropical seas could only have a degree of cold equal to the lowest temperature of their surface water. The apparent anomaly offered in the Mediterranean, whose waters are found of a higher temperature than is usual at great depths, has been explained by Arago, and results from the fact that, at the Straits of Gibraltar, where the surface water of the Atlantic flows in as a westerly current, a counter current prevails beneath, and prevents the influx from the ocean of the cold current from the pole.

The Ocean is the great moderator and equalizer of terrestrial climates; at a distance from coasts, and where its surface is not disturbed by currents of cooled or heated water, it maintains throughout the tropics, (and especially in the equatorial region from 10° N. to 10° S.) a wonderful uniformity and constancy of temperature [363] over spaces of many thousand square miles. It has, therefore, been well said [364], that an exact and long-continued investigation of the thermic relations of the tropical seas might afford, in the simplest manner, a solution of the great and long-contested problem of the invariability of climates and of the temperature of the globe. Great changes of solar heat, of considerable duration, would be reflected in the altered mean temperature of the sea, with even greater certainty than in the mean temperature of the land.

Those zones in which the waters of the ocean attain in the one case their maximum of temperature, and in the other their maximum of density resulting from their saline contents, do not coincide with each other, or with the geographical equator. The waters of highest temperature appear to form two bands not quite parallel, one on either side of the equator. Lenz, in his voyage of circumnavigation, found in the Pacific the maxima of saltness in 22° N. and in 17° S. lat., and between them a minimum zone was interposed a few degrees south of the equator. Within the region of calms the solar heat has little effect in augmenting the relative proportion of the saline particles by the process of evaporation, because the stratum of air saturated with aqueous vapour which rests on the surface of the water is rarely disturbed by the action of the wind.

The surfaces of all connected seas must be regarded as possessing a *general* equality of mean level, although local causes (such as prevailing winds or currents) may produce small permanent differences in particular seas which form deep gulfs or inlets. An instance of this occurs in the Red Sea, whose level, near its northern extremity at the Ísthmus of Suez, is, at different hours of the day, twenty-four and thirty French feet above that of the neighbouring part of the Mediterranean : the form of the Straits of Bab el Mandeb, which is more favourable to the ingress of the waters of the Indian ocean than to their egress, being probably the cause of this remarkable fact, which was not unknown to the ancients [365.] The excellent geodesic operations of Coraboeuf and Delcros have shewn that at the two extremities of the Pyrenean Chain, as well as at Marseilles and the Northern coast of Holland, there is no sensible difference between the level of the Atlantic and of the Mediterranean Seas [366].

Disturbances of equilibrium, and consequent movements of the waters, are of three kinds : 1st, irregular and transitory, occasioned by winds, and producing waves which sometimes during storms, and at a distance from coasts, attain a height of more than 37 English feet; 2d, regular and periodic, dependent on the position and attraction of the Sun and Moon; and, 3d, the less considerable but permanent phæ-nomena of oceanic currents. The tides extend over all seas (except those small inland seas where the ebb and flow is scarcely if at all perceptible) ; they have received a complete explanation by the Newtonian doctrine, and have been thus brought "within the domain of necessary facts." The duration of each of these periodical oscillations of the sea is rather more than half a day; their height in the open ocean is not more than a few feet; but when the configuration of a coast opposes the progress of the wave, they may reach upwards of 50 feet, as at St. Malo, or even of 70 feet, as in the Bay of Fundy. The great geometer, Laplace, has shewn that, regarding the depth of the sea as inconsiderable in relation to the semi-diameter of the earth, the stability of equilibrium of the ocean requires that the density of its fluid mass should be less than the mean density of the earth. We have already seen that the mean density of the earth is actually five times that of water. Tides, therefore, caused by the action of the Sun and Moon can never overflow the elevated portions of the dry land; nor can they have transported the remains of marine animals to the summits of the mountains where they are now found ([367]). It is no small testimony of the value of analysis, sometimes so contemptuously regarded in the unscientific circles of civil life, that by his complete theory of tides, Laplace has enabled us to predict in our astronomical ephemerides the height of spring tides at the periods of new and full moon.

Oceanic currents, which exercise an important influence on the climate of neighbouring coasts, and on the intercourse of nations, depend concurrently on a variety of causes of unequal magnitude and operation. Amongst these we may reckon the propagation of the tide-wave in its progress round the globe, the duration and strength of prevailing winds, the variations of density which sea-water undergoes in different latitudes and depths by changes of temperature and of the relative quantity of its saline contents, (368) and, finally, the horary variations of the atmospheric pressure, so regular in the tropics, and propagated successively from east to west. The currents of the ocean present a remarkable spectacle; maintaining a nearly constant breadth, they cross the sea in different directions, like rivers of which the adjacent undisturbed masses of water form the banks. The line of demarcation between the parts in motion, and those in repose, is most strikingly shewn in places where long bands of sea-weed, borne onward by the current, enable us to estimate its velocity. Analogous phænomena are sometimes presented to our notice in the lower strata of the atmosphere, when, after a violent storm, the path of a limited aerial current may be traced through the forest by long lanes of overthrown trees, whilst those on either side remain unscathed.

The general movement of the sea from east to west between the tropics, known by the name of the equatorial or rotation current, is regarded as the joint effect of the trade winds, and of the progressive propagation of the tide-wave. Its direction is modified by the resistance which it

experiences from the eastern coasts of continents. The velocity of this current, computed by Daussy from data supplied by bottles, purposely thrown overboard and subsequently picked up in different localities, agrees within
one eighteenth part with the velocity of ten French nautical
miles in twenty-four hours, which I had previously deduced
from comparing the experience of different navigators ([369]).
Columbus was aware of the existence of this current at
the period of his third voyage, (the first in which he sought to
enter the tropics in the meridian of the Canary Islands), since
we find in his journal the following passage :—" I regard
it as proved that the waters of the sea move from east
to west as do the heavens" (or the apparent motion of the
sun, moon, and stars), " las aguas van con los cielos" ([370]).
Of the narrow currents, or true oceanic rivers, of which
we have spoken, some carry warm water into higher, and
others cold water into lower latitudes.   To the first class
belongs the celebrated gulf stream ([371]), the existence of
which was recognised by Anghiera ([372]), and more particularly by Sir Humphry Gilbert, as early as the sixteenth
century, and of which the first origin and impulse is to be
sought to the south of the Cape of Good Hope.   After a
wide circuit, it pours itself from the Caribbean Sea and the
Mexican Gulf through the channel of the Bahamas, and,
following a direction from S.S.W. to N.N.E., deviates more
and more from the coast of the United States, until, deflected still further to the east by the banks of Newfoundland,
it crosses the Atlantic, and casts an abundance of tropical
seeds (Mimosa scandens, Guilandina bonduc, Dolichos
urens) on the coasts of Ireland, of the Hebrides, and of
Norway.   Its north-easternmost prolongation mitigates the

cold of the ocean, and exercises a beneficent influence on the climate of the northernmost point of Scandinavia. At the point where the stream is deflected to the east by the banks of Newfoundland, it sends off an arm toward the south, not far from the Azores ([373]). This is the situation of the Sargasso Sea, or that great sea of weed, or bank of fucus, which made so lively an impression on the imagination of Columbus, and which Oviedo calls sea-weed meadows,—"praderias de yerva." These evergreen masses of Fucus natans (one of the most widely distributed of the social sea plants), driven gently to and fro by mild and warm breezes, are the habitation of a countless number of small marine animals.

This great current, which, in the Atlantic valley, between Africa, America, and Europe, belongs almost entirely to the northern hemisphere, has its counterpart in the Southern Pacific Ocean, in a current the effect of whose low temperature on the climate of the adjacent coasts was first brought into notice by myself in the autumn of 1802. This current brings the cold water of high southern latitudes to the coast of Chili, and follows its shores and those of Peru northward to the bay of Arica, and thence north-westerly to the neighbourhood of Payta, where the most westerly projection of the American coast deflects the stream, and causes it suddenly to quit the shore, taking a due west direction. Here the boundary is so sharply marked, that a ship sailing northwards finds itself passing suddenly from cold to warm water. At certain seasons of the year this cold current brings into the tropics water of 15°.6 Cent. (60° Fah.), the temperature of the undisturbed masses of water in the vicinity being from 27°.5 to 28°.7 Cent. (81°.5 to 83°.7 Fah.)

We do not know the depth to which oceanic currents
(warm or cold) extend : the deflection of the South
African current by the Lagullas bank, which is from
sixty to seventy fathoms deep, would indicate it to be
considerable. The presence of sand banks and shoals, not
situated in any line of current, may be recognised by the
low temperature of the water over them ; and the knowledge
of this fact may often conduce to the safety of navigation.
It was first discovered by the justly celebrated Benjamin
Franklin, who may be said to have thereby transformed the
thermometer into a sounding line; and its explanation appears to be, that when in the general movement of the
water forming the current, the deeper situated and colder
particles strike upon a bank, their motion is inclined upwards, and they mingle with and chill the upper stratum of
water.   My late illustrious friend, Sir Humphry Davy,
attributed the phænomenon rather to the descent of the
surface particles, cooled by nocturnal radiation, and to their
being prevented from sinking deeper by the shoal, which
thus retained them in closer proximity to the surface.   Mists
are frequently met with over shoals, from the influence of
the cooled water in condensing the vapour in the atmosphere.   I have seen such mists to the south of Jamaica,
and in the Pacific, which have shewn the outline of the
shoals beneath so well defined, as to be distinctly recognised from a distance ; thus forming to the eye aerial
images reflecting the form of the bottom of the ocean.   A
still more remarkable effect of the cooling influence of shoals
is shewn sometimes in the higher regions of the atmosphere.
At sea and in very clear weather, clouds are often seen suspended above the site of sand banks or shoals, as well as
over low coral or sandy islands.   Their bearings may be

taken from a distant ship by the compass, precisely as that of a high mountain or a solitary peak.

Although the surface of the ocean is less rich in animal and vegetable forms than that of continents, yet when its depths are searched, perhaps no other portion of our planet presents such fulness of organic life. Charles Darwin, in the agreeable journal of his extensive voyages, justly remarks, that our land forests do not harbour so many animals as the low wooded regions of the ocean, where the sea-weed rooted to the shoals, or long branches of fuci detached by the force of waves and currents, and swimming free upborne by air-cells, unfold their delicate foliage. The application of the microscope still farther increases our impression of the profusion of organic life which pervades the recesses of the ocean, since throughout its mass we find animal existence, and at depths exceeding the height of our loftiest mountain chains, the strata of water are alive with polygastric worms, cyclidiæ, and ophrydinæ. Here swarm countless hosts of minute luminiferous animals, mammaria, crustacea, peridinea, and ciliated nereides, which, when attracted to the surface by particular conditions of weather, convert every wave into a crest of light. The abundance of these minute creatures, and of the animal matter supplied by their rapid decomposition, is such, that the sea water itself becomes a nutritious fluid to many of the larger inhabitants of the ocean.

If all this richness and variety of animal life, containing some highly organised and beautiful forms, is well fitted to afford not only an interesting study, but also a pleasing excitement to the fancy, the imagination is yet more deeply, I might say more solemnly, moved by the impressions of the boundless and immeasurable which every sea-voyage affords.

He who, awakened to the inward exercise of thought, delights to build up an inner world in his own spirit, fills the wide horizon of the open sea with the sublime image of the infinite; his eye dwells especially on the distant sea line where air and water join, and where stars arise and set in ever renewed alternation: in such contemplations there mingles, as with all human joy, a breath of sadness and of longing.

A peculiar predilection for the sea, and a grateful recollection of the impressions which, when viewed within the tropics, either in the calm of peaceful nights or in moments of tumultuous agitation, it has left upon my mind, have alone induced me to allude to the individual enjoyment afforded by its contemplation, before I proceed to other and more general considerations. Contact with the ocean has unquestionably exercised a beneficial influence on the cultivation of the intellect and formation of the character of many nations, on the multiplication of those bonds which should unite the whole human race, on the first knowledge of the true form of the earth, and on the pursuit of astronomy and of all the mathematical and physical sciences. This beneficial influence, enjoyed by the dwellers on the Mediterranean and on the shores of South-Western Asia, was long limited to them; but since the sixteenth century it has spread far and wide, extending to nations living even in the interior of continents. Since Columbus was " sent to unbar the gates of ocean" (as the unknown voice said to him in a dream on his sick-bed near the River Belem) ([374]), man has boldly adventured into intellectual as well as geographical regions before unknown to him.

We proceed to the second covering of our planet, its exterior and universally diffused envelope, the *aerial* ocean, at the bottom or on the shoals of which we live. The *atmosphere* presents six classes of phænomena which manifest the most intimate connection with each other. These arise from its chemical composition; the variations in its transparency, polarisation, and colour; its density or pressure; its temperature; its humidity; and its electric tension. The atmosphere contains, in the form of oxygen, the first element of the physical life of animals; and we may here notice another benefit scarcely less important, of which it is the instrument; the air is the "conveyer of sound"; the channel of the communication of ideas, the indispensable condition of all social life. The earth, deprived of its atmosphere, as we believe our moon to be, presents to our imagination the idea of a soundless desert.

Since the beginning of the present century, the relative proportion of the components of the accessible strata of the atmosphere has been the object of researches, in which Gay-Lussac and myself took an active part : but the chemical analysis of the atmosphere has only very recently attained a high degree of perfection, through the meritorious labours of Dumas and Boussingault by new and more accurate methods. According to this analysis a volume of dry air contains 20.8 parts of oxygen and 79.2 of nitrogen, besides from two to five ten-thousandth parts of carbonic acid gas, a still smaller quantity of carburetted hydrogen ([375]), and, according to the important experiments of Saussure and of Liebig, traces of ammoniacal vapours ([376]), which furnish to plants the azote they contain. Some observations of Lewy have rendered it probable, that the quantity of oxygen varies slightly, but perceptibly, in different seasons of the year, and over the sea or in the interior of continents. It is quite a conceivable case

that variations which microscopic animals may occasion in
the quantity of oxygen held in solution in the water may entail
corresponding variations in the strata of air immediately in
contact with it ([377]). The air which Martins collected on the
Faulhorn at a height of 8226 French feet had not less
oxygen than the air at Paris ([378]).

The admixture of carbonate of ammonia in the atmosphere
may have been older than the presence of organic life on the
globe. The sources from whence the atmosphere may derive
carbonic acid are manifold ([379]). We may name first the
respiration of animals, who obtain the carbon which they
exhale from vegetable food, the vegetables receiving it from
the atmosphere. Other sources are, the interior of the earth
in districts of extinct volcanoes and thermal springs; and the
decomposition of the small portion of carburetted hydrogen
existing in the air by the electric discharges of clouds, which
are much more frequent within the tropics than in our
climates. With the above-mentioned substances, which we
find to be proper to the atmosphere at all elevations within
our reach, are accidentally associated others, especially
in close vicinity to the ground, which, as miasma and pesti-
lential emanations, sometimes exercise a dangerous influence
on animal organization. Although the chemical nature of
these has not yet been ascertained by direct analysis, yet the
existence of such deleterious local admixtures may be
inferred, both from a consideration of the processes of decay
which the vegetable and animal substances with which our
globe is covered are perpetually undergoing, and from combinations and analogies belonging to the domain of pathology.
Ammoniacal and other vapours containing azote, sulphuretted
hydrogen, and even combinations analogous to the ternary

and quaternary bases of the vegetable kingdom ([380]), may produce miasma which, under many forms and conditions (and by no means exclusively on wet marshy grounds, or on coasts covered with decaying mollusca and low bushes of mangrove and avicennia) may generate agues and typhus fever. At certain seasons of the year, fogs having a peculiar smell remind us of such adventitious and unwelcome mixtures in the lower portions of the atmosphere. Winds and ascending currents, caused by the heating of the ground, may sometimes carry up even solid substances when reduced to fine powder. The sand which creates an obscurity in the air over a wide area, and falls on the Cape Verd Islands, to which attention was directed by Darwin, was found by Ehrenberg to contain innumerable siliceous-shelled infusoria.

As principal features of a general descriptive picture of the atmosphere, we may distinguish—

1. *Variations of atmospheric pressure ;* those horary fluctuations which within the tropics occur with such regularity, and are so distinctly marked : they form a kind of atmospheric tide, which, however, cannot be attributed to the attraction of the Moon ([381]), and which differs greatly at different seasons, latitudes, and elevations.

2. *Climatic distribution of heat,* dependent on the relative position of transparent and opaque masses (the fluid and solid portions of the surface of the globe), and on the hypsometric configuration of continents. These determine the geographical position and the curvature of the isothermal lines (or lines of equal mean annual temperature), both in a horizontal and a vertical direction, or on uniform or different levels.

3. *The humidity of the atmosphere.* The quantitative relations of the humidity depend on the proportion between the terrestrial and oceanic surfaces; on the distance from the Equator, and the height above the sea; on the form in which the aqueous vapour in the atmosphere is precipitated; and on the connection of these precipitations with changes of temperature and with the direction and succession of winds.

4. *The electric tension of the atmosphere;* of which the primary source, when the sky is serene, is still much contested. Under this head we have to consider the relation of ascending vapour to the electric charge and form of clouds at different periods of the day and year; the influence of cold and warm zones of the earth, and of low or elevated plains; the frequency or infrequency of thunder-storms, their periodicity, and their formation in summer and winter; and the causal connection of electricity with the exceedingly rare occurrence of night hail, and with the phænomena of water-spouts and of columns of sand, investigated with much ingenuity by Peltier.

The horary variations of the barometer, which under the tropics present two maxima (at 9 or $9\frac{1}{4}$ A.M. and at $10\frac{1}{2}$ or $10\frac{3}{4}$ P.M.) and two minima (at 4 or $4\frac{1}{4}$ P.M. and 4 A.M., which are nearly the hottest and the coldest hours of the day), were long the object of my most careful daily and nightly observation ([382]). Their regularity is such, that, in the day time especially, we may infer the hour from the height of the column of mercury, without being in error on an average more than fifteen or seventeen minutes. In the torrid zone of the New Continent, I have found the regu-

larity of this ebb and flow of the aerial ocean undisturbed
either by storm, tempest, rain, or earthquake, both on the
coasts, and at elevations of nearly 13000 English feet above
the sea, where the mean temperature sinks to 7° Cent.
(44°.6 Fah.)   The amount of horary oscillation decreases
from the Equator to 70° N. lat., (where we have very accurate observations made by Bravais [383] at Bosekop), from
1.32 to 0.18 French lines (0.117 to 0.016 English inches.)
It has been supposed that much nearer to the pole the mean
height of the barometer is really less at 10 A.M. than at
4 h. P.M., so that the hours of maximum and minimum are
inverted; but this can by no means be concluded from Parry's
observations at Port Bowen in 73° 14′ N.

Owing to the effect of the ascending current, the mean height
of the barometer is rather less at the equator and generally
under the tropics, than in the temperate zone [384], and it appears to reach its maximum in Western Europe in the
parallels of 40° to 45°.   If with Kämtz we combine, by
isobarometric lines, those places which present the same
mean difference between the monthly extremes of the barometer, we obtain curves whose inflections and geographic
positions furnish important conclusions respecting the influence which the form of the land and the distribution of
land and sea exercise on the oscillations of the atmosphere.
Hindostan, with its lofty mountain chains and triangular
peninsulas,—the eastern coast of the New Continent,
where the warm current of the Gulf Stream is deflected
to the east by the banks of Newfoundland,—shew greater
isobarometric fluctuations than do the West Indies and
Western Europe.   Prevailing winds exercise a principal

influence on the diminution of the pressure of the atmosphere; in accompaniment with which, according to Daussy, as we have already noticed, the mean level of the sea is raised ([385]).

As the most important fluctuations of atmospheric pressure, whether regular and periodical, or irregular or accidental and then often violent and dangerous ([386]), have for their principal cause, like all other phænomena of weather, the heating power of the sun's rays,—it has long been customary (partly as proposed by Lambert) to compare the direction of the wind with the height of the barometer, and with changes of temperature and the increase or decrease of moisture. Tables of atmospheric pressure accompanying different winds, which have received the name of *barometric windroses*, afford a deep insight into the mutual relation of meteorological phænomena ([387]). Dove, with admirable sagacity, has recognised in the " law of rotation" in both hemispheres, which he has propounded, the cause of many important processes and extensive movements in the aerial ocean ([388]). The difference of temperature between the equatorial and the polar regions of the globe produces two opposite currents; one in the higher portion of the atmosphere, the other near the surface of the earth. The difference between the rotatory velocity at the poles and at the equator causes the air flowing from the poles to undergo an easterly, and the equatorial current a westerly deflection: on the opposition of these two currents, on the place where the upper one descends to the surface of the earth, and on the alternate displacement of one by the other, depend the principal phænomena of atmospheric pressure, of the heating and cooling of different strata of the air, of aqueous precipitations, and even, as Dove has truly represented, the formation of clouds and their shape and aspect.   Thus the particular form of cloud, which is often a characteristic and animating feature of the landscape, announces the processes which are taking place in the upper regions of the air; and when the air is calm, the clouds on a hot summer's day will sometimes present a "projected image" of the diversities of form of the heat-radiating surface of the Earth beneath them.

In parts of the globe where radiation acts on very extensive continental and oceanic surfaces in certain relative positions to each other, as between the east coast of Africa and the west coast of the Peninsula of India, its effects are shewn in the monsoons of the Indian Seas ([389])—the Hippalos of the Greek navigators : as the direction of the monsoons varies with the declination of the Sun, their periodical character was early recognised and turned to the use of man. In the knowledge of the monsoons,—which has doubtless extended for ages over both Hindostan and China, and among the Arabs to the west and the Malays to the east,—as well as in the still more ancient and more general knowledge of land and sea breezes, there lay, as it were enveloped and concealed, the hidden germ of that meteorological science which is now making such rapid progress.   The long chain of *magnetic stations* extending from Moscow to Pekin through the whole of Northern Asia, and which have also for their object the investigation of *meteorological* relations, will furnish data of great importance towards the investigation of the *Law of the Winds*.   For example, the comparison of stations so many hundred miles apart will determine, whether the same east wind blows from the elevated desert plains of Gobi to the interior of Russia, or whether the direction of the current originates in the middle of the chain of stations by a descent of the air from the upper regions. We shall thus learn in the strictest sense "where the wind comes from." If at the present moment we take a result on this point exclusively from stations where observations have now been made for above twenty years, we find (according to Wilhelm Mahlmann's most recent and careful calculation), that in the middle latitudes of the temperate zone, and in both continents, the prevailing direction of the wind is west south-west.

We have gained a clearer insight into the distribution of heat in the atmosphere, since the attempt has been made to connect graphically by lines those points, where accurate observations have indicated equality of mean annual, mean summer, or mean winter temperature. The system of Isothermal, Isotheral, and Isochimenal lines, which I first employed in 1817, if gradually perfected by the united efforts of investigators, may perhaps prove one of the chief foundations of a *Comparative Climatology*. It is thus that researches in terrestrial magnetism have become a *science*, from the period when partial results were combined in a graphic form in lines of equal Declination, equal Inclination, and equal Intensity.

The expression *climate*, taken in its most general sense, signifies all those states and changes of the atmosphere which sensibly affect our organs : temperature, humidity, variation of barometric pressure, a calm state of the air or the effects of different winds, the amount of electric tension, the purity of the atmosphere or its admixture with more or less deleterious exhalations, and lastly the degree of habitual transparency of the air and serenity of the sky, which has an important influence not only on the organic development of plants and the ripening of fruits, but also on the feelings and the whole mental disposition of man.

If the whole surface of the earth consisted either of a homogeneous fluid mass, or of strata of rock perfectly alike in colour, density, and smoothness, and of equal capacity both for the absorption of the sun's rays, and for radiating heat through the atmosphere into space, the isothermal, isotheral, and isochimenal lines would all be parallel to the equator. In this hypothetical condition of the earth's surface, the power of absorption and emission of light and heat would, under equal latitudes, be every where the same. From this mean, and, as it were, primitive condition, which neither excludes currents of heat whether in the interior of the globe or in its external covering, nor the propagation of heat by atmospheric currents, the mathematical consideration of climate proceeds. Whatever alters the power of absorption and of radiation, in any of the several parts of the earth's surface under the same parallel of latitude, causes inflections in the isothermal lines.

The progress of *climatology* has been remarkably favoured by the circumstance, that European civilization has extended over two opposite coasts, having passed from our western shores to a coast on the other side of the Atlantic valley having an eastern aspect. When, after the ephemeral colonization from Iceland and Greenland, the British formed the first permanent settlements on the coast of the United States of America, and the colonists (whose numbers were rapidly augmented by the effects of religious persecution, fanaticism, and the love of liberty), spread over the whole extent from North Carolina and Virginia to the banks of the St. Lawrence, they were astonished at the degree of winter cold which they had to endure, when compared with the climates of Italy, France, and the British Islands in corresponding parallels of latitude. But, however well fitted to awaken attention, the comparison bore no fruit until it could be based on numerical results of *mean annual temperature*. If between 58° and 30° of north latitude, we compare Nain on the coast of Labrador with Gottenburg, Halifax with Bordeaux, New York with Naples, and St. Augustin in Florida with Cairo, we find the differences of mean annual temperature under equal latitudes, in Eastern America and Western Europe, commencing with the northernmost pair of stations, and proceeding southwards, successively 11°.5, 7°.7, 3°.8, and almost 0° Cent. (20°.7, 13°.8, 6°.8, and almost 0° Fah.) The gradual diminution of the differences in this series, which extends over twenty-eight degrees of latitude, is striking. Farther to the south, under the tropics, the isothermal lines in both parts of the world are every where parallel to the equator. We see from the above examples, that the questions often asked in society,—how many degrees America (without distinction of its eastern or western coast), is colder than Europe?—or how much the mean annual temperature of Canada and the United States is lower than that of the corresponding latitudes in Europe?—have, when thus *generally expressed,* no definite meaning. The difference is not the same under different parallels; and unless we compare separately the winter and the summer temperatures of the opposite coasts, we shall not be able to form any clear idea of their climatic relations, as influential on agriculture and other industrial pursuits, and on the comfort and well-being of man.

In enumerating the causes which exercise a disturbing influence on the form of the isothermal lines, I have distinguished between those which *raise* and those which *depress* the temperature. To the first belong,—the vicinity of a western coast in the temperate zone; a divided or intersected configuration of the land, with projecting peninsulas, and deep re-entering bays and inland seas; aspect, or the position of the land relatively to a sea free from ice extending within the polar circle, or to a considerable mass of continental land situated beneath the equator, or at least within part of the torrid zone; prevalence of southerly and westerly winds on the western side of a continent, in the temperate zone of the northern hemisphere; chains of mountains acting as screens or protecting walls against winds from colder regions; infrequency of swamps or marshes, which retain the ice in spring and early summer; absence of woods on a dry sandy soil; constant serenity of sky during the summer months; and lastly, the vicinity of an oceanic current, bringing water of a higher temperature than that of the surrounding sea.

Among the *cooling causes* which modify the mean annual temperature, I consider elevation above the sea level, especially when not forming an extensive table land; the vicinity of an eastern coast in the higher and middle latitudes; the compact and massive form of a continent, having a coast line little varied by indentations; an extension of the land in the direction of the pole, far into the frozen regions, (there being no intervening sea free from ice during

the winter); a geographical position in which the tropical
portions of the same meridians are occupied by sea, implying
the absence under those meridians of extensive tropical land
powerfully heated by the sun's rays, and giving out great
heat by radiation; chains of mountains which, by their
direction and precipitous form, impede the access of warm
winds; the neighbourhood of isolated peaks, causing the
descent of currents of cold air on their declivities; extensive
forests, preventing the heating of the ground by the direct effect of the sun's rays, and, by means of the vital organic action
of their leafy appendages causing great evaporation, while,
by the extension of those organs, they increase the quantity
of surface cooled by radiation, thus operating in a threefold
manner, by shade, evaporation, and radiation; extensive
marshes, which, in the north, form a kind of subterranean
glacier in the plains, lasting until the middle of summer;
a cloudy summer sky, or frequent mists, which impede the
action of the sun's rays; and lastly, a very serene and clear
winter sky, favouring the escape of heat by radiation ([390]).

The inflections of the isothermal lines are determined by
the joint action, or total effect of the simultaneous operation,
of all the disturbing causes, whether belonging to those
which raise or to those which depress the temperature, but
especially by the relative extent and configuration of the
continental and oceanic masses. Local perturbations occasion the convex and concave summits of the isothermal
curves. Since there are different orders of disturbing causes,
each should first be viewed singly; and afterwards, in order
to learn their total effect on the form of the isothermal line
in the part of the earth under consideration, we must consider
their mutual influence; and the operation of each, in modi-

fying, destroying, or enhancing the effect of others, as small undulatory movements which encounter and intersect each other are known to do. Such is the spirit of the method by which I persuade myself it will some day be possible to combine by empirical laws, numerically expressed, vast series of apparently insulated facts, and to manifest their reciprocal dependence.

The trade winds (which are easterly winds blowing within the tropics), are the occasion of the west and west-south-west winds which prevail in both the temperate zones. These are of course land winds to eastern coasts, and sea winds to western coasts. The surface of the ocean not being susceptible of being cooled in the same degree as that of the land, by reason of the mass of its waters, and the tendency of its particles to sink immediately they are cooled, and to be replaced by cooler from below, it follows that, where easterly winds prevail, western coasts should be warmer than eastern coasts, except a counteraction exists in oceanic currents. Cook's young companion on his second voyage of circumnavigation, the ingenious George Forster, to whom I am indebted for the lively interest which prompted me to undertake distant travels, was the first who distinctly called attention to the difference of temperature between the eastern and western coasts of the two continents; and to the similarity of temperature, in the mean latitudes, of the west coast of America and the west coast of Europe [391]. Even in more northern latitudes, exact observations shew a striking difference between the *mean annual temperatures* of the east and the west coasts of America. Nain, in Labrador, in 57° 10′ N. lat. has a mean annual temperature of —3°.8 Cent. (25°.2 Fah.), or 6°.8 Fah. *below* the freezing

point; while Sitka, on the west coast of America, in
57° 3′ N. lat., has a mean temperature of 6°.9 Cent. (44°.4
Fah.), being 12°.4 Fah. *above* the freezing point. At Nain
the mean summer temperature hardly attains 6°.2 Cent.
(43°.2 Fah.), while at the Sitka it is 13°.8 Cent. (56°.8 Fah.)
Pekin, on the east coast of Asia, in lat. 39° 54′, has a mean
annual temperature of 11°.3 Cent. (52°.3 Fah.), being more
than 9° Fah. lower than that of Naples, which is in a rather
more northern latitude. The mean winter temperature of
Pekin is at least 3° Cent. (5°.4 Fah.) *below* the freezing
point; while in Western Europe, the winter temperature of
Paris, in lat. 48° 50′, is fully 3°.3 Cent. (6°.0 Fah.) *above*
the freezing point. The mean winter temperature of Pekin
is 2°.5 Cent. (4°.5 Fah.) lower than that of Copenhagen,
situated seventeen degrees further to the north.

I have already alluded to the slowness with which the
great mass of water in the ocean follows the variations of
temperature in the atmosphere, and the consequent influence
of the sea in equalizing temperatures; it moderates both the
asperity of winter and the heat of summer: hence arises a
second important contrast,—that between insular or littoral
climates (enjoyed also in some degree by continents whose
outline is broken by peninsulas and bays), and the climate
of the interior of great masses of solid land. Leopold von
Buch was the first writer who entered fully into the subject
of this remarkable contrast, and the varied phænomena
resulting from it; its influence on vegetation and agriculture,
—on the transparency of the atmosphere and serenity of the
sky,—on the radiation from the surface,—and on the height
of the limit of perpetual snow. In the interior of the Asiatic
continent, Tobolsk, Barnaul on the Obi, and Irkutsk, have

summers which, in mean temperature, resemble those of
Berlin and Munster, and that of Cherbourg in Normandy,
and during this season the thermometer sometimes remains for weeks together at 30° and 31° Cent. (86° or
87°.8 Fah.); but these summers are followed by winters
in which the coldest month has the severe mean temperature
of --18 to —20 Cent. (—0°.4 to ⚹4°.0 Fah.)   Such continental climates do indeed deserve the name of *excessive*
which they received from Buffon, and the inhabitants
of those countries almost seem condemned in Dante's
words ([392]),

" A sofferir tormenti caldi e geli."

I have in no part of the earth, not even in the Canary
Islands, in Spain, or in the south of France, seen more
magnificent fruit, especially grapes, than at Astrachan,
near the shores of the Caspian, in lat. 46° 21'.   With
a mean annual temperature of about 9° Cent. (48°
Fah.), the mean summer temperature rises to 21°.2 Cent.
(70°.2 Fah.), which is that of Bordeaux; while not only
there, but also still more to the south, at Kislar at the
mouth of the Terek (in the latitude of Avignon and Rimini),
the thermometer sometimes falls in winter to —25° or —30°
Cent. (—13° to —22° Fah.)

Ireland, Guernsey, and Jersey, the peninsula of Brittany,
the coast of Normandy and that of the south of England, all
present, by the mildness of their winters, and by the low temperature and clouded skies of their summers, the most
striking contrast to the continental climate of the interior of
eastern Europe.   In the north-eastern part of Ireland, in lat.
54° 56', under the same parallel as Königsberg, the myrtle

flourishes as luxuriantly as in Portugal. The mean temperature of the month of August, in Hungary, is 21° Cent.
(69°.8 Fah.); in Dublin, which is situated on the same
isothermal line (or line of equal mean *annual* temperature)
of 9½° Cent. (49°.2 Fah.), it is barely 16° Cent. (60°.8 Fah.);
the mean winter temperatures of the two stations being
2°.4 Cent. (27°.7 Fah.) at Bude, and 4°.3 Cent. (39°.8 Fah.)
at Dublin. The winter temperature of Dublin is 2° Cent.
(3°.6 Fah.) higher than that of Milan, Pavia, Padua, and of
the whole of Lombardy, although they enjoy, on the mean
of the whole year, a temperature of at least 12°.7 Cent.
(54°.8 Fah.) Stromness, in the Orkneys, not half a degree south of Stockholm, has a winter temperature of 4°
Cent. (39°.2 Fah.), being nearly as mild as London, and
milder than Paris. Even in the Feroe Islands, in lat. 62°,
under the favouring influences of the sea and of westerly
winds, the inland waters never freeze. On the lovely coast
of Devonshire, where Salcombe Bay has been called, on account of its mild climate, the Montpellier of the North, the
Agave Mexicana has been seen to blossom in the open air,
and orange trees trained against espaliers, and only slightly
protected by mats, have borne fruit. There, and at Pen-
zance and Gosport, as well as at Cherbourg in Normandy, the
mean winter temperature is above 5°.5 Cent. (41°.8 Fah.),
that is, only 1°.3 Cent. (2°.4 Fah.) lower than that of
Montpellier and Florence ([393]). Hence we perceive in what
a variety of ways the same mean annual temperature may be
distributed in the different seasons of the year, and the important influence of this distribution, whether considered in
reference to vegetation, to agriculture, to the ripening of
fruits, or to the comfort and well-being of man.

We have just seen, that the lines which I have called
isochimenals and isotherals (or lines of equal *winter* and
equal *summer* temperature), are by no means parallel with
the isothermals, or lines of equal *annual* temperature. If,
however, in countries where the myrtle grows wild, and the
snow does not continue on the ground during winter, the
temperature of summer and autumn is barely sufficient to
ripen apples thoroughly,—and if the vine (to produce drinkable wine) avoids islands, and in almost all cases proximity
to coasts,—the reason is by no means exclusively the low
summer temperature of such situations, shewn by the
thermometer suspended in the shade: it is also to be
sought in a difference which has been hitherto but little
considered, although known to be most actively influential in
other classes of phænomena (for example, in the bursting into
flame of a mixture of hydrogen and chlorine),—I mean the
difference between direct and diffused light; or that which
prevails when the sky is clear, and when it is veiled by cloud
or mist. I long since ([394]) attempted to call the attention
of physicists and vegetable physiologists to this difference,
and to the heat, unmeasured by thermometers, which is
locally developed in the vegetable cells by the action of
direct light.

If we form a thermic scale of different kinds of cultivation ([395]), beginning with that which requires the hottest
climate, and proceeding successively from vanilla, cacoa,
spices, and cocoa nuts, to pine apples, sugar cane, coffee, fruit-
bearing date trees, cotton, citrons, olives, sweet chestnuts, and
vines producing drinkable wine, an exact consideration of
their various limits, both on plains and on the declivities of
mountains, will teach that in this respect other climatic re-

lations than those of mean annual temperature must be sought.
Taking only one example, the cultivation of the vine,—
the production of *drinkable* wine ([396]) requires not only a
mean annual temperature of above $9\frac{1}{2}°$ Cent. (or 49°.2 Fah.),
but also a winter temperature of above 0°.5 Cent. (32°.8
Fah.), followed by a mean summer temperature of at least
18° Cent. (64°.4 Fah.)   At Bordeaux, in the valley of the
Garonne, in lat. 44° 50′, the mean temperature of
the year, the winter, the summer, and the autumn, are
respectively 13°.8, 6°.2, 21°.7, and 14°.4 Cent. (56°.8,
43°.2, 71°.0, and 58°.0 Fah.)   On plains in the vicinity of
the Baltic in lat. $52\frac{1}{2}°$, where a wine is produced which
though it is used can scarcely be called drinkable, these numbers are respectively 8°.6, —0°.7, 17°.6, and 8°.6 Cent.
(47°.5, 30°.8, 63°.7, and 47°.5 Fah.)   If it should appear
strange that these great differences in the influence of climate on the production of wine do not shew themselves still
more markedly in the indications of thermometers, it should
be remembered that an instrument suspended in the shade,
and carefully protected from the direct rays of the sun, and
from nocturnal radiation, cannot shew at all seasons of the
year, and during all the periodical changes of temperature,
the true heat of the surface of the ground, which receives
the whole effect of the sun's rays.

The relation of the mild, equable, littoral climate of the
peninsula of Brittany, to the colder winters and hotter summers of the more compact mass of the rest of France, resembles, to a certain degree, the relation of the climate of
Europe generally to that of the great continent of Asia,
of which it is the western peninsula.   Europe owes its milder
climate to its intersected form and deeply indented coast,—

to its exposure to the prevailing west winds which have blown across the ocean,—to the sea free from ice which separates it from the polar regions,—and lastly, to the existence and position of Africa, with its wide extent of tropical land favourable to the ascending current, while the equatorial region to the south of Asia is for the most part covered by the ocean. The European climate would therefore become colder ([397]), if Africa were to be overflowed by the ocean,—or if the fabulous Atlantis were to rise from the waves, and connect the two continents,—or, finally, if either the Gulf stream were to cease to extend its warming influence to the Northern Sea, or if a tract of land were to be elevated by volcanic forces between the Scandinavian peninsula and Spitzbergen. If in Europe the mean annual temperature decreases as we proceed easterly on a parallel of latitude, from the Atlantic coast of France through Germany, Poland, and Russia, to the chain of the Ural;—the principal cause of this phænomenon is to be sought in the form of the continent being gradually less intersected, and becoming more compact and extended,—in the increasing distance from the sea,—and in the feebler influence of westerly winds. Beyond the Ural, westerly winds blowing over wide expanses of land covered during several months with ice and snow, become cold land winds. It is to such circumstances of configuration and of atmospheric currents that the cold of western Siberia is due, and by no means to a great elevation of the surface above the level of the sea, as anciently assumed by Hippocrates and Trogus Pompeius, and still related by travellers of some celebrity in the eighteenth century.

If from differences of temperature at a uniform level, we

proceed to the inequalities of the form of the surface of
our planet, we have to consider mountains either in reference to their influence on the climate of neighbouring
lowlands, or to the climatic effects of their hypsometric relations on their own summits, which often spread out into
elevated plains.   Mountain chains divide the surface of the
earth into different basins, which are sometimes narrow and,
as it were, walled in, forming circular valleys or calderas, which
(as in Greece and in parts of Asia Minor) occasion climates
locally individualised in respect to warmth, humidity,
atmospheric transparency, and frequency of winds and
tempests.   These circumstances have at all times exercised a powerful influence on natural products, and on
cultivation ; as well as on manners, institutions, and the
feelings with which neighbouring tribes mutually regard each
other.   The character of *geographical individuality*, if we
may be permitted to use the expression, reaches its maximum
in countries where the variety of the form of the ground is
the greatest possible, both in the vertical and the horizontal
direction, both in relief and in the configuration of the coast
line.   The greatest contrasts to this variety of ground are
found in the steppes of Northern Asia, in the grassy plains
(savannahs, llanos, and pampas) of the new continent, in the
heaths of Europe, and in the sandy and stony deserts of Africa.

   The law of the decrement of heat with increasing elevation
in different latitudes has a most important bearing on
meteorological processes, on the geography of plants, on the
theory of terrestrial refraction, and on the different hypotheses on which the height of our atmosphere is estimated.
In the many mountain journeys which I have undertaken,
both within and beyond the limits of the tropics, the investiga-

tion of this law has always been an especial object of my researches ([399]).

Since we have obtained a somewhat more exact knowledge of the distribution of heat on the surface of the earth,—namely, of the inflections of the isothermal, isotheral, and isochimenal lines, and of their unequal distances apart, in the different systems of temperature of Eastern and Western Asia, of central Europe, and of North America,—it is no longer admissible to ask as a general question, to what fraction of the mean annual or mean summer temperature a change of one degree of geographical latitude, taken in a particular meridian, corresponds. In each system of isothermal lines of equal curvature, there reigns an intimate and necessary connection between three elements ;—the decrease of heat in a perpendicular direction from below upwards,—the variation of temperature corresponding to a change of one degree of geographical latitude,—and the relation which exists between the mean temperature of a station on a mountain, and the latitude of a point situated at the level of the sea.

In the system of Eastern America, the mean annual temperature varies from the coast of Labrador to Boston 0°.88 Cent. (1°.58 Fah.) for each degree of latitude ; from Boston to Charleston 0°.95 Cent. (1°.71 Fah.) ; from Charleston to the tropic of Cancer in the island of Cuba, the variation diminishes, being only 0°.66 Cent. (1°.19 Fah.) Within the tropic the diminution decreases to such a degree, that from Havannah to Cumana, the variation is not more than 0°.20 Cent. (0°.36 Fah.) for each degree of latitude.

In the system of isothermal lines of central Europe, the case is quite otherwise. Between the parallels of 38° and 71°, I find the decrease of temperature to be with considera-

ble uniformity, 0°.5 Cent. (0°.9 Fah.) for one degree of
latitude. But as a decrease of temperature of 1°.0 Cent.
(1°.8 Fah.) corresponds to an increase of vertical elevation
of 480 to 522 French feet (512 to 556 English), it follows
that a difference of elevation of 240 to 264 French
feet (256 to 278 English) produces the same effect on the
annual temperature as a change of one degree of latitude.
The mean annual temperature at the Convent on the great
St. Bernard, at the elevation of 7668 French (8173 English)
feet in lat. 45° 50', should therefore be found at the level of
the sea in lat. 75° 50'.

In that portion of the chain of the Andes which falls
within the tropics, observations, made by myself at various elevations from the sea-level to a height of 18000 French feet
(19185 English), gave a decrease of temperature of 1° Cent.
(1°.8 Fah.) for 576 French feet (614 English). My friend
Boussingault found, as a mean result, thirty years afterwards,
540 French feet (575 English). By a comparison of places
in the Cordilleras, at equal elevations above the level of the sea,
but situated some on the declivities of mountains, and others
on extensive table lands, I found that the latter class of stations
shewed an higher annual temperature varying from 1°.5 Cent.
to 3° Cent. (2°.7 to 4°.2 Fah.) ; and the difference would be
still greater if it were not for the cooling effect of nocturnal
radiation. As the various climates are here placed successively
stage above stage, from the cacao plantations of the lowlands
to the limit of perpetual snow, and as the differences of temperature in the course of the year are very small, we may
obtain a tolerably accurate representation of the climates
experienced by the inhabitants of the large towns in the
Andes, by comparing them with the temperatures of parti-

cular months in the plains of France and Italy. While the
heat which prevails daily on the forest-covered banks of the
Orinoco is 4°.0 Cent. (7°.2 Fah.) greater than that of the
month of August at Palermo, we find on ascending the
chain of the Andes, at Popayan, (at an elevation of
5826 English feet), the mean temperature of the three
summer months at Marseilles; at Quito, (at a height of
9541 English feet), that of the end of the month of May
at Paris; and, on the Paramos, (at an altitude of 11510
English feet), where only dwarf Alpine bushes grow
though flowers are still numerous, that of the beginning of
April at Paris.

The ingenious Peter Martyr de Anghiera, one of the
friends of Christopher Columbus, seems to have been the
first to recognize (in the expedition undertaken in October
1510 by Rodrigo Enrique Colmenares) that the snow line
becomes always higher as the equator is approached. We
read in the fine work, entitled *De rebus Oceanicis* ([400]) :—
" The river Gaira comes from a mountain (in the Sierra Ne-
vada of Santa Martha), which, according to the report of
Colmenares's travelling companions, is higher than any
mountain hitherto discovered ; doubtless it must be so, if in
a zone distant not more than 10° from the equinoctial line,
it yet retains snow permanently." The lower limit of perpetual snow in a given latitude is the boundary line of the
snow which resists the effect of the summer ; it is the highest
elevation to which the snow line recedes in the course of
the whole year. We must distinguish between the limit
thus defined, and three other phænomena; viz. the annual
fluctuation of the snow line ; the phænomenon of sporadic
falls of snow ; and the existence of glaciers, which appear to be

peculiar to the temperate and cold zones, and on which, since the immortal work of Saussure, a new light has of late years been thrown by Venetz and Charpentier, and by the meritorious and intrepid perseverance of Agassiz.

We know only the *lower* and not the *upper* limit of perpetual snow, for the highest mountains of the earth are far from attaining to those strata of highly rarefied and excessively dry air, concerning which we may suppose, with Bouguer, that they no longer contain vesicular vapour capable of being converted into crystals of snow, and of thus becoming visible. The lower limit of perpetual snow is not, however, a mere function of the geographical latitude, or of the mean annual temperature, nor is it even under the equator as was long supposed, or even within the tropics, that the snow line reaches its greatest elevation above the level of the sea. The phænomenon is a very complicated one, depending generally on relations of *temperature* and *moisture*, and on the peculiar shape of the mountains : if these relations are subjected to a still more particular analysis, which a great number of recent determinations renders possible (⁴⁰¹), we shall recognize, as concurrent causes, the differences of temperature in the different seasons of the year,—the direction of the prevailing winds, and whether they have blown over land or sea,—the degree of dryness or of moisture of the upper strata of air,—the absolute thickness of the accumulated mass of fallen snow,—the relation of the height of the snow line to the total height of the mountain,—the position of the latter in the chain of which it forms a part,—the steepness of its declivities,—the vicinity of other summits also covered with perpetual snow,—the extent, position, and elevation of the plain from

which the snow-capped mountain rises either solitarily, or as
a portion of a group or chain,—and whether this plain may
be part of the sea coast, or of the interior of a continent,
whether wooded or grassy, of arid sand or rock, or, on the
contrary, wet and marshy.

Under the equator in South America, the height of the
snow line is equal to that of the summit of Mount Blanc in
Europe, and descends, according to recent measurements,
960 French feet (1023 English) lower in the highlands of
Mexico, in lat. 19° North; in the southern tropic, on the
contrary, in $14\frac{1}{2}$° to 18° S. lat., it *ascends* according to
Pentland, to an elevation of 2500 French (2665 English)
feet above that which it attains under the equator not far
from Quito, on Chimborazo, Cotopaxi, and Antisana; this
ascent of the snow line does not take place in the eastern
part of the Cordilleras, but in the western and maritime
chain of the Andes of Chili.   Dr. Gillies affirms that
even much further to the south, on the declivity of the
volcano of Peuquenes (lat. 33°), he found the snow line
reach an elevation of between 2270 T. and 2350 T.
(14520 and 15030 English feet).   The evaporation of the
snow in the excessively dry air of summer and under a
cloudless sky is so powerful, that the volcano of Aconcagua,
north-east of Valparaiso, in S. lat. $32\frac{1}{2}$° (the elevation of
which was found by the Expedition of the Beagle to be
1400 feet above that of Chimborazo), was once seen free from
snow [402].   In an almost equal northern latitude (30° 45′
to 31°) the snow line on the southern declivity of the
Himalaya has an elevation of 12180 French feet (12982
English), which is nearly that which might have been con-
jecturally assigned to it from many combinations and com-

parisons with other mountain chains ; but on the northern declivity, under the influence of the high lands of Thibet, whose mean elevation appears to be about 10800 French feet (11510 English), the snow limit attains a height of 15600 French feet (16630 English).   This phænomenon, was long contested both in Europe and in India, and I have developed my views respecting its causes on several occasions since the year 1820 ([403]) : it is interesting in another point of view besides the purely physical one, for it has exercised an important influence on the mode of life of numerous tribes ; meteorological processes in the atmosphere either favour or forbid an agricultural or a pastoral life to the inhabitants of extensive tracts of continent.

As the quantity of moisture in the atmosphere increases with the temperature, this element, so important for the whole organic creation, varies with the hour of the day, the season of the year, and the degree of latitude and of elevation.   Our knowledge of the hygrometric relations of the atmosphere has been materially augmented of late years by the method now so generally and extensively employed, of determining the relative quantity of vapour, or the condition of moisture of the atmosphere, by means of the difference of the *dew point* and of the temperature of the air, according to the ideas of Dalton and of Daniell, and by the use of the wet bulb thermometer.   Temperature, atmospheric pressure, and the direction of the wind, have all a most intimate relation to the atmospheric moisture so essential to organic life.   The influence, however, of humidity on organic life, is less a consequence of the quantity of vapour held in solution under different zones, than of the nature and frequency of the aqueous precipitations which refresh the ground in the form of dew, mist, rain, or snow. According to Dove ([404]), in our northern zone, " the elastic force of the vapour is greatest with south-west, and least with north-east winds; it diminishes on the western side of the windrose, and rises on the eastern side. On the western side of the windrose, the cold, dense, and dry current represses the warm and light current containing an abundance of aqueous vapour, while, on the eastern side, the contrary takes place. The south-west is the equatorial current which has descended through the lower current to the surface of the earth; the north-east is the polar current prevailing undisturbed."

The agreeable and fresh verdure which many trees preserve, in districts within the tropics where for from five to seven months no cloud is seen on the vault of heaven, and no perceptible dew or rain falls, proves that their leaves are capable of drawing water from the atmosphere by a vital process of their own, which perhaps is not simply that of producing cold by radiation. The absence of rain in the arid plains of Cumana, Coro, and Ceara (in North Brazil), is contrasted with the abundance of rain which falls in other places within the tropics; for example, at the Havannah, where, by the average of six years of observation by Ramon de la Sagra, the mean annual quantity of rain is 102 Parisian inches, which is four or five times as much as falls at Paris or at Geneva ([405]). On the declivities of the chain of the Andes, the quantity of rain as well as the temperature decreases with increasing elevation ([406]). My South American travelling companion, Caldas, found that the annual quantity of rain at Santa Fe de Bogotá, at an elevation of nearly 8700 English feet, did not exceed 37 Parisian inches, little more than on some of the western coasts of Europe. At Quito, Boussingault sometimes

saw Saussure's hygrometer recede to 26°, with a temperature
of 12° to 13° C. (53°.6 to 55°.4 Fah.)   Gay-Lussac,
in his great aerostatic ascent, in a stratum of air 6600
French feet (7034 English) high, with a temperature of
4° C. (39°.2 Fah.) saw the same hygrometer at 25°.3.   The
greatest degree of dryness which has yet been observed on
the surface of the globe in the lowlands is probably that
which Gustav Rose, Ehrenberg, and myself, found in
Northern Asia, between the valleys of the Irtysch and the
Obi.   In the steppe of Platowskaia, after south-west winds
had blown for a long time from the interior of the continent,
with a temperature of 23.7 C. (74°.7 Fah.) we found the dew
point —4.3 C. (24°. Fah.) ; the air, therefore, contained only
sixteen parts in a hundred of the quantity of vapour required
for saturation([407]).   Accurate observers, Kämtz, Bravais, and
Martins, have in the last few years raised doubts concerning
the greater dryness of mountain air, which appears to follow
from the hygrometric measurements made by De Saussure and
myself in the higher regions of the Alps and of the Cordilleras.
The more recent observations referred to furnish a comparison of the strata of air at Zurich, and on the Faulhorn ;
but the latter scarcely deserves the name of an elevated
mountain station ([408]).   The moisture by which, in the
tropical region of the Paramos (near the part where snow
begins to fall, between 12000 and 14000 English feet of elevation), some kinds of large-flowered myrtle-leaved alpine shrubs
are almost perpetually bathed, does not necessarily imply
the presence of a large absolute quantity of aqueous vapour
at that height ; it only proves, as do the abundant mists on
the fine plateau of Bogotá, the frequency of aqueous precipitations.   At such elevations, with a calm state of the atmosphere, mists arise and disappear several times in the course

of an hour.   These rapid alternations characterise the elevated plains and Paramos of the chain of the Andes.

The electricity of the atmosphere, whether considered in the lower regions of the air or in the canopy of the clouds, whether studied in its silent problematical diurnal march, or in explosions in the lightning and thunder of the tempest, shews itself in manifold relation to the phænomena of the distribution of heat, — of the pressure of the atmosphere and its disturbances,—of hydro-meteoric phænomena (or watery precipitations),—and probably also of the magnetism of the outer crust of the globe.   It acts powerfully on the whole animal and vegetable world, not only indirectly through the agency of meteorological processes, the precipitations of aqueous vapour, and the acid or ammoniacal combinations occasioned by it, but also in a direct manner as an electric force stimulating the nerves and promoting the circulation of the organic juices.   This is not the place to renew the discussion concerning the proper source of atmospherical electricity when the sky is clear.   Its existence has been ascribed, sometimes to the evaporation of impure fluids (impregnated with earths or with salts) ([409]), sometimes to the growth of plants ([410]), or to chemical decompositions taking place at the surface of the earth, sometimes to the unequal distribution of heat in the atmospheric strata ([411]), and, lastly, sometimes, according to Peltier's ingenious investigations, to the influence of a constant charge of negative electricity in the terrestrial globe ([412]).   The physical description of the universe, limiting itself to the results given by electrometric observations, (particularly those made with the ingenious electromagnetic apparatus, first proposed by Colladon), has only to notice the incontestable increase of the tension of positive electricity accompanying the increased elevation of the station and the absence of neighbouring trees ([413]), its daily variations, which take place, according to Clarke's experiments at Dublin, in more complicated periods than those found by Saussure and myself, and its variations at different seasons of the year, at different distances from the equator, and under different relations of continental or oceanic surface.

The electric equilibrium is more rarely disturbed where the aerial ocean rests on a liquid base, than where it rests on land; but it is very striking to notice in extensive seas the influence which small groups of islands exercise on the state of the atmosphere in occasioning the formation of storms. In fogs, and in the commencement of falls of snow, I have seen the previously permanent vitreous electricity pass rapidly into resinous, and the two alternate repeatedly, during long series of experiments made both in the plains of the colder latitudes and in the Paramos of the Cordilleras, at elevations from 11000 to 15000 English feet under the tropics.   The alternate transition was quite similar to that shewn by the electrometer a short time before and during a thunder-storm ([414]). When the vesicular vapour has condensed into clouds with definite outlines, the electricity of the separate vesicles of vapour passes to the surface of the cloud, and contributes to increase the general electric tension of the exterior surface ([415]). According to Peltier's experiments at Paris, slate-grey clouds have resinous, and white, rose, and orange coloured clouds have vitreous electricity.   Thunder clouds not only envelop the highest summits of the Andes,—(I have myself seen the vitrifying effects of lightning on one of the rocky pinnacles which rise above the crater of the volcano of Toluca, at an elevation of nearly 1530 English feet),—but they have also been seen over the low lands in the temperate zone at a measured vertical elevation of 25000 (26650 English) feet. On the other hand, the stratum of cloud in which thunder is taking place sometimes sinks to a height of only 5000 or even 3000 feet above the plain.

According to Arago's investigations, which are the most complete and comprehensive that we yet possess on this difficult portion of meteorology, the disengagement of light (lightning) is of three kinds : the zig-zag form, sharply defined at the edges ; lightning without definite form, illuminating at the same instant the whole cloud, which appears as it were to *open* and display its inner recesses ; and lightning in the form of balls of fire ([417]). The duration of the two first kinds is scarcely the thousandth part of a second ; but the globular lightning moves far more slowly, its appearance lasting for several seconds. Recent observations confirm the phænomenon described by Nicholson and Beccaria, in which, without any audible thunder or any indication of storm, isolated clouds remain stationary for some time high above the horizon, and lightning proceeds uninterruptedly both from their interior and from their margins : hail-stones, drops of rain, and flakes of snow, have also been seen to appear luminous with electric light when there has been no thunder. In the geographical distribution of thunder-storms, the Peruvian coast, where thunder and lightning are unknown, presents the most striking contrast to the rest of the torrid zone, in which at certain seasons of the year storms take place almost daily, four or five hours after the sun has reached

its meridian altitude.   According to the testimony of navigators (Scoresby, Parry, Ross, and Franklin), as collected
by Arago, it is undoubted that in general in the high northern
regions, between 70° and 75° of latitude, electric explosions
are exceedingly rare ([418]).

The meteorological portion of the description of nature,
which we are now concluding, shews that the various processes
which the vast aerial ocean presents,—the absorption of
light, the disengagement of heat, the variation of elastic force,
the hygrometric condition, and the electric tension,—are all so
intimately connected, that each separate meteorological process is simultaneously modified by all the rest.   This complexity of disturbing causes, (which reminds us involuntarily
of those which the near, and especially the small, cosmical
bodies, the satellites, comets, and shooting stars encounter in
their course through space,) makes it very difficult to give
the full interpretation of meteorological phænomena ; and
the same cause greatly limits or wholly precludes the possibility of that prediction of atmospheric changes, which would
be so important for agriculture and horticulture, for navigation, and for the convenience and pleasures of life.   Those
who place the value of meteorology not in the knowledge of
the phænomena themselves, but in this problematical power
of prediction, are imbued with a firm persuasion that this
branch of science, for the sake of which so many journeys
in distant mountain regions have been undertaken, has for
centuries achieved no progress whereof to boast.   The confidence which they refuse to physical philosophers they bestow
on changes of the moon, and on certain days long marked
in the calendar.

Great deviations from the mean distribution of temperature are seldom local in their occurrence; they are for the most part distributed in a uniform manner over extensive districts. The amount of deviation has its maximum at some one determinate place, in receding from which it decreases gradually until its limits are reached ; and when these limits are passed, great deviations in the opposite direction are met with. Similar relations of weather extend more often from south to north than from west to east. At the end of 1829 (when I had just completed my Siberian journey) the maximum of cold was at Berlin, while North America enjoyed unusual warmth. The assumption that a severe winter will be followed by a hot summer, or a mild winter by a cool summer, is a wholly arbitrary one, resting on no foundation. Opposite conditions of weather in adjacent countries, or in two corn-producing continents, are the means of effecting a beneficent equalisation in the prices of many products of cultivation.

It has been justly remarked that it is the barometer alone which indicates to us what is taking place in all the strata of air above the place of observation (419), while the thermometer and hygrometer are purely local, and can only inform us concerning the warmth and moisture of the lowest stratum in close proximity to the ground. The simultaneous thermic and hygrometric modifications of the upper regions of the atmosphere (when direct observations on mountains or in aerostatic ascents are wanting), can be sought only by hypothetical combinations, whereby the barometer may indeed serve also to determine temperature and moisture. Important changes of weather do not usually arise from a local cause situated at the place of observation itself: their origin is to be looked for in a disturbance of the equilibrium

of the currents of the atmosphere which has begun afar
off, and generally not at the surface of the earth but in the
higher regions, and which, bringing with it either warm or
cold, dry or moist, air, either renders the sky and the
atmosphere cloudy and thick, or serene and clear, by transforming the towering masses of cumuli into light feathery
cirrous clouds. As, therefore, the inaccessibility of the
primary phænomena is added to the multiplicity and complication of disturbances, it has always appeared to me that
meteorology must seek its foundations and its advance first
in the torrid zone ; in those more favoured regions where
the same breezes always blow, and where the ebb and flow of
atmospheric pressure, the course of hydro-meteors, and the
phænomena of electric explosions, all recur periodically.

Having now passed through the entire circle of terrestrial
inorganic nature,—having considered our planet in respect to
its form, its internal heat, its electro-magnetic charge, its
polar luminous effusions, the reaction of its interior on its
variously composed crust, and finally the phenomena of its
oceanic and atmospheric envelopes,—the view which we have
essayed to trace in broad and general outlines might be regarded as complete, and would be so according to the limitation formerly adopted in physical descriptions of the
globe. But the plan which I have proposed to myself has
a more elevated aim, and I should regard the contemplation
of nature as deprived of its most attractive feature, were
it not also to include the sphere of organic life with
its many gradations of development. The idea of life is
so intimately connected with the moving, combining,
forming, and decomposing forces which are incessantly in

action in the globe itself, that the oldest mythical representations of many nations ascribe to these forces the production of plants and animals, and represent the epoch in which
the surface of our planet was unenlivened by animated forms,
as that of a primeval chaos of conflicting elements. But
investigations into primary causes, or into the mysterious
unresolvable problems of origin, do not enter into the domain
of experience and observation; nor has the obscure commencement of the history of organisation a place in the description of the actual condition of our planet ([420]). These
reservations once made, it should still be noticed in the
physical description of the world, that all those substances
which compose the organic forms of plants and animals are
also found in the inorganic crust of the earth; and that the
same forces which govern inorganic matter are seen to prevail in organic beings likewise, combining and decomposing
the various substances, regulating the forms and properties
of organic tissues, but acting in these cases under conditions
yet unexplained, to which the vague term of "vital phænomena" has been assigned, and which have been systematically grouped according to analogies more or less happily
imagined. Hence has arisen a tendency of the mind to trace
the action of physical forces to their extremest limits, in the
development of vegetable forms, and of those organisms which
are endowed with powers of voluntary motion : and here,
also, the contemplation of inorganic nature becomes connected with the distribution of organic beings over the surface of the globe, *i. e.* the *geography of plants and animals*.

Without attempting to enter on the difficult question of
"spontaneous motion," or the difference between vegetable

and animal life, it may be remarked that if nature had endowed us with a microscopic power of vision, and if the integuments of plants had been perfectly transparent, the vegetable kingdom would be far from presenting to us that aspect of immobility and repose which our perceptions now ascribe to it. The internal parts of the cellular structure are incessantly animated by the most various currents : ascending and descending, rotating, ramifying, and continually changing their direction, they manifest themselves by the movements of a granular mucilaginous fluid in water plants (naiades, characeæ, hydrocharideæ) and in the hairs of phenogamous land plants. Such is the peculiar molecular movement discovered by the great botanist Robert Brown (which is indeed perceptible, not only in vegetables, but also in all matter reduced to an extreme state of division); such is the gyratory current (cyclose) of globules of cambium; and, lastly, such are the articulated filamentary cells which unroll themselves in the antherides of the chara, and in the reproductive organs of liverworts and algæ, and in which Meyen, too early lost to science, believed that he recognised an analogy to the spermatozoa of the animal kingdom. If we add to these various currents and molecular agitations, the phænomena of endosmose, the processes of nutrition and of growth, and internal currents of air or gases, we shall have some idea of the forces which, almost unknown to us, are incessantly in action in the apparently still life of the vegetable kingdom.

Since the time when, in an earlier work ("Ansichten der Natur," "Tableaux de la Nature"), I attempted to describe the universal diffusion of organic life on the surface of the globe, and its distribution in height and in depth, our knowledge has been wonderfully augmented by Ehrenberg's brilliant discoveries (" über das Verhalten des kleinsten Lebens in dem Weltmeere wie in dem Eise der Polar länder"), which rest not on ingenious combinations and infe- rences, but on direct and exact observation.   By these dis- coveries the sphere of animated existence—we may say the horizon of life—has expanded before our view.   " Not only is there no interruption of minute microscopic forms of animal life in the vicinity of either Pole where larger animals can- not maintain themselves, but we find among the micros- copic animals of the South Polar Seas, collected in the Antarctic Expedition of Captain James Ross, a remarkable abundance of new forms, which are often of great elegance. Even in the residuum obtained from melted ice which floats in rounded fragments in lat 78° 10' S., there have been found above fifty species of siliceous-shelled polygastrica, and even coscinodiscæ with green ovaries, which were there- fore certainly living and able to resist the.extreme severity of the cold.   In the Gulf of the Erebus and Terror, sixty- eight siliceous-shelled polygastrica and phytolitharia, together with a single calcareous-shelled polythalamia, were brought up by the lead from depths of 1242 to 1620 English feet."

By far the greater number of the oceanic microscopic forms hitherto observed belong to the siliceous-shelled in- fusoria, although the chemical analysis of sea-water has not shewn silica to be one of its essential constituents, and it could only indeed exist in water in a state of simple mixture or suspension.   It is not only in particular localities, in inland waters or in the vicinity of coasts, that the ocean

is thus thickly peopled with living atoms invisible to
the naked eye. Samples of water taken up by Schayer
in 57° S. lat., on his return from Van Diemen Island, as
well as those taken between the tropics in the middle of
the Atlantic, shew that the ocean water in its ordinary condition, without any appearance of discoloration, contains
innumerable microscopic organisms, quite distinct from
the siliceous filaments of the genus Chætoceros, floating in a
fragmentary state like the oscillatoria of our fresh waters.
Some polygastrica which have been found mixed with sand
and excrements of Penguins in the Cockburn Islands, appear
to be generally distributed over the globe; other species belong
to both the arctic and antarctic polar regions [421]. Thus we
see that animal life reigns in the perpetual night of the depths
of the ocean, while on continents, vegetable life, stimulated
by the periodical action of the solar rays, chiefly predominates.
The mass of vegetation on the earth very far exceeds that of
the animal creation ; for what, in point of bulk, would be an
assemblage of all the great cetaceæ and pachydermata living at
one time, compared to the thickly-crowded colossal trunks of
trees of 8 and 12 feet in diameter from the tropical forests which
cover only one region of the Earth, namely that comprised
between the Orinoco, the Amazons, and the Rio da Madeira?
If the characteristic aspect of different portions of the Earth's
surface depend conjointly on all external phænomena,—if
the contours of the mountains, the physiognomy of plants
and animals, the azure of the sky, the form of the clouds,
and the transparency of the atmosphere, all combine in forming that general impression which is the result of the whole,
yet it cannot be denied that the vegetable covering with

which the earth is adorned is the principal element in the impression. Animal forms are inferior in mass, and their individual power of motion often withdraws them from our sight: vegetable forms, on the contrary, produce a greater effect by reason of their amplitude and of their constant presence. The age of trees is announced by their magnitude, and the union of age with the manifestation of constantly renewed vigour,—the ancient trunk with the fresh verdure of spring,—is a charm peculiar to the vegetable creation([422]). In the animal kingdom (and this knowledge is also a result of Ehrenberg's discoveries) it is precisely the minutest forms which, owing to their prodigious fecundity ([423]), occupy the greatest space. The minutest infusoria (the Monadines) only attain a diameter of $\frac{1}{3000}$th of a French line, and yet these siliceous-shelled animalcula form in humid districts subterranean strata of many fathoms in thickness.

Those whose minds and feelings are awakened to the influences of the contemplation of nature, are impressed, under every zone, by the diffusion of life over the surface of the globe; but this impression is most powerful in the regions where the palms, the bamboos, and the tree-ferns flourish, and where the ground rises from the margin of a sea filled with mollusca and corals to the limit of perpetual snow. The distribution of animal and vegetable life is scarcely arrested by height or depth: organic forms descend even into the interior of the earth, not only where the labours of the miner have opened extensive excavations, but also in closed natural caverns into which rain or snow-water can only penetrate through minute fissures. In such a cavern,

when opened by blasting, I have observed the snow-white
stalactitic walls covered by the delicate net-work of an Usnea.
Podurellæ are found in fissures of the glaciers of Monte
Rosa, and of those of the Grindelwald and the upper
Aar; the Chionæa araneoides, described by Dalman, and the
microscopic Discerea nivalis (formerly called Protococcus),
live in the polar snows as well as in those of our high
mountains. The red colour of long-fallen snow had been
noticed by Aristotle, who had probably observed it in the
Macedonian mountains (424). On the highest summits of
the Swiss Alps, lecideæ, parmeliæ, and umbilicariæ, scantily
tinge the surface of the rocks where they are denuded of snow;
but on the chain of the tropical Andes, beautiful phænogamous
plants, the woolly Culcitium rufescens, Sida pinchinchen-
sis, and Saxifraga boussingaulti, present isolated flowers, at
elevations of about 15000 English feet above the level of the
sea. Thermal springs contain small insects (Hydroporus ther-
malis), gallionellæ, oscillatoria, and confervæ: whilst their
waters nourish the fine fibres of the roots of phænogamous
plants. Not only are earth, air, and water, filled with life,
and that at the most different temperatures, but also the
interior of the various parts of animal bodies: there are
animalcula in the blood of frogs and of salmon: according
to Nordmann, the fluids of the eyes of fishes are often
filled with a worm which lives by suction (Diplostomum);
and the same naturalist has even discovered in the gills of
the Bleak an extraordinary double animal (Diplozoon para-
doxon) having two heads and two caudal extremities disposed in rectangular directions.

Although the existence of meteoric infusoria is more

than doubtful, it is certainly very possible that small infusoria may be passively carried up with ascending aqueous
vapour, and may float for a time in the atmosphere like the
pollen of the flowers of pines which falls every year from the
air (425). This circumstance deserves to be especially considered in any renewal of the old discussion respecting
" spontaneous generation ;" (426) and the more so, because,
as I have before noticed, Ehrenberg has discovered the remains of eighteen species of siliceous-shelled polygastric
infusoria, in the dust or sand which often falls on ships
navigating the ocean near the Cape Verd Islands at a
distance of 380 geographical miles from the African coast.

The geography of plants and animals may be considered
either with reference to the differences and relative numbers
of typical forms, (the distribution of genera and species
in different localities,) or to the number of individuals
of each species on a given surface.   Both in plants and in
animals it is essential to distinguish between social or grega-
rious, and solitary species.   Those species of plants which I
have termed " social" (427) spread a uniform covering over
large extents of country : to this class belong many kinds of
sea weeds,—cladoniæ and mosses which spread over the waste
plains of Northern Asia,—grasses, and cacti growing like
the pipes of an organ,—avicenniæ and mangroves in the
tropics,—and forests of coniferæ and of birches in the
plains of the Baltic and of Siberia.   It is especially by
this particular mode of geographical distribution, that certain vegetable forms constitute the leading feature in the
physiognomy of a country, imparting to it a character

determined by their general aspect and magnitude, and the form and colour of their leaves and flowers ([428]). The animal creation, from its smaller mass, and from its mobility, is far less influential in this respect, notwithstanding its variety and interest, and its greater aptitude to excite in us feelings either of sympathy or of aversion. Agricultural nations enlarge artificially the domain of social plants, and thus give to extensive districts in the temperate zone a character of greater uniformity than would belong to them in a state of nature; cultivation extirpates, and causes gradually to disappear, many species of wild plants, whilst others, without being purposely conveyed, follow man in his most distant migrations. The luxuriance of nature in the tropics offers a more powerful resistance to the changes which human efforts thus have a tendency to introduce in the aspect of creation.

Observers, who in short intervals of time traversed extensive regions, and ascended lofty mountains in which different climates are found stage above stage in close proximity, must have been early impressed with the character of regularity in the distribution of vegetable forms : those who recorded their observations were unconsciously collecting the raw materials of a science of which the name had not yet been spoken. The same zones or regions of vegetation which, in the sixteenth century, Cardinal Bembo, when a youth, observed and described on the acclivities of Etna ([429]), were found on those of Ararat by Tournefort, who compared the alpine flora with the flora of the plains in different latitudes, and was the first to remark that the distribution of vegetation is similarly influenced, by the elevation of the ground above the level of the sea in mountains, and by the distance from the equator in the plains. Menzel, in an unedited flora of Japan, used almost accidentally the expression " Geography of plants," and the same expression is found in the fanciful but graceful work of Bernardin de St. Pierre, entitled " Etudes de la Nature." A scientific treatment of the subject, however, only commenced when the geography of plants was brought into close connection with the study of the distribution of heat over the surface of the earth ; and when, by a classification into natural families, it had become possible to distinguish numerically the forms which increase or decrease in frequency in receding from the equator towards the poles, and to assign the numerical proportion which in different regions of the earth each family bears to the entire mass of the phænogamous flora of the same region. I regard it as a happy circumstance in my life, that at a time when my views were almost exclusively turned to botanical studies, I was led, by the favouring influence of the grand spectacle presented by the mountainous regions of the tropics, where the most varied climates and vegetations are brought into close proximity and contrast, to those subjects of investigation of which I have here spoken.

The geographical distribution of animals, on which Buffon first put forward general views and in many instances just ones, has of late years benefitted greatly by the progress made in the study of the geography of plants. The isothermal curves, and particularly the isochimenal curves, whether of latitude or of elevation, coincide with the limits which species of plants, and of animals which do not wander far from their

fixed habitations, rarely pass. The elk, for example, lives
in the Scandinavian peninsula almost ten degrees of latitude
farther north than in the interior of Siberia, where the isochimenal lines, or lines of equal winter temperature, present
a form so strikingly concave. Plants migrate in the germ ;
the seeds of many species are provided with appropriate
organs, by means of which they are wafted through the air;
but when once rooted they become dependent on the soil,
and on the temperature of the surrounding atmosphere ;
animals, on the contrary, having the power of locomotion,
can migrate at pleasure beyond the bounds of their usual
habitations ; and they do so more particularly where the
isotheral lines are greatly inflected, and where hot summers follow severe winters. The royal tiger, perfectly identical with the Bengal species, makes incursions every summer into the North of Asia, as far as the latitudes of Berlin
and Hamburgh,—a fact which Ehrenberg and myself have
fully established in another work ([430]).

The associations of different species of plants, to which we
are accustomed to give the name of "floras," do not appear to
me, from what I have myself seen of the surface of the
earth, to manifest that predominance of particular families,
which would justify us in distinguishing geographically the
regions of the umbellatæ, of the solidaginæ, of the la-
biatæ, or of the scitamineæ. In this respect my individual
opinion differs from the views of several of my friends who
are among the most distinguished botanists of Germany.
It appears to me, that the character of the floras of the
highlands of Mexico, of New Granada and Quito, of
the plains of European Russia, and of Northern Asia,

is given not so much by the greater number of the species which constitute one or two natural families, as by
the far more complicated relations which result from the
coexistence of a great number of families, and the relative
proportions of their respective species.   Doubtless the gramineæ and the cyperaceæ predominate in the prairies and
the steppes, as do coniferæ, cupuliferæ, and betulinæ in our
northern forests; but this predominance of certain forms is
only apparent, produced by the particular aspect given
by the social plants.   The north of Europe, and the zone
of Siberia which is situated to the north of the Altai,
are not more deserving of the title of regions of gramineæ or
of coniferæ, than are the vast llanos of South America, or the
pine forests of Mexico.   It is by the association of forms
which may partially replace each other, and by their relative
abundance and their groupings, that the vegetation of a
country produces its impression of luxuriance and variety,
or of poverty and uniformity.

In this brief and fragmentary consideration of the
phænomena of organisation, I have ascended from the
simplest cell (431), or as it were the first manifestation
of life, progressively to higher forms.   " Mucilaginous
granules produce by their juxtaposition a *cytoblast* of
definite form, around which a vesicular membrane forms
a closed cell;" and this first germ of organisation is
either occasioned by a pre-existing cell (432),—so that
cell produces cell,—or the origin and evolution of the
cell is concealed in the obscurity of a chemical process,
analogous to the fermentation which produces the fungus in
yeast.   It is not, however, the province of this work to do

more than touch very lightly on the mysterious subject of modes of origin; the geography of organic forms treats rather of germs already developed—of their habitations, their migrations voluntary or passive, and their distribution over the surface of the Earth.

The general view of nature which I have endeavoured to present would be incomplete, were I to close it without attempting to trace, by a few characteristic traits, a corresponding sketch of *man*, viewed in respect to physical gradations, to the geographical distribution of contemporaneous types, to the influences which terrestrial forces exercise on him, and to the reciprocal but less powerful action which he in turn exerts on them. Subject, though in a less degree than plants and animals, to the circumstances of the soil and the meteorological conditions of the atmosphere, and escaping from the control of natural influences by the activity of mind and the progressive advance of intelligence, as well as by a marvellous flexibility of organisation which adapts itself to every climate, man forms everywhere an essential portion of the life which animates the globe. It is by these relations that the obscure and much contested problem of community of origin enters into the circle of ideas comprised in the physical description of the world. Its examination will stamp (if I may so express myself) with that nobler interest which attaches itself to all that belongs to mankind, the termination of my work. The immense domain of the science of languages, in whose varied structure the aptitudes of nations are mysteriously reflected, borders closely on that which treats of the parentage and affinity of races: and if we would know what even slight diversities of race may be

capable of producing, we have a striking example in the
Hellenic nations when in the flower of their intellectual culture. The most important questions in the history of civilisation are connected with the descent of races, the community of language, and the greater or less persistency in
the original direction of the intellect and disposition.

Whilst attention was exclusively directed to the extremes
of colour and of form, the result of the first vivid impressions
derived from the senses was a tendency to view these differences as characteristics, not of mere varieties, but of originally
distinct species. The permanence of certain types ([433]) in the
midst of the most opposite influences, especially of climate,
appeared to favour this view, notwithstanding the shortness
of the time to which the historical evidence applied : but in
my opinion more powerful reasons lend their weight to the
other side of the question, and corroborate the unity of the
human race. I refer to the many intermediate gradations ([434])
of the tint of the skin and the form of the skull, which have been
made known to us by the rapid progress of geographical science
in modern times ; to the analogies derived from the history of
varieties in animals, both domesticated and wild ; and to the
positive observations collected respecting the limits of fecundity in hybrids ([435]). The greater part of the supposed contrasts, to which so much weight was formerly assigned, have
disappeared before the laborious investigations of Tiedemann
on the brain of negroes and of Europeans, and the anatomical
researches of Vrolik and Weber on the form of the pelvis.
When we take a general view of the dark-coloured African
nations, on which the work of Prichard has thrown so much
light, and when we compare them with the natives of the

Australasian Islands, and with the Papuas and Alfourous (Ha-
rofores, Endamenes), we see that a black tint of skin, woolly
hair, and negro features, are by no means invariably associated (⁴³⁶). So long as the western nations were acquainted
with only a small part of the Earth's surface, partial views
almost necessarily prevailed; tropical heat and a black colour
of the skin appeared inseparable. "The Ethiopians," said
the ancient tragic poet Theodectes of Phaselis (⁴³⁷), "by the
near approach of the Sun-god in his course, have their bodies
coloured with a dark sooty lustre, and their hair curled and
crisped by his parching rays." The campaigns of Alexander,
in which so many subjects connected with physical geography
were originally brought into notice, occasioned the first discussion on the problematical influence of climate on nations
and races. "Families of plants and animals," says one of
the greatest anatomists of our age, Johannes Müller, in his
comprehensive work entitled *Physiologie des Menschen,* "in
the course of their distribution over the surface of the Earth
undergo modifications within limits prescribed to genera and
species, which modifications are afterwards perpetuated orga-
nically in their descendants, forming types of varieties of the
same species: the present races of animals have been produced by a concurrence of causes and conditions, internal as well as external, which it is impossible to follow
in detail; but the most striking varieties are found in
those families which are susceptible of the widest geographical extension. The different races of mankind are
forms or varieties of a single species: their unions are
fruitful, and the descendants from them are so likewise;
whereas if the races were distinct species of a genus, the

descendants of mixed breed would be unfruitful : but whether the existing races of men are descended from one or from several primitive men is a question not determinable by experience ([438])."

Mankind are therefore distributed in *varieties,* which we are often accustomed to designate by the somewhat vague appellation of " *races.*" As in the vegetable kingdom, and in the natural history of birds and fishes, an arrangement into many small families proceeds on surer grounds than one which unites them into a few sections embracing large masses; so also, in the determination of races, it appears to me preferable to establish smaller families of nations. In the opposite mode of proceeding, whether we adopt the old classification of my master, Blumenbach, into *five* races (Caucasian, Mongolian, American, Ethiopian, and Malay), or that of Prichard into *seven* races ([439]) (Iraunian, Turanian, American, Hottentots and Bushmen, Negroes, Papuas and Alfourous), it is impossible to recognise in the groups thus formed any true typical distinction, any general and consistent natural principle. The extremes in colour and form are separated indeed, but without regard to nations which cannot be made to arrange themselves under any of the above-named classes, and which have sometimes been called Scythian, and sometimes Allophyllic races. Iraunian is indeed a less objectionable name for the European nations than Caucasian ; but it may be affirmed generally, that geographical denominations, designed to mark the points of departure of races, are exceedingly vague and undetermined, especially when the place which is to give its name to the race has been inhabited at different epochs ([440]) by very different races—as, for example, Turan (or Mawerannahr) by Indo-Germanic and Finnish, but not by Mongolian races.

Languages, as intellectual creations of man, and closely entwined with his whole mental development, bear the stamp of national character, and as such are of the highest importance in the recognition of similarity or diversity of race: the descent of languages from a common origin is the conducting thread which enables us to tread the labyrinth, in which the connection of physical and mental powers and dispositions presents itself under a thousand varied forms. The brilliant progress which the philosophical study of languages has made within the last half century in Germany, is favourable to researches on their national character ([441]), or on that which they appear to have derived from the influence of race. But here, as in all fields of ideal speculation, there are many illusions to be guarded against, as well as a rich prize to be attained. Positive ethnographical studies, supported by profound historical knowledge, teach us that a great degree of caution is required in these investigations concerning nations, and the languages spoken by them at particular epochs. Subjection to a foreign yoke, long association, the influence of a foreign religion, a mixture of races even when comprising only a small number of the more powerful and more civilised immigrating race, have produced in both continents similarly recurring phenomena; viz. in one and the same race, two or more entirely different families of languages; and in nations differing widely in origin, idioms belonging to the same linguistic stock. Great Asiatic conquerors have been most powerfully instrumental in the production of striking phenomena of this nature.

But language is an integral part of the natural history of the human mind; and notwithstanding the freedom with which the mind pursues perseveringly in happy independence its self-chosen direction under the most different physical conditions—notwithstanding the strong tendency of this freedom to withdraw the spiritual and intellectual part of man's being from the power of terrestrial influences—yet is the disenthralment never completely achieved. There ever remains a trace of the impression which the natural disposition has received from climate, from the clear azure of the heavens, or from the less serene aspect of a vapour-loaded atmosphere. Such influences have their place among those thousand subtle and evanescent links in the electric chain of thought, from whence, as from the perfume of a tender flower, language derives its richness and its grace. Seeing, then, how close is the bond which unites the physical world with the world of the intellect and of the feelings, we are unwilling altogether to deprive this general sketch of nature of those brighter lights and tints, which might be imparted to it by considerations, however lightly touched, on the mutual relations of races and of languages.

By maintaining the unity of the human species, we at the same time repel the cheerless assumption ([442]) of superior and inferior races of men. There are families of nations more readily susceptible of culture, more highly civilized, more ennobled by mental cultivation than others; but not in themselves more noble. All are alike designed for freedom;

for that freedom which in ruder conditions of society belongs to individuals only, but, where states are formed, and political institutions enjoyed, belongs of right to the whole community. " If," in the words of Wilhelm von Humboldt, " we would point to an idea which all history throughout its course discloses as ever establishing more firmly and extending more widely its salutary empire—if there is one idea which contributes more than any other to the often contested, but still more often misunderstood, perfectibility of the whole human species—it is the idea of our common humanity ; tending to remove the hostile barriers which prejudices and partial views of every kind have raised between men; and to cause all mankind, without distinction of religion, nation, or colour, to be regarded as one great fraternity, aspiring towards one common aim, the free development of their moral ·faculties. This is the ultimate and highest object of society ; it is also the direction implanted in man's nature, leading towards the indefinite expansion of his inner being. He regards the earth and the starry heavens as inwardly his own, given to him for the exercise of his intellectual and physical activity. The child longs to pass the hills or the waters which surround his native dwelling ; and his wish indulged, as the bent tree springs back to its first form of growth, he longs to return to the home which he had left; for by a double aspiration after the unknown future and the unforgotten past—after that which he desires, and that which he has lost—man is preserved, by a beautiful and touching instinct, from exclusive attachment to that which is present. Deeply rooted in man's inmost nature, as well as commanded by his highest tendencies, the full recognition of the bond of humanity, of the community of the whole human race, with the sentiments and sympathies which spring therefrom, becomes a leading principle in the history of man." (443)

With these words—which derive their charm from the depth of the feelings from whence they sprang—let a brother be permitted to close the general description of the phenomena of the universe. From the remotest nebulæ, and from the revolving double stars, we have descended to the minutest animal forms of sea and land, and to the delicate vegetable germs which clothe the naked precipice of the ice-crowned mountain summit. Laws partially known have enabled us in some degree to arrange these phenomena; other laws of a more mysterious nature prevail in the highest sphere of the organic world, in that of man with his varied conformation, the creative intellectual energies with which he is endowed, and the languages which have sprung therefrom. We have thus reached the point at which a higher order of being is presented to us, and the realm of *mind* opens to the view: here, therefore, the *physical* description of the universe terminates; it marks the limit, which it does not pass.

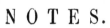

N O T E S.

NOTES.

(¹) page 7.—This expression is taken from a beautiful description of tropical forest scenery by Bernardin de St.-Pierre, in Paul and Virginia.

(²) p. 9.—The comparisons in the text are only approximate. The several elevations above the level of the sea are more exactly as follows :—The Schnee or Riesenkoppe, in Silesia, 824 toises, according to Hallaschka. Rigi, 923 toises, taking the height of the lake of Lucerne at 223 toises from Eschmann, (Results of the Trigonometrical Measurements in Switzerland, 1840, p. 230). Athos, 1060 toises, according to Captain Gaultier. Pilatus, 1180 toises. Etna, 1700.4 toises, or 10874 English feet, according to Captain Smyth; 1700.7 toises, or 10876 English feet, by barometrical measurement by Sir John F. W. Herschel, which he communicated to me in manuscript in 1825 ; 1704 toises, or 10896 English feet, by angles of altitude taken by Cacciator at Palermo, and assuming the terrestrial refraction at 0·076. Schreckhorn, 2093 toises. Jungfrau, 2145 toises, according to Tralles. Mont Blanc, 2467 toises, according to the results discussed by Roger in the Bibliothèque Universelle, May 1828, pp. 24—53 ; 2460 toises, according to Carlini's determination, taken from Mont Colombier ; 2463 toises, measured by the Austrian engineers, from Trelod and the Glacier d'Ambin. [It should be observed that the real height of the Swiss snow-capped mountains fluctuates, according to Eschmann, as much as 3½ toises, from the varying thickness of the snow covering]. Chimborazo, 3350 toises, according to my trigonome‑ trical measurements, (Humboldt, Recueil d'Obs. Astro. Tome i. p. 73). Dhawalagiri, 4390 toises. All these elevations have been given in toises of 6 Parisian feet to a toise. As Blake's and Webb's determinations differ

70 toises, the elevation assigned to the Dhawalagiri (or white mountain, from the Sancrit *dhwala*, white, and *giri*, mountain), cannot lay claim to as great exactness as the determination of the height of the Jawahir, 4027 toises, = 24160 Parisian feet, = 25749 English feet, = 7848 metres, which rests on a complete trigonometrical measurement, (vide Herbert and Hodgson, in Asiat. Researches, Vol. xiv. p. 189 ; and Supplement to Encyclopædia Brit. Vol. iv. p. 643). I have shewn elsewhere (Ann. des Sciences Nat., Mars 1825), that the measurement of the Dhawalagiri (4391 toises, = 28078 English feet), depends on several elements which have not been ascertained with full certainty, *i. e.* geographical positions and azimuths (Humboldt, Asie Centrale, T. iii. p. 282). There is still less foundation for the conjecture that, in the Tartaric chain (north of Thibet, towards the Kuen-lun mountains), there are single summits which attain an elevation of 30000 English feet, (4691 toises, almost double the height of Mont Blanc), or at least 29000 English feet, or 4535 toises, (vide Captain Alexander Gerard and John Gerard's Journey to the Boorendo Pass, 1840, Vol. i. pp. 143 and 311). In the text, the Chimborazo is only called " one of the highest summits of the chain of the Andes ;" for, in the year 1827, the highly-informed traveller, Pentland, in his memorable expedition to Upper Peru (Bolivia), measured the elevation of two mountains to the east of the lake of Titiaca, *viz.* the Sorata, 3948 toises, and the Illimani, 3753 toises, both being greatly higher than the Chimborazo (which is only 3350 toises), and approaching the elevation of the Jawahir (4027 toises), which is the highest of the Himalaya mountains of which the elevation has been accurately measured. Mont Blanc (which is 2467 toises, = 14802 Parisian feet, = 4808 metres), is therefore 883 toises lower than the Chimborazo ; the Chimborazo 598 toises lower than the Sorata; the Sorata 79 toises lower than the Jawahir, and probably 443 toises lower than the Dhawalagiri. By a more recent measurement (1838), the elevation of the Illimani was found by Pentland 7275 metres, or 3732 toises—a difference of only 21 toises from the measurement taken in 1827. In this note the elevations have been given with minute exactness, incorrect numbers having been introduced into many recent maps and profiles by erroneous reductions of the original measures.

(³) p. 10.—The paucity of palms and of tree ferns on the temperate portions of the projecting spurs of the Himalaya is shewn in Don's Flora Nepalensis, 1825, and in the remarkable lithographed catalogue of Wallich's Flora Indica, containing 7683 species. From Nepaul (lat. 26½° to 27¼°), we know as yet only one species of palm, Chamærops martiana, Wall. (Plantæ

Asiat. L. iii. p. 5) : it was found in the shady valley of Bunipa, at a height of
5000 French feet above the sea.   The magnificent tree fern Alsophila
brunoniana, Wall. (of which a stem 48 feet long has been in the possession
of the British Museum since 1831), is not from Nepaul, but from the mountains of Silhet, north-east of Calcutta, in lat. 24° 50'. The Nepaul fern,
Paranema cyathoides, Don, previously Sphæropteris barbata, Wall. (Pl. Asiat.
L. i. pp. 42—48), is, indeed, nearly related to the Cyathea, of which, in the
South American Missions of Caripe, I have seen a species 32 feet high, but
it is not properly a tree.

(⁴) p. 10.—Ribes nubicola, R. glaciale, R. grossularia.   The character
of the Himalaya vegetation is given by 8 species of Pine (notwithstanding an assertion of the ancients concerning Eastern Asia ; Strabo, Lib.
xi. p. 510, Cas), 25 Oaks, 4 Birches, 2 Beeches, (the wild Chesnut tree
of Kashmir, which grows to a height of 100 feet, is inhabited by a
large white ape with a black face, as far north as 33° N. lat.—Carl von
Hügel, Kaschmir, 1840, Th. ii. S. 249), 7 Maples, 12 Willows, 14
Roses, 3 species of Strawberry, 7 Rhododendrons (one of which is 20 feet
high), and many other northern forms. Among the Coniferæ, the Pinus
deodwara, or deodara (in Sanscrit dèwa-dùru, the timber of the gods), is
nearly allied to Pinus cedrus.   Near the limit of perpetual snow are found
the large and showy flowers of the Gentiana venusta, G. moorcroftiana,
Swertia purpurescens, S. speciosa, Parnassia armata, P. nubicola, Pæonia
emodi, Tulipa stellata, and, besides species of European genera peculiar to
these Indian Alps, true European species, as Leontodon taraxacum, Prunella
vulgaris, Galium aparine, and Thlaspi arvense. The heath, which was mentioned by Saunders in Turner's Travels, and which has been confounded
with Calluna vulgaris, is an Andromeda,—a fact of great importance in the
geography of Asiatic plants. If in this note I have made use of the unphi-
losophical expressions of *European* forms, or *European* species, *growing wild.
in Asia*, it has been a consequence of the old botanical language, which has
dogmatically substituted, for the idea of distribution in space or coexistence
of organic forms, the hypothesis of migration, and, from predilection for
Europe, migration from west to east.

(⁵) p. 11.—The limit of perpetual snow, on the southern declivity of the
Himalaya chain, is 2030 toises above the level of the sea ; on the northern
declivity, or rather on the peaks which rise above the Tartarian plateau, this
limit is 2600 toises, from 30½° to 32° of latitude, while under the Equator, in
the Andes of Quito, it is 2470 toises. I have deduced this result from the

collection and combination of many data furnished by Webb, Gerard, Herbert, and Moorcroft, (vide my two Mémoires sur les Montagnes de l'Inde, in 1816 and 1820, in the Annales de Chimie et de Physique, T. iii. p. 303, and T. xiv. pp. 6, 22, 50). The greater elevation to which the snow line recedes on the Thibetian declivity is the result conjointly of the radiation of heat from the neighbouring elevated plains, the serenity of the sky, and the infrequent formation of snow in very cold and dry air (Humboldt, Asie Centrale, T. iii. pp. 281—326). The result which I have given as the most probable for the elevation of the snow line on the two sides of the Himalaya, had Colebrooke's great authority in its favour. He wrote to me in June, 1824 :—" I also find, from the materials in my possession, 13000 English feet (2033 toises) for the elevation of the line of perpetual snow, on the southern declivity, under the parallel of 31°. Webb's measurements would give me 13500 English feet (2111 toises), being 500 feet higher than by Captain Hodgson's observations. Gerard's measurements fully confirm your statement that the snow line is higher on the northern than on the southern side." It has not been until the present year (1840), that we have obtained, through Mr. Lloyd, the publication of the collected journals of the two brothers, Gerard, (Narrative of a Journey from Caunpoor to the Boorendo Pass in the Himalaya, by Captain Alexander Gerard and John Gerard, edited by George Lloyd, Vol. i. pp. 291, 311, 320, 327, and 341). Much information respecting different localities is brought together in a " Visit to Shatool, for the purpose of determining the Line of Perpetual Snow on the Southern Face of the Himalaya, in August, 1822 ;" but, unfortunately, the travellers always confound the elevation at which sporadic falls of snow take place with the highest elevation to which the snow line recedes above the Thibetian plateau. Captain Gerard distinguishes between the summits in the *midst of the elevated plateau*, where he gives the limit of perpetual snow at from 18000 to 19000 English feet (2815 to 2971 toises), and that portion of the northern declivities of the Himalaya chain which abut on the deep valley of the Sutlej, where the plateau, being interrupted, can radiate but little heat. The elevation of the village of Tangno is given at only 9300 English feet, or 1454 toises ; while that assigned to the plateau round the sacred lake Manasa is 17000 English feet, or 2658 toises. Where the chain of the Himalaya is broken through, Captain Gerard finds the snow line on the northern declivity 500 feet lower than on the side facing India, where he estimates it at 15000 English feet, (2346 toises). As respects vegetation, the most striking differences are presented between the Thibetian plateau and the Indian declivity of the

Himalaya. On the latter, the cultivation of grain only ascends to the height
of 1560 toises, (or 9912 English feet); and even there the corn has often
to be cut green, or even while still in the blade. The woody region, comprising tall oaks and deodars, only attains 1870 toises; and dwarf birches,
2030 toises. On the plateau, Captain Gerard saw pastures up to 2660 toises;
grain succeeds up to 2200 toises, and even up to 2900 toises; birch trees,
with tall stems, reach 2200 toises; and low bushes, 2660 toises,—*i. e.* 200
toises higher than the limit of perpetual snow in the province of Quito under
the Equator. It is exceedingly desirable that both the *mean* elevation of the
table land of Thibet, which, between the Himalaya and the Kuen-lun, I
assume at only about 1800 toises, and the relative height of the snow line on
the northern and southern faces of the Himalaya, should be investigated anew
by travellers accustomed to general views. Hitherto, estimations have often
been confounded with actual measurements, and the elevations of summits
rising out of the table land with that of the surrounding plateau, (compare
Carl Zimmermann's Hypsometric Remarks, in his Geographischen Analyse
der Karte von Inner-Asien, 1841, S. 98). Lord calls attention to a contrast
between the relative heights of the line of perpetual snow on the two sides of
the Himalaya and on those of the Hindu Coosh. " The latter chain," he
says, " has the table land to the south, and, therefore, the snow line is higher
on its southern face, which is the contrary case to that of the Himalaya, which
has low plains to the south as the Hindu Coosh has to the north." However
much the hypsometric data here treated of may require critical correction when
taken separately, yet the fact is sufficiently established that the remarkable
form of a portion of the earth's surface, in the interior of Asia, renders it
possible for men to dwell and to find food and firing at an elevation which,
in almost all other parts of the two continents, is covered with perpetual
snow. I except the exceedingly dry mountains of Bolivia, which are so
deficient in snow, and where, in 1838, Pentland found the mean height of the
snow limit 2450 toises, in 16° to 17¾° S. lat. What appeared to me to be
the probable difference of the height of the snow line on the northern and
southern faces of the Himalaya has been confirmed by the barometric measurements of Victor Jacquemont, who fell an early sacrifice to his noble and
unwearied activity, (vide his Correspondance pendant son Voyage dans
l'Inde, 1828 à 1832, Livr. 23, pp. 290, 296, 299). He says, " Les neiges
perpétuelles descendent plus bas sur la pente meridionale de l'Himalaya que
sur les pentes septentrionales, et leur limite s'élève constamment à mésure que
l'on s'éloigne vers le nord de la chaine qui borde l'Inde. Sur le col de

Kioubrong à 5581 métres de hauteur selon le Capitaine Gerard je me trouvai encore bien au dessous de la limite des neiges perpétuelles qui dans cette partie de l'Himalaya je croirais," (much too high an estimate), " de 6000 métres, ou 3078 toises." The same traveller says, that, " on the southern declivity, the climate preserves the same character at all heights, the distribution in the different seasons of the year being the same as in the Indian plains. The summer solstice there brings the same torrents of rain, which last, without interruption, to the autumnal equinox. Kashmir, at an elevation of 5350 English feet (837 toises, which is nearly the height of the towns of Popayan and Merida), is the first place where a new and wholly different kind of climate begins," (Jacquemont, Corresp. T. ii. pp. 58 and 74). As Leopold von Buch remarks, the monsoons cannot carry the warm and moist sea air of the plains of India beyond the rampart of the Himalaya into the Thibetian districts of Ladak and Lhassa. Carl von Hügel, from a determination of the boiling point of water, estimates the elevation of the valley of Kashmir above the level of the sea at 5818 English feet, or 910 toises (Theil ii. S. 155; and Journal of the Geogr. Soc. Vol. vi. p. 215). In this calm and sheltered valley, scarcely ever visited by tempests, in latitude 34° 7', the snow, from December to March, is found several feet in thickness.

(⁶) p. 11. — Vide, generally, my Essai sur la Géographie des Plantes et Tableau physique des Régions Equinoxiales, 1807, pp. 80—88. On the diurnal and nocturnal variations of temperature, vide Plate 9 of my Atlas géographique et physique du nouveau Continent; and the Tables in my own work, entitled, De distributione geographica plantarum secundum cœli temperiem et altitudinem montium, 1817, pp. 90—116; the meteorological portion of my Asie Centrale, Vol. iii. pp. 212—224; and, lastly, the more recent and far more exact representation of decreasing temperature with increasing elevation in the chain of the Andes, given in Boussingault's Memoire sur la profondeur à laquelle on trouve la couche de température invariable sous les tropiques (Annales de Chimie et de Physique, 1833, T. liii. pp. 225—247). This treatise contains determinations of the mean temperature and of the elevation of 128 points, from the level of the sea to the declivity of Antisana, at a height of 2800 toises (about 17900 feet), and varying between 27°·5 and 1°·7 Centigrade, (or 81°·6 and 35° Fahrenheit).

(⁷) p. 14.—On the proper Madhjadeca, see Lassen's excellent work, entitled, Indische Alterthumskunde, Bd. i. S. 92. The Chinese give the name of Mo-kie-thi to the southern Bahar, *i. e.* the part to the south of the Ganges, (Foe-koue-ki, par Chy-Fa-Hian, 1836, p. 256). Djambu-dwipa is the name given to the whole of India; it sometimes comprehends, besides, one of the four Budhistic continents.

(^{8}) p. 14.—Ueber die Kawi-Sprache auf der Insel Java, nebst einer Einleitung ueber die Verschiedenheit des menschlichen Sprachbaues und ihren Einfluss auf die geistige Entwickelung des Menschengeschlecht's von Wilhelm v. Humboldt, 1836, Bd. i. S. 5—310.

(^{9}) p. 15.—Aber im stillen Gemach entwirft bedeutende Zirkel
  Sinnend der Weise, beschleicht forschend den schaffenden Geist
  Prüft der Stoffe Gewalt, der Magnete Hassen und Lieben,
  Folgt durch die Lüfte dem Klang, folgt durch den Aether dem Strahl,
  Sucht das vertraute Gesetz in des Zufall's grausenden Wundern—
  Sucht den ruhenden Pol in der Erscheinungen Flucht.—*Schiller*, 1795.

Science the while, deep musing in cell over circle and figure,
  Knows and adores the Power which through creation it tracks,
  Measures the forces of matter—the hates and loves of the magnets—
  Sound through its wafting breeze, light through its æther pursues,
  Seeks in the marvels of chance the law which pervades and controls it—
  Seeks the reposing pole fixed in the whirl of events.

From a translation by Sir John Herschel, Bart. printed for private circulation.

(^{10}) p. 19.—Arago's "micrométre oculaire," a happy improvement on Rochon's "micrométre prismatique," or "à double réfraction." See the note of M. Mathieu in the Hist. de l'Astr. au 18me siècle, par Delambre, 1827, p. 651.

(^{11}) p. 22.—Carus, "Von den Urtheilen des Knochen- und Schalen-Gerüstes," 1828, § 6.

(^{12}) p. 22.—Plut., in vita Alex. Magni, cap. 7.

(^{13}) p. 27. — The melting points of substances of very difficult fusion are usually given much too high. According to the very accurate researches of Mitscherlich, the melting point of granite can hardly exceed 1300° Cent.

(^{14}) p. 28.—See the classical work on the fishes of the old world by Agassiz, entitled, Recherches sur les Poissons fossiles, 1834, Vol. i. p. 38; Vol. ii. pp. 3, 28, 34; Addit. p. 6. The whole genus of Amblypterus, Ag., nearly allied to Palæoniscus (formerly Palæothrissum), lies buried beneath the oolitic formation in the old carboniferous strata. Scales, which in some situations are formed like teeth, and covered with enamel, from the family of Lepidoides (Order of Ganoides), belong to the oldest forms of ancient fishes

after the Placoides : their still living representatives are found in two genera,—
Bichir in the Nile and Senegal, and Lepidosteus in the Ohio.

([15]) p. 29.—Goethe, in the Aphorisms on Natural Science, in the small
edition of his Works, 1833, Vol. i. p. 155.

([16]) p. 37.—Arago's discoveries in the year 1811 (Delambre, Histoire de
l'Astronomie, p. 652).

([17]) p. 37.—Goethe, Aphoristiches über die Natur : Werke, B. 1. S. 4.

([18]) p. 39.—Pseudo-Plato, Alcib. ii. p. 148, ed. Steph.; Plut. Instituta
lacouica, p. 253, ed. Hutten.

([19]) p. 44.—The "Margarita philosophica" of Gregorius Reisch, prior of the
Chartreuse near Freiburg, appeared first under the title of " Æpitome Omnis
Philosophiæ, alias Margarita philosophica tractans de omni genere scibili."
This title was retained in the Heidelberg edition of 1486, and in the Strasburg
edition of 1504; but in the Freiburg edition of 1504, and in twelve subsequent ones which appeared in the short interval between that year and 1535,
the first part of the title was omitted. This work had a great influence on
the extension of mathematical and physical knowledge in the beginning of
the sixteenth century; and Chasles, the learned author of the Aperçu historique des Méthodes en Géométrie, (1837,) has shewn the great importance
of Reisch's Encyclopædia in the mathematical history of the middle ages. I
have endeavoured, by means of a passage found in a single edition of the
Margarita philosophica, (that of 1513,) to elucidate the important question
of the relations between the geographer of St. Die, Hylacomilus (Martin
Waldseemüller), who first, in the year 1507, gave the new continent the name
of *America*, and Amerigo Vespucci, Réné king of Jerusalem, and the celebrated editions of Ptolemy of 1513 and 1522. See my Examen critique
de la Géographie du Nouveau Continent, et des Progrès de l'Astronomie
nautique aux 15e et 16e Siècles, T. iv. p. 99—125.

([20]) p. 45.—Ampère, Essai sur la Philosophie des Sciences, 1834, p. 25 ;
Whewell, Philosophy of the Inductive Sciences, Vol. ii. p. 277 ; Park,
Pantology, p. 87.

([21]) p. 45.—He reduces all changes of condition in the material world to
motion. Aristot. Phys. ausc. iii. 1 and 4, p. 200 and 201. Bekker, viii.
1, 8, and 9, p. 250, 262, and 265. De Gener. et Corr. ii. 10, p. 336.
Pseudo-Aristot. de Mundo, cap. 6, p. 398.

([22]) p. 50.—On the question already raised by Newton himself, of the
difference between the attraction of masses and molecular attraction, see Laplace,
Exposition du Système du Monde, p. 384 ; and Supplement au Livre X. de

la Mécanique céleste, pp. 3 and 4. Kant's Metaphysical Principles of Natural Science, coll. works, 1839, Vol. v. p. 309. Peclet's Physique, 1838, T. i. p. 59—63.

(²³) p. 52.—Poisson, in Conn. des Tems pour l'Année 1836, p. 64—66. Bessel, in Poggendorff's Annalen der Physik, Vol. xxv. p. 417. Encke, in Abhandlungen der Berliner Academie, 1826, p. 257; Mitscherlich, Lehrbuch der Chemie, 1837, Vol. i. p. 352.

(²⁴) p. 53.—Compare Ottfried Müller, Dorier, Bd. i. S. 365.

(²⁵) p. 54.—Geographia generalis in qua affectiones generales telluris explicantur. The oldest Amsterdam (Elzevir) edition dates in 1650; the second and third, in 1672 and 1681, were prepared at Cambridge at Newton's suggestion. This exceedingly important work of Varenius is a Physical Geography in the proper sense of the term. Telluric phænomena had not been treated with such generality, since the excellent natural description of the new continent, traced by the Jesuit, Joseph de Acosta, in his work, entitled, Historia natural de las Indias, 1590. Acosta is richer in observations of his own: Varenius embraces a greater range of ideas; his life in Holland, the center of the commerce of the world at that period, having brought him into contact with many well-informed travellers. "Generalis sive universalis Geographia dicitur, quæ tellurem in genere considerat atque affectiones explicat non habita particularium regionum ratione." The General Geography of Varenius (Pars absoluta, cap. 1—22) is in the full import of the term a comparative geography, although the author uses the term, Geographia comparativa (cap. 33—40), in a much more restricted sense. I may cite as remarkable parts the enumeration of systems of mountains, with the relation of their directions to the form of the continents (pp. 66—70, ed. Cantabr. 1681); the list of active and extinct volcanoes; the assemblage of notices on the distribution of single islands and groups of islands (p. 220); the considerations on the depth of seas compared with the heights of neighbouring coasts (p. 103); on the equal level of the surface of all open seas (p. 97); on currents as depending on prevailing winds; on the unequal saltness of the sea, and the configuration of coasts (p. 139); the directions of winds as consequences of diversity of temperature, &c. The considerations, in page 140, on the general Equinoctial current from east to west as a cause of the Gulf stream, which begins at Cape San Augustin and issues forth between Cuba and Florida, are also excellent. The directions of the current along the west coast of Africa, between Cape Verd and the island of Fernando Po, in the Gulf of Guinea, are described with extreme exactness. Sporadic islands are regarded, by Varenius, as the "raised bottom of the sea;"—magna spirituum inclusorem vi, sicut aliquando montes e terrâ protrusos esse quidam scribunt (p. 215). The edition of 1681, by Newton ("auctior et emenda-tior"), contains unfortunately no additions from the pen of the great master. The spheroidal figure and compression of the earth are no where mentioned, although Richer's Pendulum Experiments are nine years older than the Cambridge edition: Newton's Principia were first communicated to the Royal Society of London in manuscript, in April, 1686.—There is much uncertainty as to the native country of Varenius. According to Jöcher, he was born in England; according to the Biographie Universelle (T. xlvii. p. 495), in Amsterdam; but both suppositions are shewn to be erroneous by the dedication of his General Geography to a Burgomaster of Amsterdam, in which he says expressly, "that he had sought refuge in that city, his own native town having been laid in ashes and entirely destroyed during the long war." These words seem to indicate Northern Germany, and the devasta-tions of the Thirty Years' war. Moreover, Varenius remarks, in the dedi-cation of his Descriptio Regni Japoniæ, (Amst. 1649) to the Senate of Hamburgh, that he had made his first mathematical studies in the Hamburgh Gymnasium. There is little or no reason to doubt that this admirable geographer was a German, and a native of Lüneburg. (Witten, Mem. Theol. 1685, p. 2142; Zedler, Universal-Lexikon, Th. xlvi. 1745, S. 187.)

(²⁶) p. 54.—Carl Ritter's Erdkunde im Verhaltniss zur Natur und zur Geschichte des Menschen, oder allgemeine vergleichende Geographie.

(²⁷) p. 56.—Κοσμος, in its most ancient and proper signification, was merely "ornament," (as belonging either to the dress of men and women, or to the caparison of horses); figuratively, it implied "order," for ευταξια and *ornament of speech*. The ancients are unanimous in affirming that Pythagoras was the first who employed the word to signify "order of the universe," or World or Universe itself. As he left no writings, the oldest passages in evi-dence of this are the fragments of Philolaus (Stob. Eclog. pp. 360 and 460. Heeren. Philolaos von Böckh, S. 62 und 90). I do not, with Näke, adduce Timæus of Locris, because his authenticity is doubtful. Plutarch (de plac. phil. ii. 1) says decidedly that Pythagoras first called the Universe *Kosmos*, because of the order which reigns throughout it; (also Galen, hist. phil. p. 429.) The word with its new signification passed from the philosophic school into the language of poets of nature and of prose writers. Plato continues to designate celestial bodies as Uranos, but he too calls the order of the universe Kosmos; and in Timæus (p. 30, B,) the universe is called an animal endowed with soul (κοσμος ζωον εμψυχον). Compare Anaxag. Claz. ed. Schaubach. p. 111; and Plut. de plac. phil. ii. 3, on the immaterial ordering spirit of the universe. In Aristotle (de cœlo, i. 9) Kosmos is "universe" and "order of the universe." It was also considered as divided in space into the sublunary world and the higher world above the moon (Meteor. i. 2, 1, and i. 3, 13, p. 339, *a*. and 340, *b*. Bekk.) The definition of the Kosmos cited by me in the text from the Pseudo-Aristoteles de Mundo cap. 2, (p. 391), runs thus in the original:—κοσμος εστι συσ]ημα εξ ουρανου και γης και των εν τουΊοις περιεχομενον φυσεων. Δεγεΐαι δε και ετερως κοσμος η των ολον ταξις τε και διακοσμησις, υπο θεων τε και δια θεων φυλαττομενη. I find most of the passages from Greek writers on the subject of Kosmos assembled—1st, in the controversial writings of Richard Bentley against Charles Boyle (Opuscula philologica, 1781, pp. 347, 445; Dissertation upon the Epistles of Phalaris, 1817, p. 254), on the historic existence of Zaleucus, the Locrian legislator; 2d, in Näke's excellent Sched. crit. 1812, pp. 9—15; and 3d, in Theoph. Schmidt ad Cleom. cycl. theor. met. i. 1, p. ix. 1 und 99. Kosmos was also used in a more restricted sense in the plural (Plut. 1, 5), either as applied to each separate star, or heavenly body, or "world" (Stob. i. p. 514; Plut. ii. 13), or to separate systems of worlds, or world-islands, each having a sun and a moon assumed to exist in infinite space (Anaxag. Claz. fragm. pp. 89, 93, 120; Brandis, Gesch. der Griechisch-Römischen Philosophie, Bd. i. S. 252). As each group thus became a Kosmos, the universe, το παν, was a higher and more comprehensive idea, different from Kosmos (Plut. ii. 1). It was not till long after the time of the Ptolemies that the word Kosmos was first applied to the earth. Böckh has made known inscriptions to the praise of Trajan and Adrian (Corpus Inscr. Græc. T. i. N. 334 and 1306), in which κοσμος is substituted for οικουμενη, just as we often say "world," meaning only the earth. The triple division of universal space into "Olympus," "Kosmos," and "Uranos," alluded to in the text, (Stob. i. p. 488; Philolaos, S. 94—102,) relates to the different regions which surround the hearth or focus of the universe, the Pythagorean εστια του παντος. The innermost region, between the earth and the moon, the domain of the "variable," is termed in the Fragments "Uranos." The middle, or that of the unvarying, well-ordered revolving planets, is, in accordance with a peculiar view of the universe, exclusively termed "Kosmos." The outer region is a fiery one, and is "Olympus." Bopp, the profound investigator into the affinities of languages, remarks, that "if we derive κοσμος from the Sanscrit root 'sud', purificari, as Pott had previously done. Etymol.

Forschungen, Th. i. S. 39 und 252), then we shall have to consider, in phonic relations,—1st, that the Greek κ (in κοσμος) has proceeded from the palatial *s*, which Bopp expresses by *ś*, and Pott by ç, as δεκα, dccem, Gothic *taihun*, from the Indian *dasán*; 2d, that the Indian *d'* corresponds regularly (Vergleichende, Gramm. § 99) to the Greek θ, whereby we perceive the relation of κοσμος (for κοθμος) to the Sanscrit root 'sud'; whence also καθαρος. Another Indian expression for world is gágat (pronounced dschagat), of which the proper signification is, "the going," as the participle of "gágâmi," I go (from the root, gâ)." In the inner circle of Hellenic etymology, κοσμος connects itself, according to Etym. M. p. 532, 12, most directly with καζω, or rather with καινυμαι, whence κεκασμενος or κεκαδμενος. Welcker also combines with this (Eine kretische col. in Theben, S. 23) the name κιδμος, as in Hesychius καδμος signifies a Cretan suit of armour. When the Romans adopted the technical language of the philosophy of the Greeks, they appropriated in a similar manner the word mundus, (which, like κοσμος, had originally signified female ornament,) to the "universe" or "world." Ennius appears to have been the first who ventured on this novelty : according to a fragment preserved by Macrobius, he says in his dispute with Virgil (Sat. vi. 2) : — " Mundus cœli vastus constitit silentio ;" as Cicero, " quem nos lucentem mundum vocamus," (Timæus, s. de univ. cap. 10). The Sanscrit root *mand*, from which Pott (Etym. Forsch. Th. i. S. 240) derives the Latin mundus, combines the two significations of shining and adorning. " *Löka*," in Sanscrit, signifies both " world " and " people," like the French " monde," and is derived, according to Bopp, from lök, to see and to shine : in a similar manner the Slavonian *swjet* is both light and world, *Licht und Welt* (Grimm, Deutsche Gramm. Bd. iii. S. 394). The original meaning of the latter word, Welt, of which we now make use, weralt in the old High German, worold in old Saxon, and veruld in Anglo-Saxon, referred rather to time, sæculum, than to space. Amongst the Tuscans, the " mundus " was an inverted dome, an imitation of the dome of the sky, or the vault of the heavens (Ottfried Müller, Etrusker, Th. ii. S. 96, 98, und 143). In its more restricted telluric meaning, the word appears in the Gothic language as the disk of earth girt by sea (marei, meri), or as " merigard," a sea-garden.

(28) p. 57.—Respecting Ennius, see the ingenious investigations of Leopold Krahner, in his Grundlinien zur Geschichte des Verfall's der römischen Staats-Religion, 1837, S. 41—45. It is probable that Ennius did not quote from writings of Epicharmus himself, but from poems written under his name and according to the views of his system.

(29) p. 59.—Aul. Gell. "Noct. Att." V. xviii.

(30) p. 65.—Schelling's "Bruno on the Divine and Natural Principles of Things," p. 181.

(31) p. 76.—The optical considerations relative to the difference in the intensity of light presented by a single luminous point and by a disk subtending an appreciable angle, are explained in Arago's Analyse des travaux de Sir William Herschel (Annuaire du Bureau des Longitudes, 1842, pp. 410—412, and 441.)

(32) p. 76.—"The two Magellanic clouds, Nubecula Major and Minor, are extremely remarkable objects. The greater is a clustering collection of stars, and consists of clusters of irregular form, of globular clusters, and nebulæ of various magnitudes and degrees of condensation; among these there occur large nebulous spaces not resolvable into stars, but which yet are probably *star-dust* (composed of very minute stars), and which, even in the twenty-feet reflector, appear only as a general illumination of the field of view, and form a bright ground upon which other objects of very remarkable and mysterious characters are scattered. In no other part of the heaven are so many nebulæ and clusters of stars crowded together as in this Nubecula. The Nubecula Minor is less striking. It exhibits more nebulous, irresolvable light, and the clusters which are scattered over it are fewer in nebula and less brilliant." (From a letter written by Sir John F. W. Herschel, from Feldhausen at the Cape of Good Hope, June 13, 1836.)

(33) p. 77.—I should have introduced the fine expression χορϙος ουρανου, (which Hesychius borrowed from some unknown poet,) when speaking in an earlier page of the "Garden of the Universe," if χορϙος had not rather signified more generally an *enclosed space*. The connection with the Germanic *Garten*, garden, the Gothic *gards* (derived, according to Jacob Grimm, from *gairdan*, cingere, Engl. *gird*), cannot, however, be mistaken, any more than the affinity to the Slavonian *grod, gorod*, and, as noticed by Pott in his Etymol. Forschungen, Th. 1, S. 144, to the Latin *chors*, (whence *corte, cour*, and the Ossetic *khart*). We may perceive a further connection with the Northern *gard, gerd* (enclosure or place hedged round as a court, or as a country seat), and the Persian *gerd, gird*, circle, district, as applied to a princely country seat, castle, or town, as in ancient names of places in Firdusi's Shahnameh, Siyawakschgird, Darabgird, &c.

(34) p. 79.—Respecting α Centauri, see Maclear's results in 1839 and 1840, in the Trans. of the Astronomical Society, vol. xii. p. 370: probable mean error 0″·0640. For 61 Cygni, see Bessel, in Schumacher's Jahrbuch. 1839,

S. 47—49, and in the Astronomische Nachrichten, Bd. 17, S. 401, 402: mean error 0″0141. Respecting the relative distances of stars of different magnitudes, how those of the third magnitude may probably be three times more distant, and in what manner we may represent to ourselves the arrangement of the sidereal strata, I find in Kepler's "Epitome Astronomiæ Copernicanæ," 1618, T. i. lib. 1, p. 34—39, a remarkable passage: "Sol hic noster nil aliud est quam una ex fixis, nobis major et clarior visa, quia propior quam fixa. Pone terram stare ad latus, uno semidiametro viæ lacteæ, tunc hæc via lactea apparebit circulus parvus, vel ellipsis parva, tota declinans ad latus alterum; eritque simul uno intuitu conspicua, quæ nunc non potest nisi dimidia conspici quovis momento. Itaque fixarum sphæra non tantum orbe stellarum, sed etiam circulo lactis versus nos deorsum est terminata."

(^{35}) p. 82.—"Si dans les zones abandonnées par l'atmosphère du soleil il s'est trouvé des molecules trop volatiles pour s'unir entre elles ou aux planètes, elles doivent en continuant de circuler autour de cet astre, offrir toutes les apparences de la lumière zodiacale, sans opposer de résistance sensible aux divers corps du système planétaire, soit à cause de leur extrême rareté, soit parceque leur mouvement est à fort peu près le même que celui des planètes qu'elles rencontrent."—Laplace, "Exp. du Syst. du Monde" (ed. 5), p. 415.

(^{36}) p. 82.—Laplace, "Exp. du Syst du Monde" (ed. 5), pp. 396 and 414.

(^{37}) p. 82.—Littrow's "Astronomie," 1825, Bd. ii. S. 107. Mädler, Astr., 1841, S. 212. Laplace, "Exposition du Système du Monde," p. 210.

(^{38}) p. 84.—Kepler on the increasing volume and density of the planets, with the increase of their distance from the sun or central body, itself described as the densest of all the heavenly bodies (Epitome Astron. Copern. in vii. libros digesta, 1618—1622, p. 420). Leibnitz was also inclined to the opinion of Kepler and Otto von Guericke, $i.\ e.$ that the planets increase in volume in proportion to their distance from the sun. See his letter to the Magdeburg Burgomaster (Mainz, 1671), in Leibnitz, deutschen Schriften herausg. von Guhrauer, Th. I. S. 264.

(^{39}) p. 84.—For a tabular statement of the masses, see Encke, in Schumacher's Astr. Nachr., 1843, No. 488, p. 114.

(^{40}) p. 87.—Taking the semi-diameter of the moon at 0.2725, according to Burckhardt's determination, and its volume at $\frac{1}{49.09}$, its density is found 0.5596 or $\frac{5}{9}$ths nearly. Compare also Wilhelm Beer and H. Mädler, der Monde, S. 2, und 10, as well as Mädler, Astr. S. 157. The material contents are, according to Hansen, nearly $\frac{1}{54}$th, and according to Mädler, $\frac{1}{49.6}$ of those of the earth, and its mass $\frac{1}{87.73}$ of that of the earth. In the largest of Jupiter's Satellites (the third) its ratios of volume and mass to the central planet are, of volume $\frac{1}{18870}$, of mass $\frac{1}{11800}$. On the ellipticity of Uranus, see Schumacher's "Astron. Nachr." 1844, No. 493.

([41]) p. 90.—Beer und Mädler, der Mond., § 185, S. 208, and § 347, S. 332; and Phys. Kentniss der himml. Körper., S. 4 und 69, Tab. I.

([42]) p. 92.—The paths of the four first comets which it has been possible to investigate, have been computed from Chinese observations. They are those of 240 (in the reign of Gordian III.), of 539 in that of Justinian, of 565 and of 837. According to Du Sejour, the last of these must have been for four-and-twenty hours 2000000 geographical miles only from the earth. It appeared during the reign of Louis-le-Débonnaire (the son of Charlemagne), and was thought to prognosticate disasters which that monarch sought to avert by founding conventual establishments, whilst the Chinese astronomers were scientifically engaged in tracing its path. Its tail was 60 degrees in length, and appeared sometimes simple and sometimes divided. The first comet computed exclusively from European observations, is that of 1456, or Halley's comet, so named on the occasion which was long, though erroneously, supposed to have been its first well-ascertained appearance. Arago, Annuaire, 1836, p. 204; Laugier, Comptes-rendus des Séances de l'Académie, 1843, T. xvi. p. 1006.

([43]) p. 93.—Arago, "Annuaire," 1832, p. 209—211. As the tail of the comet of 1402 was seen in bright sunshine, so also in the recent great comet of 1843 the nucleus and tail were visible on the 28th of February in North America, between one and three o'clock in the afternoon. (J. G. Clarke, of Portland, in the State of Maine.) The distance of the very dense nucleus from the sun's limb could be measured with great exactness. The nucleus and tail appeared as a very pure white cloud; between the tail and the nucleus there was one darker part. (American Journal of Science, vol. xlv. No. 1, p. 229; Schum. Astr. Nachr. 1843, N. 491, S. 175.)

([44]) p. 93.—"Phil. Trans." for 1808, Pt 2, p. 155; and for 1812, Pt 1, p. 118. Herschel found the diameters of the nuclei 538 and 428 English miles. For the dimensions of the comets of 1798 and 1805, vide Arago, in the "Annuaire" for 1832, p. 203.

([45]) p. 95.—Arago, des changemens physiques de la comète de Halley du 15-23 Oct. 1835, in the "Annuaire" for 1836, p. 218—221. The usual direction of the emanations had been remarked in the time of Nero "Comæ radios solis effugiunt," Seneca, Nat. Quaest. vii. 20.

([46]) p. 95.—Bessel, in Schumacher's "Astronomische Nachrichten," 1836,

N. 300—302, S. 188, 192, 197, 200, 202, and 230; and in Schum.
"Jahrbuch." 1837, S. 149—168. William Herschel considered that in his
observations of the fine comet of 1811, he found evidences of the rotation of
the nucleus and tail. Phil. Trans. 1812, part 1, p. 140; Dunlop, at
Paramatta, thought the same respecting the third comet of 1825.

(⁴⁷) p. 96.—Bessel, in the "Astr. Nachr." 1836, N. 302, S. 231; Schum.
"Jahrb." 1837, S. 175. Also compare Lehmann, "über Cometen schweife,"
in Bode's "Astron. Jahrb." für 1826, S. 168.

(⁴⁸) p. 96.—Aristot. "Meteor." I. 8, 11—14 and 19—21 (ed. Ideler, t. 1,
p. 32—34). Biese, "Phil. des Aristoteles," Bd. ii. S. 86. The great influence
which the writings of Aristotle exercised on the whole of the middle ages,
renders it a cause of extreme regret that he should have been so opposed to
the grander and juster views of the fabric of the universe entertained by the
more ancient Pythagorean school. He pronounces comets to be transitory
meteors belonging to our atmosphere, in the same book in which he mentions
the Pythagorean opinion, that they were planets of long revolution (Aristot.
I. 6, 2). This Pythagorean doctrine, which, according to the testimony of
Apollonius Myndius, had been still more anciently held by the Chaldeans,
descended to the Romans, who here, as elsewhere, merely repeated the lesson
learnt from others. The Myndian describes the path of comets as extending
far into the upper celestial spaces. Hence Seneca says, in the "Nat. Quaest."
vii. 17: "Cometes non est species falsa, sed proprium sidus sicut solis et
lunæ: altiora mundi secat et tunc demum apparet quum in imum cursum sui
venit;" and (vii. 27): "Cometas æternas esse et sortis ejusdem, cujus cœtera
(sidera), etiamsi faciem illis non habent similem." Pliny (ii. 25) likewise obviously refers to Appollonius Myndius, when he says: "Sunt qui et hæc sidera
perpetua esse credant suoque ambitu ire, sed non nisi relicta a sole cerni."

(⁴⁹) p. 96.—Olbers, in the "Astr. Nachr." 1828, S. 157 und 184. Arago,
de la constitution physique des cometes, "Annuaire," 1832, p. 203—208.
The ancients had been struck by the circumstance that it is possible to see
through comets as through flame. The oldest testimony to stars having been
seen through them is that of Democritus (Aristot. Meteor. I. 6, 11); and it
led Aristotle to make the not unimportant remark that he had himself observed
one of the stars of Gemini occulted by Jupiter. Seneca only speaks decidedly
of the transparency of the tail. He says (Nat. Quaest. vii. 18) that stars are
seen through comets as through a cloud; not indeed through the body of the
comet, but through the rays of the tail: "non in ea parte qua sidus ipsum est
spissi et solidi ignis, sed qua rarus splendor occurrit et in crines dispergitur.

Per intervalla ignium, non per ipsos, vides" (vii. 26). The addition is superfluous, for we certainly can see through flame if it has not too great a thickness ; as shewn by Galileo in the Saggiatore (Lettera a Monsignor Cesarini,
1619).

(^{50}) p. 96.—Bessel, "Astr. Nachr." 1836, N. 301, S. 204—206.   Struve,
"Recueil des Mem. de l'Acad. de St.-Pétersbourg," 1836, p. 140—143 ; and
"Astr. Nachr." 1836, N. 303, S. 238.   At Dorpat the star was in conjunction
only 2."2 from the brightest point in the comet.   The star remained constantly
visible, and its light was not perceptibly weakened, whereas the nucleus of the
comet seemed to fade even to extinction before the light of this small star of
the ninth or tenth magnitude.

(^{51}) p. 97.—Arago's first attempt to analyse the light of comets by polarisation was on the 3d of July, 1819, on the evening of the sudden appearance of
the great comet.   I was present at the Observatory, and satisfied myself, as did
Mathieu and the since deceased astronomer Bouvard, of the dissimilarity in
brightness of the images in the polariscope when the instrument received the
cometary light.   When it received the light from Capella, which was near
the comet and at the same altitude, the images were equal in intensity.   When
Halley's comet appeared in 1835, the apparatus was altered, so as to give, according to Arago's " chromatic polarisation," two images of complementary
colours (green and red).   "Annales de Chimie," vol. xiii. p. 108.   " Annuaire," 1832, p. 216.   " On doit conclure," says Arago, " de l'ensemble de
ces observations, que la lumière de la comete n'était pas en totalité composée
de rayons doués des propriétés de la lumière directe, propre ou assimilée : il
s'y trouvoit de la lumière réfléchie spéculairement et polarisée, c'est à dire
venant du soleil.   On ne peut décider par cette méthode, d'une manière
absolue, que les cométes brillent seulement d'un eclat d'emprunt.   En effet en
dévenant lumineux par eux-mêmes, les corps ne perdent pas pour cela la
faculté de réflechir des lumières étrangères."

(^{52}) p. 98.—Arago, in the "Annuaire" for 1832, p. 217—220.   Sir John
Herschel's Astronomy, § 488.

(^{53}) p. 99.—Encke, in the "Astr. Nachr." 1843, N. 489, S. 130—132.

(^{54}) p. 100.—Laplace, "Exp. du Syst. du Monde," p. 216 and 237.

(^{55}) p. 100.—Littrow, " Beschreibende Astr." 1835, S. 274.   Respecting
the comet of short period recently discovered by Faye, of the Paris Observatory, of which the eccentricity is 0.551, its solar distance at its perihelion
1.690 and at its aphelion 5.832, see "Schum. Astr. Nachr." 1844, N. 495 ;
see also N. 239 of the "Astr. Nachr." 1833, on the supposed identity of the

comet of 1766 with the third comet of 1819; and N. 237 of the same work, on the identity of the comet of 1743 and the fourth comet of 1819.

([56]) p. 102.—Laugier, in the "Comptes-rendus des Séances de l'Académie," 1843, t. xvi. p. 1006.

([57]) p. 105.—Fries, "Vorlesungen über die Sternkunde," 1833, S. 262—267. A not very happy instance of a comet of "good omen" is met with in Seneca, "Nat. Quaest." vii. 17 and 21; he says of comets, "quem nos Neronis principatu lætissimo vidimus et qui cometis detraxit infamiam."

([58]) p. 107.—A friend of mine, accustomed to exact trigonometrical measurements, in the year 1788, at Popayan, a town situated in 2° 26′, N. lat. and at an elevation of 5520 feet (about 5880 English), saw at noon, with the sun shining brightly in a cloudless sky, his whole room illuminated by a ball of fire. He was standing with his back to the window, and on turning round great part of the track left by the meteor was still brilliantly marked. These phænomena among different nations and tribes have been connected with very different names and associations. In the Lithuanian Mythology a fanciful but graceful and noble symbolical meaning has been attached to them. It was said that when a child was born, the "Verpeja" began to spin the thread of the infant's destiny, that each of these threads was attached to a star, and that when death approached the person, the thread broke and the star fell glimmering to the earth and was extinguished. (Jacob Grimm, Deutsche Mythologie, 1843, S. 685.)

([59]) p. 107.—From the account of Denison Olmsted, Professor at Yale College, Newhaven, Connecticut. Vide "Poggend. Annalen der Physik," Bd. xxx. S. 194. Kepler, who excluded balls of fire and shooting stars from the domain of astronomy, considering them as "meteors produced by terrestrial exhalations mixing with the higher ether," expresses himself on the whole with great care respecting them. He says, "Stellæ cadentes sunt materia viscida inflammata. Earum aliquæ inter cadendum absumuntur, aliquæ verè in terram cadunt, pondere suo tractæ. Nèc est dissimile vero, quasdam conglobatas esse ex materiâ fœculentâ, in ipsam auram æthceream immixta : exque ætheris regione, tractu rectilineo, per aërem trajicere, ceu minutos cometas, occultâ causa motus utrorumque. (Kepler, Epit. Astron. Copernicanæ, t. i. p. 80.)

([60]) p. 107.—Relation Historique, t. i. p. 80, 213, and 527. If in shooting stars, as in comets, we distinguish between the head or nucleus and the tail or train, we shall recognise that the greater length and brilliancy observed in the train in tropical countries, is to be attributed to the greater transparency of the atmosphere; the phænomenon itself is the same, but is more easily and longer visible. This influence of the condition of the atmosphere sometimes shews itself even in the temperate zone, and in a difference between places at very small distances apart; Wartmann mentions that on one occasion of the periodical November phænomenon, the number of shooting stars observed at Geneva and aux Planchettes (places very near to each other), were as 1 : 7. (Wartmann, Mém. sur les étoiles filantes, p. 17). The tail or train of a shooting star, which Brandes has made the subject of so many exact and delicate observations, is by no means to be ascribed to the prolongation of impressions on the retina. Its visibility sometimes lasts an entire minute, in rare cases even longer than the light of the head or nucleus of the shooting star. The luminous path in such cases remains motionless (Gilb. Ann. Bd. xiv. S. 251). This circumstance, too, shews the analogy between large shooting stars and fire balls. Admiral Krusenstern, in a voyage round the world, saw the train of a fire-ball which had long disappeared continue to shine for the space of an hour, during which time it changed its place exceedingly little. (Reise, Th. i. S. 58.) Sir Alexander Burnes gives a charming description of the transparency of the atmosphere in Bokhara as favourable to a love for astronomy; the latitude is 39°43′, and the elevation above the level of the sea about 1200 (1280 English) feet. "There is a constant serenity in the atmosphere, and an admirable clearness in the sky. At night the stars have an uncommon lustre, and the milky way shines gloriously in the firmament. There is also a never-ceasing display of the most brilliant meteors, which dart like rockets in the sky : ten or twelve of them are sometimes seen in an hour, assuming every colour, fiery, red, blue, pale and faint. It is a noble country for astronomical science, and great must have been the advantage enjoyed by the famed observatory of Samarkand." (Burnes, Travels into Bokhara, vol. ii. 1834, p. 158). A solitary traveller must not be reproached for calling ten or twelve shooting stars in an hour many; it has only been by very careful observations in Europe directed to this particular subject, that it has been found that, for the range of vision of a single individual, eight is the mean number of meteors that may be seen per hour (Quetelet, Corresp. Mathém. Nov. 1837, p. 447), while so diligent an observer as Olbers limited the number to five or six. (Schum. Jahrb. 1838, S. 325.)

([61]) p. 109.—On meteoric dust. (See Arago, Annuaire, 1832, p. 254). I have very recently endeavoured to show in another work, (Asie Centrale, t. i. p. 408), the probability that the Scythian tradition of the sacred gold which fell glowing from Heaven, and remained in the possession of the Paralatæ, (Herodot. iv. 5—7), arose from the obscure recollection of a fall of aerolites. The ancients had also a strange fable (Dio Cassius, lxxv. 1259) of silver which fell from heaven, and with which it was attempted, under the Emperor Severus, to silver over bronze coins: the presence of metallic iron in meteoric stones was, however, known (Plin. ii. 56). The frequent expression, " lapìdibus pluit," must not, however, be always interpreted to mean falls of aerolites. In Liv. xxv. 7, it probably refers to erupted pumice (rapilli), from the then not quite extinct volcano Mons Albanus (Monte Cavo); see Heyne, " Opuscula Acad." T. iii. p. 261 ; and my " Relat. Hist." T. i. p. 394, The conflict of Hercules with the Lygians, on the way from the Caucasus to the Hesperides, belongs to a different set of ideas : it is an attempt to explain mythically the origin of the round quartz blocks in the Lygian field of stones at the mouth of the Rhone, which Aristotle supposed to have been ejected from a fissure during an earthquake, and Posidonius ascribes to the action of the waves of an inland sea. In the fragment of the " Prometheus Freed" of Æschylus, there is a proceeding which closely resembles a fall of aerolites. Jupiter draws together a cloud, and " covers the ground with rounded stones for rain." Posidonius allowed himself to laugh at the geological mythus of stones and blocks. The Lygian field of stones is, however, very naturally and faithfully described. The district is now called La Crau. (Vide Guerin, Mesures barométriques dans les Alpes et Météorologie d'Avignon, 1829, chap. xii. p. 115).

(⁶²) p. 109.—The specific gravity of aerolites varies from 1.9 (Alais) to 4.3 (Tabor). The most usual density is 3, water being 1. In regard to the actual diameters of fire-balls, the numbers in the text refer to the few tolerably certain measurements which can be collected. These give for the fire-ball of Weston, in Connecticut, 14th of December, 1807, only 500 feet ; for the one observed by Le Roi, 10th of July, 1771, about 1000 feet ; and for the one of the 18th of January, 1783, (estimated by Sir Charles Blagden), 2600 feet diameter. Brandes gives to shooting stars a diameter of 80—120 feet, with luminous trains of 3 or 4 (12 or 16 Engl.) miles in length, (Unterhalt. Bd. i. S. 42). There are not, however, wanting optical reasons which render it probable that the apparent diameters of fire-balls and shooting stars have been greatly over-estimated. The volume of the largest which has been seen cannot properly be compared to the volume of Ceres, (should we even assign to that planet a diameter of only 70 English miles). Vide the generally so exact and excellent treatise on the " Connexion of the Physical Sciences," 1835, p. 411. To elucidate what I have said, in page 110, of the large aerolite which fell in the bed of the river near Narni, but which has not been again found there, I subjoin the passage which Pertz has made known from the "Chronicon Benedicti monachi Sancti Andreæ in Monte Soracte," a document belonging to the tenth century, and which is preserved in the Chigi Library at Rome. The barbarous Latin of the period has been left unaltered : " Anno—921—temporibus domini Johannis Decimi pape, in anno pontificatus, illius 7 visa sunt signa. Nam iuxta urbem Romam lapides plurimi de cœlo cadere visi sunt. In civitate quæ vocatur Narnia, tam diri ac tetri, ut nihil aliud credatur, quam de infernalibus locis deducti essent. Nam ita ex illis lapidibus unus omnium maximus est, ut decidens in flumen Narnus, ad mensuram unius cubiti super aquas fluminis usque hodie videretur. Nam et ignitæ faculæ de cœlo plurimæ omnibus in hac civitate Romani populi visæ sunt, ita ut pene terra contingeret. Aliæ cadentes," &c. (Pertz, Monum. Germ. hist. Scriptores, T. iii. p. 715). Respecting the aerolite at Ægos Potamos, the fall of which is placed by the Parian Chronicle in the 78.1 Olym. (Böckh, Corp. Inscr. Græc. T. ii. pp. 302, 320, and 340), compare Aristol. Meteor. i. 7 (Ideler, Comm. T. i. pp. 404—407); Stob. Ecl. Phys. i. 25, p. 508, Heeren ; Plut. Lys. c. 12 ; Diog. Laert. ii. 10. (See also in the sequel Notes 69, 87, 88, and 89.) According to a Mongolian popular tradition, in a plain near the sources of the Yellow River in Western China, there is a fragment of black rock 40 French feet high which fell from heaven. (Abel-Rémusat, in Lamétherie's Journ. de Phys. 1819, Mai, p. 264).

(63) p. 110.—Biot, Traité d'Astronomie physique (3me édition), 1841, T. i. pp. 149, 177, 238, and 312. My illustrious friend Poisson attempted to solve the difficulty, attendant on the assumption of the spontaneous ignition of meteoric stones at a height where the density of the atmosphere is almost insensible, in a very peculiar manner : " A une distance de la terre où la densité de l'atmosphère est tout-à-fait insensible, il seroit difficile d'attribuer, comme on le fait, l'incandescence des aerolites à un frottement contre les molécules de l'air. Ne pourrait-on pas supposer que le fluide électrique, à l'état neutre, forme une sorte d'atmosphère, qui s'étend beaucoup au-delà de la masse d'air ; qui est soumise à l'attraction de la terre, quoique physiquement impondérable ; et qui suit, en conséquence, notre globe dans ses mouvements ? Dans cette hypothèse, les corps dont il s'agit, en entrant dans cette atmosphère impondérable, décomposeraient le fluide neutre, par leur action inégale sur les deux électricités, et ce serait en s'électrisant qu'ils s'échaufferaient et deviendraient incandescents." (Poisson, Rech. sur la Probabilité des Jugements, 1837, p. 6).

(64) p. 111.—Phil. Trans. Vol. xxix. pp. 161—163.

382 NOTES.

([65]) p. 111.—The first edition of Chladni's important memoir " On the Origin of the Mass of Iron found by Pallas, and of other similar Masses," appeared two months before the fall of stones at Sienna, and two years before Lichtenberg stated in the " Göttingen Taschenbuch" " that stones arrive in our atmosphere from the regions of universal space." Compare also Olbers, Letter to Benzenberg, 18th of November, 1837, in Benzenberg's Memoir on Shooting Stars, p. 186.

([66]) p. 111.—Encke, in Poggend. Annalen, Bd. xxxiii. (1834), S. 213. Arago, Annuaire, 1836, p. 291. Two letters from myself to Benzenberg, 19th of May and 22d of October, 1837, on a conjectured retrogression of the nodes in the orbit of periodical streams of shooting stars, (Benzenberg, Sternschnuppen, S. 207 und 209). Olbers subsequently adopted this opinion of the gradual retardation of the November phenomenon. (Ast. Nach. 1838, N. 372, S. 180.) If I may combine two of the showers of falling stars mentioned by Arabian writers, with the epochs which Boguslawski has found for the fourteenth century, I obtain the following more or less accordant elements of the movement of the nodes :—

In October 902, on the night when King Ibrahim-ben-Ahmed died, there was a great fall of shooting stars, " like a fiery rain." The year was called, on this account, the year of stars. (Condé, Hist. de la domin. de los Arabes, p. 346).

On the 19th of October 1202, the stars were falling the whole night through. " They fell like locusts." (Comptes-rendus, 1837, T. i. p. 294 ; and Fraehn, in the Bulletin de l'Académie de St.-Pétersbourg, T. iii. p. 308).

On the 21st of October (old style), 1366, " die sequente post festum xi. millia Virginum, ab hora matutina usque ad horam primam, visæ sunt quasi stellæ de cœlo cadere continuo, et in tantâ multitudine quod nemo narrare sufficit." This remarkable notice, which will be again alluded to in the text, was found, by the younger von Boguslawski, in Benesse (de Horowic) de Weitmil or Weithmül, Chronicon Ecclesiæ Pragensis, p. 389. The chronicle is also found in the second part of the Scriptores rerum Bohemicarum, by Pelzel and Dobrowsky, 1784 (Schum. Astr. Nachr. Dec. 1839).

On the night 9—10 Nov. 1787, many shooting stars were observed in Southern Germany, and especially at Manheim, by Hemmer (Kämtz, Meteor. Th. iii. S. 237.)

After midnight, on the 12th of November, 1799, the prodigious fall of shooting stars at Cumana, which has been described by Bonpland and myself; and which was observed over a great part of the earth (Relat. hist. t. i. pp. 519—527).

In the night 12—13 Nov. 1822, shooting stars, mingled with balls of fire, were seen in great numbers by Klöden, at Potsdam (Gilbert's Ann. Bd. lxxii. S. 219).

13 Nov. 1831, at 4 A.M. a great fall of shooting stars was seen by Cap. Bérard, on the Spanish Coast near Cartagena del Levante (Annuaire, 1836, p. 297).

On the night of the 12—13 Nov. 1833, in North America, the memorable phænomenon of which Denison Olmstead has given so excellent a description.

On the night 13—14 Nov. 1834, in North America, the same stream, but less considerable in numbers (Poggend. Ann. Bd. xxxiv. S. 129).

On the 13th Nov. 1835, near Belley in the Département de l'Ain, a barn was set on fire by the fall of a sporadic fire-ball (Annuaire, 1836, p. 296).

In 1838 the stream showed itself most decidedly on the night 13—14 Nov. (Astr. Nachr. 1838, N. 372).

(⁶⁷) p. 112.—I am aware that among the 62 shooting stars simultaneously observed at the request of Professor Brandes, in Silesia, in 1823, there were a few which appeared to have an elevation of 45.7 to 60, and even 100 German miles (or 182.8 to 240, and even 400 English miles) (Brandes, Unterhaltungen für Freunde der Astronomie und Physik, Heft i. S. 48); but all determinations above 30 German, or 120 English miles, are regarded by Olbers as doubtful, on account of the smallness of the parallax.

(⁶⁸) p. 112. — The planetary velocity of translation in the orbit is, in Mercury, 26.4; in Venus, 19.2; and in the Earth, 16.4 miles in a second.

(⁶⁹) p. 113.—Chladni informs us that an Italian physicist, Paolo Maria Terzago, in 1660, was the first who noticed the possibility of aerolites being stones from the moon. This was on the occasion of a fall of aerolites in Milan, by which a Franciscan monk was killed. He says, in a writing entitled "Musæum Septalianum Manfredi Septalæ, Patricii Mediolanensis industrioso labore constructum," Tortona, 1664, p. 44, "Labant philosophorum mentes, sub horum lapidum ponderibus; ni dicere velimus, lunam terram alteram, sive mundum esse, ex cujus montibus divisa frusta in inferiorem nostrum hunc orbem delabantur." Without knowing any thing of this conjecture, Olbers was led, on the occasion of the celebrated fall of meteoric stones at Sienna (16 June, 1794), to undertake, in the following year, an investigation of the initial projectile force which would be requisite to bring to the earth masses erupted at the surface of the moon. This balistic problem occupied for ten or twelve years the attention of the geometers—Laplace, Biot, Brandes, and Poisson. The then prevailing, but since abandoned, opinion of the existence of active volcanoes in the moon, where air and water are absent, caused the public to confound two things extremely different, *viz.* a mathematical possibility and a physical probability. Olbers, Brandes, and Chladni, considered that, in the relative velocity of 4 to 8 German, or 16 to 32 English miles, with which balls of fire and shooting stars enter our atmosphere, they found a refutation of lunar origin. According to Olbers, the initial velocity required to reach the earth, without taking into account the resistance of the atmosphere, would be 7780 French feet in a second; according to Laplace, 7377; according to Biot, 7771; and according to Poisson, 7123. Laplace calls this an initial velocity only five or six times greater than that of a cannon-ball; but Olbers has shewn, "that with an initial velocity of 7500 to 8000 French feet in a second, meteoric stones would arrive at the surface of the earth with a velocity only of 35000 feet (1·53 German geographical miles): now as the mean measured velocity per second of meteoric stones is 5 German geographical miles, or above 114000 feet per second, it follows that the initial velocity at the surface of the moon should be almost 110000 feet, or 14 times greater than that assumed by Laplace." (Olbers, in Schum. Jahrb. 1837, S. 52—58; and in Gehler's neuem Physik. Wörterbuche, Bd. vi. Abth. 3, S. 2129—2136). It is true, that if we could assume volcanic forces to be active at the surface of the moon at the present time, the absence of atmospheric resistance would give to the projectile force of lunar volcanoes an advantage over that of our terrestrial volcanoes; but even in respect to a measure of the latter force, data on which we can depend are extremely deficient, and it is probable that it has been greatly over-estimated. A very accurate observer of the phænomena of Etna, Dr. Peters, found the greatest velocity of any of the stones which he saw ejected from the crater only 1250 feet in a second; observations on the Peak of Teneriffe, in 1798, gave 3000 feet. Although Laplace, at the end of his work (Expos. du Syst. du Monde, éd. de 1824, p. 399), says respecting aerolites, "que selon toutes les vraisemblances, elles viennent des profondeurs de l'espace céleste;" yet we see from another passage (Chap. vi. p. 233), that, being probably unacquainted with the enormous planetary velocity of meteoric stones, he turned with a degree of preference to the hypothesis of a lunar origin, always, however, premising the assumption that the stones projected from the moon "deviennent des satellites de la terre, décrivant autour d'elle une orbit, plus ou moins allongée, de sorte qu'ils n'atteignent l'atmosphère de la terre qu'après plusieurs et même un très grand nombre de révolutions." As an Italian at Tortona conceived that aerolites came from the moon, so some of the Greek philosophers imagined that they came from the sun. Diogenes Laertius (ii. 9) records such an opinion respecting the origin of the mass which fell near Ægos Potamos (Note 62). Pliny, who registered every thing, repeats the opinion, and derides it the more because he, with earlier writers (Diog. Laert. ii. 3 and 5, p. 99, Hübner), accuses Anaxagoras of having predicted the fall of aerolites from the sun :—"Celebrant Græci Anaxagoram Clazomenium Olympiadis septuagesimæ octavæ secundo anno prædixisse cœlestium litterarum scientia, quibus diebus saxum casurum esse e sole, idque factum interdiu in Thraciæ parte ad Ægos flumen.—Quod si quis prædictum credat, simul fateatur necesse est majoris miraculi divinitatem Anaxagoræ fuisse, solvique rerum naturæ intellectum, et confundi omnia, si aut ipse sol lapis esse aut unquam lapidem in eo fuisse credatur ; decidere tamen crebro non erit dubium." The fall of a stone of moderate size preserved in the Gymnasium of Abydos is also said to have been foretold by Anaxagoras. Probably the fall of aerolites during bright sunshine, and when the moon's disk was not visible, led to the idea of "sun stones." According to one of the physical dogmas of Anaxagoras, the sun was regarded as "a molten incandescent mass" (μυδρος διαπυρος). Following these views, the sun is called, in the Phaëton of Euripides, a "golden mass ;" meaning a brightly-shining mass, not thereby tending to any inference of aerolites being "golden sun stones" (see Note 61). Compare Valckenaer, Diatribe in Eurip. perd. dram. Reliquias, 1767, p. 30; Diog. Laert. ii. 10. We find, therefore, among the Greek naturalists, four hypotheses respecting the origin of shooting stars, two of which may be termed telluric, and two cosmical : 1. from terrestrial exhalations ; 2. from masses of stone carried up by violent tempests (Aristoteles Meteor. Lib. i. Cap. iv. 2—13, and Cap. viii. 9) ; 3. from the sun ; 4. from the regions of space, as heavenly bodies which had long been invisible on account of their distance. Respecting this latter opinion of Diogenes of Apollonia, which is in entire accordance with our own in the present day, see page 124 in the text, and Note 88. It is a curious circumstance, of which I have been assured by a learned orientalist who instructed me in Persian, M. Andrea de Nerciat, now resident in Smyrna, that in Syria, according to an old popular belief, falls of aerolites are looked for on very clear moonlight nights. The ancients, on the contrary, were particularly on the watch for such falls during lunar eclipses : vide Plin. xxxvii., 10, p. 164; Solinus, c. 37 ; Salm. Exerc. p. 531 ; and the passages collected by Ukert, in his Geogr. der Griechen und Römer, Th. ii., 1, S. 131, Note 14. On the improbability that aerolites are formed from gases holding in solution metallic substances, which, according to Fusinieri, exist in the upper strata of the atmosphere, and which suddenly aggregate from a state of extreme dispersion,—and on the mutual penetration and mixture of gases, see my Relat. Hist. T. i. p. 525.

([70]) p. 114.—Bessel, in Schum. Astr. Nachr. 1839, Nr. 380 and 381, S. 222 and 346. At the conclusion of the memoir, there is a comparison of the longitudes of the sun with the epochs of the November phænomenon, since the date of the first observation at Cumana in 1799.

([71]) p. 114.—Dr. Thomas Forster mentions in his Pocket Encyclopædia of Natural Phænomena, 1827, p. 17, that there is preserved at Christ's College, Cambridge, a manuscript supposed to have been written by a monk, and entitled, " Ephemerides rerum naturalium," in which the natural phænomena proper to each day of the year are indicated; such as the first blossoming of plants, arrival of birds, &c. The 10th of August is marked by the word meteorodes. It was this indication, combined with the tradition of the fiery tears of St. Lawrence, which were the immediate occasion of Dr. Forster's zealous inquiry into the August phænomenon. (Quetelet, Corresp. Mathématique, Série iii., T. i., 1837, p. 433).

([72]) p. 115.—Humboldt, Rel. Hist. T. i. pp. 519—527. Ellicot, in the Transactions of the American Soc. 1804, Vol. vi. p. 29. Arago says of the November phænomenon, "Ainsi se confirme de plus en plus l'existence d'une zone composée de millions de petits corps dont les orbites rencontrent le plan de l'écliptique, vers le point que la terre va occuper tous les ans, du 11 au 13 Novembre. C'est un nouveau monde planétaire qui commence à se révéler à nous." (Annuaire, 1836, p. 296).

([73]) p. 115.—Compare Muschenbroek, Introd. ad Phil. Nat. 1762, T. ii. p. 1061. Howard, Climate of London, Vol. ii. p. 23 ; observations of 1806 ; seven years, therefore, after the earliest observations of Brandes, in Benzenberg über Sternschnuppen, S. 240—244. August Observations of Thomas Forster, in Quetelet's Corr. Math., pp. 438—453. Observations by Adolph Erman, Boguslawski and Kreil in Schumacher's Jahrbuch, 1838, pp. 317—330. Respecting the point of origin in Perseus on the 10th of August, 1839, see the exact measurements of Bessel and Erman, (Schum. Astr. Nachr. Nr. 385 und 428); but on the 10th of August, 1837, the path does not appear to have been retrograde. See Arago, in the Comptes-rendus, 1837, T. ii. p. 183.

([74]) p. 115.—On the 25th of April, 1095, " innumerable eyes saw in France the stars fall from heaven as thick as hail," (ut grando, nisi lucerent, pro densitate putaretur, Baldr. p. 88) ; and this event was spoken of at the Council of Clermont as prognosticating the great movement in Christendom, (Wilken, Gesch. der Kreuzzüge, Bd. i. S. 75). On the 22d of April, 1800, a great fall of shooting stars was seen in Virginia and Massachusetts : it was " a fire of rockets which lasted two hours." Arago was the first to call attention to this (April) " traînée d'astéroides" as a recurring phænomenon, (Annuaire, 1836, p. 297). The fall of aerolites in the beginning of the month of December is also deserving of notice. In favour of their periodical recurrence as a meteoric stream, we have the early observation of Brandes in the night 6—7 December, 1798, when he counted 2000 shooting stars, (Brandes, Unterhalt für Freunde der Physik, 1825, Heft 1, S. 65) ; and perhaps the immense fall of aerolites on the 11th of December, 1836, in Brazil, by the River Assu, near the village of Macao, (Comptes-rendus, T. v. p. 211). Capocci, in the interval from 1809 to 1839, has made out twelve actual falls of aerolites between the 27th and 29th of November, as well as others on the 13th of November, 10th of August, and 17th of July, (Comptes-rendus, T. ii. p. 357). It is remarkable that hitherto no *periodical* falls of shooting stars, or streams of aerolites, have been observed in that part of the Earth's orbit to which the months of January and February, and, perhaps, also the month of March, correspond, although I myself witnessed a striking display of shooting stars on the 15th of March, 1803, in the Pacific ; and a great number had been seen in the city of Quito a short time before the great earthquake of Riobamba (4th of February, 1797). Reviewing what has been stated, the epochs most deserving of attention appear to be—

22—25 April,

17 July (17—26 July ?) (Quetelet, Corr. 1837, p. 435),

10 August,

12—14 November,

27—29 November,

6—12 December.

The frequency of these streams ought not to astonish us, if we remember the myriads of comets which fill the regions of space, great as is the difference between insulated comets and rings composed of asteroids.

(75) p. 116.—Ferd. v. Wrangel, Reise längs der Nordküste von Sibirien in den Jahren, 1820—1824, Th. ii. S. 259. Respecting the return of the denser swarm of the November stream after a period of 34 years, see Olbers, in the Jahrbuch, 1837, S. 280. I was told at Cumana that, a short time before the terrible earthquake of 1766—33 years, therefore, before the great

shower of shooting stars of 11—12 November, 1799,—a similar display had been seen. The earthquake, however, was not in November, but on the 21st of October, 1766. It may still, perhaps, be possible for some traveller to ascertain in Quito the precise day on which, during a whole hour, the volcano of Cayamba appeared as if veiled by the number of falling stars, and the alarmed inhabitants instituted processions. (Relat. Hist. T. i. Chap. iv. p. 307 ; Chap. x. pp. 520 and 527).

(76) p. 117.—From a letter to myself, dated January 24, 1838. The prodigious fall of shooting stars of November, 1799, was seen almost exclusively in America, where it was witnessed from Neu-Herrnhut in Greenland, to the Equator. In 1831 and 1832, the phænomenon was seen in Europe, and scarcely elsewhere ; and, in 1833 and 1834, was seen only in the United States of North America.

(77) p. 118.—Lettre de M. Edouard Biot à M. Quetelet sur les anciennes apparitions d'étoiles filantes en Chine ; Bulletin de l'Académie de Bruxelles, 1843, T. x. No. 7, p. 8. On the notice from the Chronicon Ecclesiæ Pragensis, see Boguslawski (the son), in Poggend. Annalen, Bd. xlviii. S. 612. Add to Note 42, that the paths of the four comets of 568, 574, 1337, and 1385, have also been calculated exclusively from Chinese observations. See John Russell Hind, in Schumacher's Astronomische Nachrichten, 1844, Nr. 498.

(78) p. 118.—" Il paroît qu'un nombre, qui semble inépuisable, de corps trop petits pour être observés, se meuvent dans le ciel, soit autour du soleil, soit autour des planètes, soit peut-être même autour des satellites. On suppose que quand ces corps sont rencontrés par notre atmosphère, la différence entre leur vitesse et celle de notre planète est assez grande pour que le frottement qu'ils éprouvent contre l'air, les échauffe au point de les rendre incandescents, et quelquefois de les faire éclater.—Si le groupe des étoiles filantes forme un anneau continu autour du soleil, sa vitesse de circulation pourra être très différente de celle de la terre ; et ses déplacemens dans le ciel, par suite des actions planétaires, pourront encore rendre possible ou impossible, à différentes époques, le phénomène de la rencontre dans le plan de l'écliptique." (Poisson, Recherches sur la probabilité des jugements, pp. 306—307).

(79) p. 119.—Humboldt, Essai politique sur la Nouvelle Espagne, (2me édit.) T. iii. p. 310.

(80) p. 119.—Pliny had remarked the peculiar colour of the crust of aerolites, " colore adusto" (ii. 56 and 58). The expression " lateribus pluisse," also refers to the burnt appearance of their exterior.

(81) p. 120.—Humboldt, Rel. hist. T. ii. Chap. xx. pp. 299—302,

(82) p. 121.—Gustav Rose, Reise nach dem Ural, Bd. ii. S. 202.

(83) p. 121.—Gustav Rose, in Poggend. Ann. 1825, Bd. iv. S. 173—192. Rammelsberg, Erstes Suppl. zum chem. Handwörterbuche der Mineralogie, 1843, S. 102. Olbers acutely observes, that " It is a remarkable circumstance, not hitherto noticed, that no fossil meteoric stones have as yet been found, like fossil shells, in secondary and tertiary formations. Are we to infer that, previous to the last and present arrangement of the surface of our planet, no meteoric stones had fallen upon it, although, according to Schreibers, it is probable that 700 falls of aerolites now take place in each year?" (Olbers, in Schum. Jahrb. 1838, S. 329). Problematical nickeliferous masses of native iron have been found in Northern Asia (Gold-washing work of Petropawlowsk, 80 geographical miles south-east of Kusnezk), at a depth of 31 French feet, and recently among the Carpathian mountains (Magura, near Szlanicz). Both these masses are very like meteoric stones. Compare Erman, " Archiv für wissenschaftliche Kunde von Russland," Bd. i. S. 315 ; and Haidinger's " Bericht über die Slaniczer Schürfe in Ungarn."

(84) p. 121.—Börzelius, Jahresber. Bd. xv. S. 217 and 231 ; Rammelsberg, Handwörterbuch, Abth. ii. S. 25—28.

(85) p. 122.—Sir Isaac said, " he took all the planets to be composed of the same matter with this earth, viz. earth, water, and stones, but variously concocted." Turner, coll. for the hist. of Grantham, cont. authentic memoirs of Sir Isaac Newton, p. 172.

(86) p. 123.—Adolph Erman, in Poggendorf's Annalen, 1839, Bd. xlviii. S. 582—601. Biot had previously thrown some doubt on the probability of the reappearance of the November stream at the beginning of May (Comptes-rendus, 1836, T. ii. p. 670). Mädler has examined the mean depression of temperature on the three ill-reputed days in the month of May, viz. 11th, 12th, and 13th, by 86 years of observations at Berlin (Verhandl. des Vereins zur Beförd. des Gartenbaues, 1834, S. 377) ; and has found a retrogression of temperature of 1°.22 Cent. (2°.2 Fahr.) just at the season which is very nearly that of the most rapid advance of temperature. It is much to be wished that this phænomenon, which some have been inclined to attribute to the cooling effect of the melting of large masses of ice in the north-eastern part of Europe, should be examined at distant parts of the globe, as in North America, and in the southern hemisphere. Compare Bulletin de l'Acad. Imp. de St.-Pétersbourg, 1843, T. i. No. 4.

(87) p. 123.—Plut. Vitæ par. in Lysandro, cap. 22. The account given by Damachos (Daïmachos) of a fiery cloud, throwing out sparks like shooting

stars, having been seen in the sky without interruption for the space of seventy days, at the end of which it descended nearer to the earth and let fall the stone of Ægos Potamos, "which was only an inconsiderable portion of the cloud," is very improbable, because it would require the direction and velocity of the ball of fire to be for so many days the same as the Earth's ; in the case of the fire-ball of July 19, 1686, described by Halley (Phil. Trans. Vol. xxix. p. 163), this only lasted for a few minutes. It is somewhat uncertain whether this Daïmachos, the writer περι ευσεβειας, was or was not the Daïmachos of Platæa, who was sent by Seleucus to India to the son of Andracottos, and whom Strabo calls a "fabler" (p. 70, Cassaub.), but it seems not unlikely from another passage of Plut. Compar. Solonis, c. Pop. cap. 4 : at any rate it is only the account of a very late author, who wrote a century and a half after the event, and whose authenticity is doubted by Plutarch. Compt., Note 62.

(⁸⁸) p. 124.—Stob. ed. Heeren i. 25, p. 508 ; Plut. de plac. Philos. ii. 13.

(⁸⁹) p. 124.—The remarkable passage in Plut. de plac. Philos. ii. 13, is to the following effect : — "Anaxagoras teaches that the ambient ether is a fiery substance, which, by the force of its rotatory motion, has torn rocks from the earth, inflamed them, and transformed them into stars." Availing himself of an ancient fable to establish a physical dogma, the Clazomenian appears to have attributed the fall of the Nemean lion from the Moon to the Earth in the Peloponnesus to an analogous effect of the general movement of rotation, or to the centrifugal force. (Ælian. xii. 7 ; Plut. de facie in orbe lunæ, c. 24 ; Schol. ex Cod. Paris. in Apoll. Argon. lib. i. p. 498, ed Schæf. T. ii. p. 40 ; Meineke, Annal. Alex. 1843, p. 85). We have had stones from the moon ; we have here an animal fallen from the moon ! According to an ingenious remark of Böckh, the mythus of the lunar lion of Nemea may have an astronomical origin, and a symbolical connection in chronology, with the cycle of intercalation of the lunar year, the worship of the Moon at Nemea, and the games by which it was accompanied.

(⁹⁰) p. 126.—The following remarkable passage on the radiation of heat from the fixed stars—one of Kepler's many inspirations—is found in the Paralipom. in Vitell. Astron. pars Optica, 1604, Propos. xxxii. p. 25 :—Lucis proprium est calor, sydera omnia calefaciunt. De syderum luce claritatis ratio testatur, calorem universorum in minori esse proportione ad calorem unius solis, quam ut ab homine, cujus est certa caloris mensura, uterque simul percipi et judicari possit. De cincindularum lucula tenuissima negare non potes, quin cum calore sit. Vivunt enim et moventur, hoc autem non sine calefactione perficitur. Sed neque putrescentium lignorum lux suo calore destituitur ; nam ipsa putredo quidam lentus ignis est. Inest et stirpibus suus calor." (Compare Kepler, Epit. Astron. Copernicanæ, 1618, T. i. lib. 1, p. 35).

(⁹¹) p. 129.—" There is another thing, which I recommend to the observation of mathematical men: which is, that in February, and for a little before and a little after that month, (as I have observed several years together,) about 6 in the evening, when the twilight hath almost deserted the horizon, you shall see a plainly discernible way of the twilight striking up towards the Pleiades, and seeming almost to touch them. It is so observed any clear night, but it is best *illac nocte*. There is no such way to be observed at any other time of the year (that I can perceive), nor any other way at that time to be perceived darting up elsewhere. And I believe it hath been, and will be, constantly visible at that time of the year. But what the cause of it in nature should be, I cannot yet imagine, but leave it to further inquiry."— (Childrey, Britannia Baconica, 1661, p. 183.) This is the first view and simple description of the phænomenon.—(Cassini, Découverte de la lumière céleste qui paroît dans le zodiaque, Mém. de l'Acad. T. viii. 1730, p. 276. Mairan, Traité phys. de l'aurore boréale, 1754, p. 16.) I find in the singular work of Childrey referred to, very correct details respecting the epochs of the maxima and minima of annual and of diurnal temperature, and notices respect· ing the retardation in the effects of maximum and minimum in all meteorological phænomena (p. 91). It is to be regretted that we find in the same work (p. 148) that the Earth is elongated at the poles, an opinion shared by Bernardin de St.-Pierre: the author says that the globe was originally a true sphere, but the constant increase of the masses of ice at the poles gradually alters its figure, and, as the ice is formed from water, the quantity of water is every where diminishing.

(⁹²) p. 129.—Dominic Cassini (Mém. de l'Acad. T. viii. 1730, p. 188) and Mairan (Aurore bor. p. 16) maintained that the phænomenon, which was seen in Persia in 1668, was the zodiacal light. Delambre (Hist. de l'astron. modèrne, T. ii. p. 742) ascribes the discovery of the zodiacal light to the celebrated traveller, Chardin; but both in the Couronnement de Soliman, and in several passages in the narrative of his travels (éd. de Langlés, T. iv. p. 326; T. x. p. 97), Chardin notices as " niazouk" (nyzek), or " petite lance," only: " la grande et fameuse comète qui parut presque par toute la terre en 1668, et dont la tête étoit cachée dans l'occident, de sorte qu'on ne pouvoit en rien apercevoir sur l'horizon d'Ispahan."—(Atlas du Voyage de Chardin, Tab. iv. from the observations at Schiraz.) The head or nucleus of this comet was, however, seen in Brazil and in India.—(Pingré, Cométogr.

T. ii. p. 22.   Respecting the conjectured identity of this comet with the
recent great comet of 1843, see Schum. Astr. Nachr. 1843, Nr. 476 and
480.   In Persian, the expression, nîzehi âteschîn (fiery spear or lance), is also
used for the rays of the rising or setting sun, in the same way as nayâzik,
according to Freytag's Arabic Lexicon, signifies "stellæ cadentes." The comparison of comets with lances and swords was, however, very common in the
middle ages in all languages.   The great comet, which was seen from April
to June 1500, was always spoken of by the Italian writers of the day under
the title of *il Signor Astone.*—(See my Examen critique de l'histoire de la
Géographie, T. v. p. 80.)   The many conjectures, which have been made,
that Descartes (Cassini, p. 230 ; Mairan, p. 16), and even Kepler (Delambre,
T. i. p. 601), were acquainted with the zodiacal light, appear to me quite
untenable.   Descartes speaks very obscurely (Principes, iii. art. 136, 137)
of the origin of tails of comets : " par des rayons obliques qui, tombant sur
diverses parties des orbes planétaires, viennent des parties latérales à notre
œil par une réfraction extraordinaire ;" also how comets' tails might be
seen, morning and evening, " comme une longue poutre," if the sun is
between the comet and the earth.   This passage is no more to be interpreted
as referring to the zodiacal light, than is that in which Kepler says of the
existence of a solar atmosphere (limbus circa solum, coma lucida), which, in
total eclipses, " prevents its being quite night."—Epit. Astron. Copernicanæ,
T. i. p. 57, and T. ii. 893.)   The statements of Cassini (p. 231, art. xxxi.)
and of Mairan (p. 15), that the " trabes quas δοκους vocant" (Plin. ii. 26
and 27) had allusion to the zodiacal light rising in the form of a tongue, are
even more uncertain, or rather erroneous.   Every where, among the ancients,
the trabes are associated with the bolides (ardores et faces), and other igneous
meteors, and sometimes even with long-bearded comets.   (Respecting δοκὸs,
δοκίas, δοκίτης, see Schäfer, Schol. par. ad Apoll. Rhod. 1813, T. ii. p.
206 ; Pseudo-Aristot. de Mundo, 2, 9 ; Comment. Alex., Joh. Philop. et
Olymp. in Aristot. Meteor. lib. i. cap. vii, 3. p. 195, Ideler ; Seneca, Nat.
Quæst. i. 1.)

(93) p. 130.—Humboldt, Monumens des peuples indigènes de l'Amérique,
T. ii. p. 301.   The curious manuscript which belonged to the Archbishop of
Rheims, Le Tellier, contains a variety of extracts from an Aztec book of rites,
an astrological calendar and historical annals from 1197—1549.   These
annals contain notices of different natural phænomena, epochs of earthquakes,
of comets—as those of 1490 and 1529,—and of solar eclipses, which are
important to Mexican chronology.   In Camargo's manuscript, Historia de

Tlascala, the light rising in the east almost to the zenith, is strangely enough called "sparkling, and as if thick set with stars." The description of the phænomenon, which is said to have lasted forty days, can by no means be understood to apply to volcanic eruptions of the Popocatepetl, which is situated at only a small distance to the south-east.—(Prescott, History of the Conquest of Mexico, Vol. i. p. 284.) More recent commentators have confounded this phænomenon, which Montezuma regarded as a presage of misfortune, with the " estrella que humeava," (" which scintillated :" Mexican *choloa*, to scintillate). Respecting the connection of this vapour with the star Citlal Choloha, (the planet Venus) and the " mountain of the star," (Citlaltepetl, the volcano of Orizaba,) see my Monumens, T. ii. p. 303.

(⁹⁴) p. 130.—Laplace, Exp. du Syst. du Monde, p. 270; Méc. cél. T. ii. p. 169 and 171. Schubert, Ast. Vol. iii. § 206.

(⁹⁵) p. 130. — Arago, Annuaire, 1842, p. 408. Compare Sir John Herschel's considerations on the volume and faintness of the light of planetary nebulæ, in Mrs. Somerville's Connexion of the Physical Sciences, 1835, p. 108. The idea of the sun being a nebulous star, whose atmosphere presents the phænomenon of the zodiacal light, was first started, not by Dominic Cassini, but by Mairan, in 1731 (Traité de l'Aurore bor. p. 47 and 263 ; Arago, in the Annuaire, 1842, p. 412). It was a renewal of Kepler's views.

(⁹⁶) p. 130.—Dominic Cassini assumed, as did subsequently Laplace, Schubert, and Poisson, the hypothesis of a detached ring, to explain the form of the zodiacal light. He says distinctly :—" Si les orbits de Mercure et de Vénus étoient visibles (matériellement dans toute l'étendue de leur surface), nous les verrions habituellement de la même figure et dans la même disposition à l'égard du soleil et aux mêmes tems de l'année que la lumière zodiacale." (Mém. de l'Acad. T. viii. 1730, p. 218 ; and Biot, in the Comptes-rendus, 1836, T. iii. pp. 666). Cassini supposed that the nebulous ring of the zodiacal light consisted of a countless number of small planetary bodies which revolve round the sun. He was inclined to believe that the fall of the fire-balls might be connected with the passage of the earth through the zodiacal nebulous ring. Olmsted, and especially Biot, in the above-mentioned volume of the Comptes-rendus, p. 673, have attempted to connect it with the November fall of aerolites, but Olbers regarded this as very doubtful. (Schum. Jahrbuch, 1837, S. 281). On the question whether the plane of the zodiacal light coincides perfectly with the plane of the sun's equator, see Houzeau, in Schum. Astr. Nachr. 1843, Nr. 492, S. 190.

(⁹⁷) p. 131.—Sir John Herschel, Astron. § 487.

([98]) p. 131.—Arago, Annuaire, 1842, p. 246. Several physical facts appear to indicate that when a mass of matter is mechanically reduced to a state of extreme division, if the mass be very small in proportion to the surface, the electric tension may increase sufficiently for the development of light and heat. Experiments with a large concave mirror have not hitherto given any decided proof of the presence of radiant heat in the zodiacal light. (Lettre de M. Matthiessen à M. Arago, Comptes-rendus, T. xvi. 1843, Avril, p. 687.)

([99]) p. 132.—" What you tell me of the changes of light in the zodiacal light, and of the causes to which, within the tropics, you ascribe such variations, has excited my interest the more, because I have been for a long time past particularly attentive every spring to this phænomenon in our northern latitudes. I, too, have always believed the zodiacal light to rotate; but I assumed it (contrary to Poisson's opinion which you communicate to me) to extend the whole way to the sun, increasing rapidly in intensity. The luminous circle which in total eclipses shews itself round the darkened sun, I have supposed to be this brightest portion of the zodiacal light. I have satisfied myself that the light is very different in different years, sometimes for several successive years being very bright and extended, and in other years scarcely perceptible. I think I find the first trace of any notice of its existence in a letter from Rothmann to Tycho Brahe. Rothmann remarks that in spring he has observed the twilight ceased when the sun had descended 24° beneath the horizon. Rothmann must certainly have confounded the disappearance of the zodiacal light in the vapours of the western horizon with the real termination of the evening twilight. I have not myself been able to observe the sudden fluctuations in the light, probably on account of the faintness with which it appears to us in this part of the world. You are certainly right in ascribing the rapid variations in the light of celestial objects, which you have perceived in the climate of the tropics, to changes taking place in our atmosphere, and especially in its higher regions. This shews itself in a most striking manner in the tails of great comets. Often, and particularly in the clearest weather, pulsations in the tails of comets are seen to commence from the head or nucleus as the lowest part, and to run in one or two seconds through the whole extent of the tail, which in consequence appears to lengthen several degrees, and contract again. That these undulations, which engaged the attention of Robert Hooke, and in later times of Schröter and Chladni, *do not take place in the cometary tails themselves*, but are produced by our atmosphere, appears evident if we reflect that the several particles of these cometary tails (which are many millions of miles in length) are at *very different distances* from us, and that the light from them can only reach our eyes at intervals of time which differ several minutes from each other. I will not attempt to decide whether what you saw on the banks of the Orinoco, not at intervals of seconds, but of minutes, were actual coruscations of the zodiacal light, or whether they belonged solely to the upper strata of our atmosphere. Nor can I explain the remarkable lightness of entire nights, or the anomalous increase and prolongation of twilight in the year 1831, particularly if, as it has been said, the *lightest part* of these singular twilights did not coincide with the place of the sun below the horizon." (Extract from a letter from Dr. Olbers to myself, written from Bremen, March 26, 1833.)

([100]) p. 133.—Biot, Traité d'Astron. physique, 3e éd. 1841, T. i. pp. 171, 238, and 312.

([101]) p. 134.—Bessel, in Schum. Jahrb. for 1839, S. 51; perhaps four millions of geographical miles in a day, in relative velocity at least 3336000 miles, or more than double the velocity of revolution of the earth in her orbit.

([102]) p. 135.—On the proper motion of the solar system, according to Bradley, Tobias Mayer, Lambert, Lalande, and William Herschel, see Arago, Annuaire, 1842, p. 388—399; Argelander, in Schum. Astr. Nachr. Nr. 363, 364, and 398: and on Perseus as the central body of our sidereal stratum, in the treatise, Von der eigenen Bewegung des Sonnensystems, 1837, S. 43; also Otho Struve, in the Bull. de l'Acad. de St.-Pétersb. 1842, T. x. No. 9, p. 137—139. By a more recent combination, Otho Struve found, for the direction of the movement of the solar system, 261° 23' A.R., + 37° 36' Decl.; and, uniting his result with Argelander's, we find, by a combination of 797 stars, 259° 9' A.R., + 34° 36' Decl.

([103]) p. 136.—Aristot. de Cœlo, iii. 2, p. 301; Bekker, Phys. viii. 5, p. 256.

([104]) p. 137.—Savary, in the Connaissance des Tems, 1830, p. 56 and 163; Encke, Berl. Jahrb. 1832, S. 253 ff.; Arago, Annuaire, 1834, p. 260—295; John Herschel, in Mem. of the Astron. Soc. Vol. v. p. 171.

([105]) p. 137.—Bessel, Untersuchung des Theils der planetarischen Störungen welche aus der Bewegung der Sonne entstehen, in Abh. der Berl. Akad. der Wissensch. 1824 (Mathem. Classe) S. 2—6. The question has been raised by Johann Tobias Mayer, in Comment. Soc. Reg. Gotting. 1804, 108; Vol. xvi. p. 31—68.

([106]) p. 138.—Phil. Trans. 1803, p. 225. Arago, Annuaire, 1842, p. 375. —Some idea of the distance assigned in the text to the nearest fixed stars may be obtained by considering, that if we take one French foot as the Earth's solar distance, the planet Uranus will be 19 feet, and α Lyræ 138 geographical miles from the Sun.

(^{107}) p. 138.—Bessel, in Schum. Jahrb. 1839, S. 53.

(^{108}) p. 138.—Mädler, Astr. S. 476.   The same author, in Schum. Jahr. 1839, S. 95.

(^{109}) p. 140.—Sir William Herschel, Phil. Trans. 1817, Pt. 2, p. 328.

(^{110}) p. 140.—Arago, Annuaire, 1842, p. 459.

(^{111}) p. 141.—Sir John Herschel, in a letter from Feldhausen, Jan. 13, 1836.   Nicholl, Archit. of the Heavens, 1838, p. 22.   See also some scattered notices by Sir William Herschel on the starless space which separates us from the Milky Way, in the Phil. Trans. for 1817, Pt. 2, p. 328.

(^{112}) p. 141.—Sir John Herschel, Astron. § 624. The same author in Observations of Nebulæ and Clusters of Stars (Phil. Trans. 1833, Pt. 2, p. 479, fig. 25): "we have here a brother system, bearing a real physical resemblance and strong analogy of structure to our own."

(^{113}) p. 141.—Sir Wm. Herschel, in the Phil. Trans. for 1785, Pt. 1, p. 257. Sir John Herschel, Astron. § 616; and in a letter, addressed to myself, in March 1823, he says :—" The nebulous region of the heavens forms a nebulous milky way, composed of distinct nebulæ as the other of stars."

(^{114}) p. 142.—Sir John Herschel, Astron. § 585.

(^{115}) p. 142.—Arago, Annuaire, 1842, pp. 282—285, 409—411, and 439—442.

(^{116}) p. 142.—Olbers on the Transparency of Celestial Spaces, in Bode's Jahrbuch, 1826, S. 110—121.

(^{117}) p. 143.—" An opening in the heavens," William Herschel, in the Phil. Trans. for 1785, Vol. lxxv. Pt. 1, p. 256.   Le François Lalande, in the Connaiss. des Tems pour l'An VIII. p. 383.   Arago, Annuaire, 1842, p. 425.

(^{118}) p. 143.—Aristot. Meteor. ii. 5, 1.   Seneca, Natur. Quæst. i. 14, 2. " Cœlum discessisse," in Cic. de Divin. i. 43.

(^{119}) p. 143.—Arago, Annuaire, 1842, p. 429.

(^{120}) p. 144.—In December, 1837, Sir John Herschel saw the star η Argûs, which, till that time, had always appeared of the second magnitude and invariable, increase rapidly in brightness until it became of the first magnitude. In January, 1838, its brightness was already equal to that of α Centauri.   In March, 1843, it appeared to Maclear as bright as Canopus; and " even α Crucis looked faint by the side of η Argûs."

(^{121}) p. 144.—" Hence it follows that the rays of the light of the remotest nebulæ must have been almost two millions of years on their way, and that

consequently so many years ago this object must already have had an existence in the sidereal heavens, in order to send out those rays by which we now
perceive it."—William Herschel, in the Phil. Trans. for 1802, p. 498.   Sir
John Herschel, Astron. § 590.   Arago, Annuaire, 1842, p. 334, 359, and
382—385.

(^{122}) p. 145.—From the beautiful sonnet, " Freiheit und Gesetz," by my
brother, Wilhelm von Humboldt: Gesammelte Werke, Bd. iv. S. 358, No. 25.

(^{123}) p. 145.—Otfried Müller, Prolegomena, S. 373.

(^{124}) p. 149.—In speaking of the greatest depths reached by mining and
boring operations, we must distinguish between the *absolute* depth, or that
below the surface of the earth at the point where the work began, and the
relative depth, or that below the level of the sea.   The greatest relative depth
which has yet been attained is probably that of the salt-works of Neu-Salzwerk,
near Minden, in Prussia: in June, 1844, its exact depth was 607·4 mètres
(1993 English feet) ; the absolute depth was 680 mètres (2231 English feet).
The temperature of the water at the bottom was 32·7° Cent. or 90·8 Fah. ;
which, assuming the mean temperature of the air at 9·6° Cent. (49° Fah.),
gives an increase of 1° Cent. for 29·6 mèt., or 1° Fah. for 53·8 feet.   The
absolute depth of the Artesian well of Grenelle is only 1683 feet (1794 Engl.
feet).   According to the accounts of the missionary, Imbert, the depth of our
Artesian wells is much exceeded by that of the fire-springs (Ho-tsing) in
China, which are sunk for the purpose of obtaining hydrogen gas to be employed in salt-boiling.   In the Chinese province Szu-tschuan the fire-springs
very ordinarily reach a depth from 1800 to 2000 French feet : it is even said
that at Tseu-lieu-tsing (place of continual flow), there is a Ho-tsing 3000 feet
deep (3197 Engl.), which was bored with a rod in the year 1812 (Humboldt,
Asie centrale, T. ii. pp. 521 and 525.   Annales de l'Association de la Propagation de la Foi, 1829, No. 16, p. 369).   The relative depth reached at
Monte Massi, in Tuscany, south of Volterra, amounts, according to Matteucci,
only to 382 mètres (1253 Engl. feet).   The boring at Neu-Salzwerk has probably very nearly the same relative depth as the coal mine at Apendale, near
Newcastle-under-Line (Staffordshire), where men work at a depth of 725
yards below the surface.   (Thomas Smith, Miner's Guide, 1836, p. 160.)
Unfortunately, I do not know the exact height above the level of the
sea.   The relative depth of the Monkwearmouth mine, near Newcastle, is
1496·5 Engl. feet (Phillips, Phil. Mag. Vol. v. 1834, p. 446) ; that of
the Liège coal mine of Espérance, at Séraing, is 1271 (1355 Engl.)
feet, according to M. von Dechen, the director; and the old coal mine

of Marihaye, near Val-St.-Lambert, in the valley of the Meuse, is
1157 French feet (1233 Engl.), according to M. Gernaert, Ingénieur des
Mines.   The mines of greatest absolute depth are for the most part situated
in such elevated mountain valleys or plains, that their deepest portions either
do not descend to the level of the sea, or reach but a little way below it.
Thus the Eselschacht, at Kuttenberg, in Bohemia, which is now inaccessible,
had the prodigious absolute depth of 3545 (3778 Engl.) feet (Fr. A. Schmidt,
Berg gesetze der österr. Mon. Abth. i. Bd. i. S. 32).   At St.-Daniel and
at Rörerbühel (Landgericht kitzbühl) there were, in the 16th century, excavations 2916 French feet deep.   The plans of the works, dated 1539, are still
preserved (Joseph von Sperges, Tyroler Bergwerksgeschichte, S. 121).   Compare also Humboldt, Gutachten über Herantreibung des Meissner Stollens in
de Freiberger Erzrevier," printed in Herder über den jetzt begonnenen Erb-
stollen, 1838, S. 124.   It may be supposed that the knowledge of the extraordinary depth of the Rörerbühel had reached England at an early period, for
I find it stated in Gilbert de Magnete, that men had penetrated into the crust
of the earth to depths from 2400 to 3000 feet.   " Exigua videtur terræ
portio quæ unquam hominibus spectanda emerget aut eruitur : cum profundius
in ejus viscera, ultra eflorescentis extremitatis corruptelam, aut propter aquas
in magnis fodinis, tanquam per venas scaturientes, aut propter aëris salubrioris
ad vitam operariorum sustinendam necessarii defectum, aut propter ingentes
sumptus ad tantos labores exantlandos, multasque difficultates, ad profundiores
terræ partes penetrare non possumos ; adeo ut quadringentas aut (quod ra-
rissime) quingentas orgyas in quibusdam metallis descendisse stupendus
omnibus videatur conatus."—Gulielmi Gilberti Colcestrensis, de Magnete
Physiologia nova. Lond. 1600, p. 40.

   The absolute depth of the mines in the Saxon Erzgebirge, near Freiberg,
are : in the Thurmhofer Zug, 1824 (1944 Eng.) feet ; in the Hohenbirker
Zug, 1714 (1827 Eng.) feet : the relative heights are nearly 626 and 620
(677 and 277 Eng.) feet, assuming the elevation of Freiberg above the level
of the sea, according to Reich's recent determination, to be 1191 (1269 Eng.)
feet.   The absolute depth of the rich and celebrated mine of Joachimsthal in
Bohemia (Verkreuzung des Jung Hauer Zechen- und Andreas ganges) is 1989
(2120 Eng.) feet ; so that assuming from Herr von Dechen's measurements
the ground at the surface to be about 2250 (2388 Eng.) feet above the level
of the sea, it follows that the excavations have not yet reached that level.   In
the Harz, the Samson mine as Andreasberg has an absolute depth of 2062
(2197 Eng.) feet. In what was Spanish America, the deepest mine with

which I am acquainted is the Valenciana, near Guanaxuato in Mexico, where I found the absolute depth of the Planes de San Bernardo 1582 (1686 Eng.) feet ; but this is still 5592 (5960 Eng.) feet above the level of the sea. If we compare the depth of the Kuttenberger mine (a depth greater than the height of the Brocken, and only 200 feet less than the height of Vesuvius, with the loftiest buildings erected by man, with the Pyramid of Cheops and with the Cathedral of Strasburg, we find the proportion of 8 to 1. Our geological writings contain so many statements either vague or disfigured by erroneous reduction to Parisian feet, that I have thought it desirable to bring together in this note all the certain information which I could collect respecting the greatest absolute and relative depths of artificial excavations. In descending eastward from Jerusalem to the Dead Sea and the valley of the Jordan, a view is enjoyed, which, according to our present hypsometric knowledge of the surface of the earth, has no parallel in any other region. The rocks on which the traveller treads, with the open sky over his head, are, according to the barometric measurements of Berton and Russegger, 1300 (1388 Engl.) feet below the level of the Mediterranean. (Humboldt, Asie centrale, T. ii. p. 323.)

([125]) p. 150.—Basin-shaped strata, which sink and reappear at distances which can be measured, although their deepest portions may be inaccessible to the miner, yet afford sensible evidence of the constitution of the crust of the earth at great depths beneath the surface. Facts of this kind possess, therefore, great geological interest : I am indebted to the excellent geologist Herr von Dechen for those subjoined :—" The depth of the Liège coal basin at Mont-St.-Gilles I infer from the joint investigation of our friend Herr von Oeynhausen and myself to be 3650 feet (3809 Engl.) below the surface, or 3250 (3464 Engl.) feet below the level of the sea, the elevation of Mont-St.-Gilles being certainly under 400 French feet; the coal basin of Mons must be fully 1750 (1865 Engl.) feet deeper still. But all these depths are small compared to that which may be deduced from the superposition of the coal strata of the Saar-Revier (Saarbrücken). I infer from repeated surveys that the lowest coal strata which we know in the district of Duttweiler, near Bettingen, north-east of Saarlouis, descend to a depth of 19406 (20682 Engl.) and 20656 (21358 Engl.) feet below the level of the sea, or 3·6 geographical miles." This result exceeds by 8000 (8526 Engl.) feet the assumption made in the text for the depth of the basin of the devonian strata : it is a depth below the level of the sea equal to the height of the Chimborazo above it, and at which we should infer the temperature to be as high as 224° Cent. (467°

Fah.)   We have, therefore, from the highest summits of the Himalaya to the lowest portions of the basins which contain the fossil floras or vegetable remains of an earlier state of the globe, a vertical distance of 45000 (or about 48000 Engl.) feet, or $\frac{1}{485}$th of the earth's semi-diameter.

(^{126}) p. 154.—Plato, Phædo, p. 97 (Aristot. Metaph. p. 985).   Compare Hegel, Philosophie der Geschichte, 1840, S. 16.

(^{127}) p. 155.—Bessel, Allgemeine Betrachtungen über Gradmessungen nach astronomisch-geodätischen Arbeiten, at the close of Bessel und Baeyer, Grad-messung in Ostpreussen, S. 427.   Respecting the accumulation of matter on the side of the moon which is turned towards the earth, see Laplace, Expos. du Syst. du Monde, p. 308.

(^{128}) p. 155.—Plin. ii. 68; Seneca, Nat. Quæst. Præf. c. ii.   "El mundo es poco" (the earth is small), said Columbus, in a letter to Queen Isabella, written from Jamaica, July 7, 1503; using the expression, however, rather from his desire to shew that the passage from Spain is not long "when we seek the east from the west."   Compare my Examen crit. de l'hist. de la géeogr. du 15me siècle, T. i. p. 83, and T. ii. p. 327; where I have shewn that the opinion supported by Delisle, Fréret, and Gosselin, of the extravagant differences in the estimates of the circumference of the earth in Greek writers being merely apparent, and caused by the different values of the stadia em-ployed, had been put forward as early as 1495, by Jaime Ferrer, in a propo-sition having for its object the determination of the Papal line of demarcation.

(^{129}) p. 155.—Brewster, Life of Sir Isaac Newton, 1831, p. 162:—"The discovery of the spheroidal form of Jupiter, by Cassini, had probably directed the attention of Newton to the determination of its cause, and consequently to the investigation of the true figure of the earth."   It is true that it was in 1691 that Cassini first announced the amount of the compression of Jupiter ($\frac{1}{15}$th) (Anciens Mémoires de l'Acad. des Sciences, T. ii. p. 108); but we know, through Lalande (Astron. 3me éd. T. iii. p. 335), that Maraldi was in possession of some printed sheets of a Latin work on the spots of the planets, commenced by Cassini, from which it appeared that he was acquainted with the ellipticity of Jupiter even before 1666, or 21 years before the publication of Newton's Principia.

(^{130}) p. 157.—Bessel's investigation of ten measurements of degrees, in which the error discovered by Puissant in the calculation of the French arc is taken into account, and allowed for (Schumacher, Astr. Nachr. 1841, N^r 438, S. 116), gives the semi-major axis of the elliptical spheroid of revolution, which the irregular figure of the earth most nearly resembles,

3272077·14 T.; the semi-minor axis, 3261139·33 T.; the ellipticity, $\frac{1}{299\cdot152}$; and the length of a mean degree of the meridian, 57013·109 T. with a probable error of 2·8403 T.; whence the length of a German geographical mile, 15 to a degree, is 3807·23 T. [In British measures, the semi-major axis is 20924774 feet; the semi-minor axis, 20854821 feet; the length of a mean degree of the meridian 364596·0 feet, with a probable error of 18·16 feet; and that of a geographical mile, 60 to a degree, 6086·76 feet. ED.] Previous combinations of measurements of degrees varied between $\frac{1}{302}$ and $\frac{1}{297}$: thus Walbeck (De forma et magnitudine telluris in demensis arcubus meridiani definiendis), in 1819, gives $\frac{1}{302\cdot78}$; Ed. Schmidt (Lehrbuch der mathem. und phys. Geographie, S. 5), in 1829, gives $\frac{1}{297\cdot48}$ from seven measures. Respecting the difference of the compression deduced from measurements in different longitudes, see Bibliothèque universelle, T. xxxiii. p. 181, and T. xxxv. p. 56; also Connaissance des tems, 1829, p. 290. From the lunar inequalities, Laplace found, by the older tables of Bürg, $\frac{1}{304\cdot3}$ (Expos. du syst. du monde, p. 229), and subsequently, by employing the observations of the moon discussed by Burckhardt and Bouvard, $\frac{1}{299\cdot1}$ (Mécan. céleste, T. v. pp. 13 and 43).

(^{131}) p. 157.—The values of the compression deduced by means of the pendulum are as follows:—The general result of Sabine's great expedition (1822 and 1823, from the equator to 80° N. lat.), $\frac{1}{288\cdot7}$; Freycinet, excluding the experiments at the Isle of France, Guam, and Mowi, $\frac{1}{291\cdot2}$; Foster, $\frac{1}{289\cdot5}$; Duperrey, $\frac{1}{286\cdot4}$; Lütke (Partie nautique, 1836, p. 232) from 11 stations, $\frac{1}{270}$. Lesser ellipticities had been given by the pendulum observations between Formentera and Dunkirk, $\frac{1}{298\cdot2}$ according to Mathieu (Connaissance des tems, 1816, p. 330); and by those between Formentera and Unst, $\frac{1}{304}$ according to Biot. Baily, Report on Pendulum Experiments, in the Memoirs of the Royal Astr. Society, Vol. vii. p. 96; also Borenius, in the Bulletin de l'Acad. de St.-Pétersbourg, 1843, T. i. p. 25. The first proposal to adopt the length of the pendulum as a standard of measure, and to establish the third part of the seconds pendulum (supposed to be every where of equal length) as a *pes honorarius*,—the measure of a unity which might be recovered at any future age of the earth, and by nations dwelling on any part of its surface,—is found in Huygens' Horologium oscillatorium, 1673, Prop. 25. In 1742, the same wish was publicly enounced in an inscription on a monument erected at the equator by Bouguer, La Condamine, and Godin. On a handsome marble tablet, which I have seen uninjured in the old Jesuits' College at Quito, it is said:—" Penduli simplicis æquinoctialis unius minuti secundi archetypus, mensuræ naturalis exemplar, utinam universalis!" From what La Condamine has said, in his Journal du Voyage à l'Equateur, 1751, p. 163, respecting parts of the inscription which are not filled up, and a difference between Bouguer and himself about the numbers, I had conjectured that I should find considerable discrepancy between the tablet, and the inscription as published in Paris; but, after a careful comparison, I discovered only two discrepancies of little importance,—" *ex arcu graduum* 3½," instead of " *ex arcu graduum plus quam trium;*" and the date of " 1745," instead of " 1742." The date of 1745 is singular, because La Condamine returned to Europe in 1744, Bouguer in June of the same year, and Godin left South America in July 1744. The most necessary and useful amendment of the numbers in the inscription would have been the astronomical longitude of Quito. (Humboldt, Recueil d'Observ. astron. T. ii. pp. 319 — 354.) Nouet's latitudes, engraved on Egyptian monuments, offer a more recent example of the danger of perpetuating, thus solemnly, erroneous or imperfectly computed results.

([132]) p. 158.—On the increased intensity of gravitation in volcanic islands (St. Helena, Ualan, Fernando de Noronha, Isle of France, Guam, Mowi, and the Galapagos,) Rawak (Lütke, p. 240) being an exception, perhaps, on account of its proximity to the high land of New Guinea, see Mathieu, in Delambre, Hist. de l'Astronomie au 18me siècle, p. 701.

[The fact of the increased intensity of gravitation in volcanic islands was, I believe, first made known by myself (Pendulum Experiments, 1825, pp. 237—341), as the result of my own experiments at the islands of Ascension and of St. Thomas (in the gulph of Guinea), and at other stations of different geological character. The comparison of the whole series furnished a numerical scale of local influence in which the volcanic islands of Ascension and St. Thomas occupied the higher extremity, and stations of alluvial soil and sand the lower extremity, whilst the intermediate gradations of local influence were seen to correspond in a remarkable manner with the density of the superficial strata. " The scale afforded by the pendulum for measuring the intensities of local attraction appears to be sufficiently extensive to render it an instrument of possible utility in inquiries of a purely geological nature. It has been seen that the rate of a pendulum may be ascertained with proper care to a single tenth of a second per diem ; whilst the variation of rate occasioned by the geological character of stations has amounted in extreme cases to nearly ten seconds per diem: a scale of 100 determinable parts is thus afforded in which the local attraction dependent on the geological accidents may be estimated." (Pend. Exp. p. 341.) The work from which this passage is taken was published before M. Mathieu's notice in Delambre's History of Astronomy, and is referred to by M. Mathieu in the passage to which M. de Humboldt alludes.—EDITOR.]

(133) p. 158.—Numerous observations shew great irregularities in the length of the pendulum in continental, as well as in insular and littoral localities, and which are also ascribed to local attraction. (Delambre, Mésure de la Méridienne, T. iii. p. 548; Biot, Mém. de l'Académie des Sciences, 1829, T. viii. pp. 18 and 23. In crossing the south of France and Lombardy from west to east, we find a minimum of intensity of gravitation at Bordeaux; thence it increases rapidly at the more easterly stations of Figéac, Clermont-Ferrand, Milan, and Padua, at which last city it reaches a maximum. In the opinion of Elie de Beaumont (Recherches sur les Révolutions de la Surface du Globe, 1830, p. 729), the influence of the Alps on the variations of gravitation on their southern side is not alone to be ascribed to their mass, but still more to the rocks of melaphyre and serpentine which have elevated the chain. On the slope of Mount Ararat, (which, with Caucasus, may be said to be situated near the centre of gravity of the old continent, consisting of Europe, Asia, and Africa,) Fedorow's very exact pendulum experiments indicate likewise the presence of dense volcanic masses in lieu of cavities (Parrot, Reise zum Ararat, Bd. ii. S. 143). In the geodesical operations of Carlini and Plana in Lombardy, differences of 20″ to 47″.8 have been found between geodesical measurements and the direct determinations of latitude; see the instances of Andrate and Mondovi, Milan and Padua, in the Opérations géodes. et astron. pour la mesure d'un arc du parallèle moyen, T. ii. p. 347; Effemeridi astron. di Milano, 1842, p. 57. It follows from the French triangulation that the latitude of Milan, deduced from that of Berne, is 45° 27′ 52″; whereas direct astronomical observation gives it 45° 27′ 35″. As the perturbations extend in the plain of Lombardy to Parma, far south of the Po, (Plana, Opérat. géod. T. ii. p. 847,) we may conjecture that there are deflecting causes in the plain itself. Struve has met with the same anomalies in the most level parts of eastern Europe, (Schumacher, Astr. Nachr. 1830, Nr. 164, S. 399). On the influence of dense masses supposed to exist at a small depth equal to the mean height of the chain of the Alps, see the analytical expressions which Hossard and Rozet have inserted in the Comptes-rendus, T. xviii. 1844, p. 292, and compare them with Poisson, Traité de Mécanique, (2me éd.,) T. i. p. 482. The earliest notices of the influence which rocks of different kinds might exercise

on the vibrations of the pendulum were given by Thomas Young, in the Phil. Trans. for 1819, pp. 70—96.   But in drawing conclusions from the length of the pendulum relatively to the curvature of the earth, it ought not to be overlooked that the crust of the earth may possibly have been hardened previous to metallic and dense basaltic masses having penetrated from great depths through open channels and clefts, and approached the surface.

([134]) p. 158.—Laplace, Expos. du Système du Monde, p. 231.

([135]) p. 159.—La Caille's pendulum experiments at the Cape of Good Hope, which Mathieu has calculated with great care (Delambre, Hist. de l'Astr. au 18me siècle, p. 479), give an ellipticity of $\frac{1}{284\cdot4}$; from several comparisons of observations in equal latitudes in the two hemispheres (New Holland and the Falkland Islands compared with Barcelona, New York, and Dunkirk,) there is as yet no ground for supposing the mean ellipticity of the southern hemisphere to be greater than that of the northern (Biot, in the Mém. de l'Acad. des Sci. T. viii. 1829, pp. 39—41.)

([136]) p. 159.—The three methods of observation give the following results: —1st, from the deflection of the plumb-line by the proximity of the Schehallien Mountain (Gaelic, Thichallin), in Perthshire, 4·713, resulting from the experiments of Maskelyne, Hutton, and Playfair 1774—1776 and 1810), according to a method which had been proposed by Newton ;—2d, by pendulum vibrations on a mountain, 4.837, (Carlini's observations on Mount Cenis compared with Biot's observations at Bordeaux, Effemer. astr. di Milano, 1824, p. 184) ; —3d, by the balance of torsion in Cavendish's experiments with an apparatus originally devised by Mitchell, 5·48 ; or, according to Hutton's revision of the calculation, 5·32, or Eduard Schmidt's revision, 5·52, (Lehrbuch der math. Geographie, Bd. i. S. 487) ; by the balance of torsion in Reich's experiments, 5·44.   In the calculation of these experiments of Professor Reich, which are a model of exactness, the original mean result (having a probable error of only 0.0233) was 5·43 ; which, being increased by the quantity by which the centrifugal force of the earth diminishes the force of gravity for the latitude of Freiberg (50° 55′), becomes 5·44.   The employment of masses of cast-iron instead of lead has not shewn any sensible difference, or none which is not well within the limits of observation error, disclosing therefore no traces of magnetic influence (Reich, Versuche über die mittlere Dichtigkeit der Erde, 1838, S. 60, 62 und 66).   By the assumption of too small an ellipticity of the earth in the calculations, and by the difficulty of forming a correct estimate of the density of rocks at the surface, the mean density of the earth, previously deduced from the experiments on or near mountains, appeared $\frac{1}{5}$th too small, *viz.* 4·761 (Laplace, Mécanique céleste, T. v. p. 46,) or 4·785 (Eduard Schmidt, Lehrb. der math. Geog. Bd. i. § 387 und 418). On Halley's hypothesis of the earth being a hollow sphere, alluded to in the following page, (and which was the germ of Franklin's notions respeating earthquakes), see Phil. Trans. for the year 1693, Vol. xvii. p. 563. "On the structure of the internal parts of the earth, and the concave habited arch of the shell."

[Carlini's result, by the second method noticed by M. de Humboldt, requires a correction which has not yet been applied to it, and which would probably alter it considerably; namely, for the *true* reduction to a vacuum of the vibrations of the pendulum observed at Bordeaux and on Mont Cenis. The beight above the sea of the station at Mont Cenis was 6374 English feet; and the difference in the density of the air, in which the pendulum was vibrated at the two stations, must have been quite sufficient to occasion a very sensible error in the difference in the lengths of the respective pendulums computed by the mode of reduction practised before Bessel's discovery alluded to in page 26 of the text of Cosmos, was made. The pendulum employed by Carlini was that of Biot (with few but valuable modifications); and the true reduction to a vacuum for pendulums of that description has not yet been experimentally investigated. As an instrument for ascertaining the density of the earth, the pendulum is so much inferior to the torsion-balance employed by Cavendish and Reich, and still more recently by Baily (whose results do not appear to have been known to M. de Humboldt when this volume of Cosmos was published), that if the true reduction to a vacuum of Biot's pendulum were required only for the correction of the mean density of the earth derived from the experiments on Mont Cenis, it might not be worth while to undertake its determination: but it should not be forgotten that, until such a determination has been made, the true results of the laborious experiments by which M. Biot himself sought to ascertain the length of the seconds pendulum at Paris, and at various other stations in France and elsewhere, are unknown: there can be little doubt that the *absolute* lengths computed at the time will have to receive very large corrections; and though the *relative* lengths will certainly be less seriously affected, they may be altered to a degree which may influence, in part at least, the conclusions which M. de Humboldt has drawn in Note 133, respecting the variations of gravity in the South of France.—EDITOR.]

([137]) p. 162.—To these belong the excellent analytical investigations of Fourier, Biot, Laplace, Poisson, Duhamel, and Lamé. Poisson, in his work entitled "Théorie mathématique de la Chaleur," 1835, p. 3, 428—430, 436, and 521—524 (see also the extract which de la Rive has made in the "Bibliothèque universelle de Genève," T. lx. p. 415), has developed an hypothesis wholly different from Fourier's views; "Théorie analytique de la Chaleur." Poisson denies the actual fluidity of the interior of the Earth; he thinks "that in the process of cooling by radiation to the medium surrounding the earth, the particles which were first solidified at the surface sunk, and that by a double current, descending and ascending, the great inequality was diminished which would otherwise take place in a solid body cooling from its surface." It seemed to this great geometer more probable that the solidification should have begun in the strata nearer the center; "that the phenomenon of the temperature increasing with increasing depth does not extend to the whole mass of the Earth, and is merely a consequence of the movement of our planetary system in space, of which some parts are of very different temperature from others, by reason of stellar heat (chaleur stellaire)." Thus, according to Poisson's views, the warmth of the waters in our Artesian wells would be merely a warmth which had penetrated into the Earth from without; and "the earth itself might be compared to a mass of rock conveyed from the equator to the pole so rapidly as not to have entirely cooled. The increase of temperature in such a block from the surface inwards would not extend to its center." The physical doubts which have been justly raised against this singular cosmical view (attributing to the regions of space that which is better explained by the transition from a primitive gaseous to a solid state) will be found collected in Poggendorff's Annalen, Bd. xxxix. S. 93—100.

([138]) p. 163.—The increase of temperature has been found in the Puits de Grenelle at Paris to be 98.4 (104.9 Eng.) feet; in the mine at Neu-Salzwerk, near Minden in Prussia, to nearly 91 (97.0 Eng.) feet; and, according to Auguste de la Rive and Marcet, to the same amount (viz. 91 French feet at Pregny near Geneva, although the mouth is situated 1510 (1609 Eng.) feet above the level of the sea. The agreement between the results derived by a method first proposed by Arago, in 1821 (Annuaire, 1835, p. 234), from mines that are severally 1683 (1794 Eng.) feet, 2094 (2232 Eng.) feet, and 680 (725 Eng.) feet, in absolute depth, is remarkable. It is probable that if there are two points on the Earth's surface situated at a small vertical distance above each other, whose annual mean temperatures are exactly determined, they are to be found at the Paris Observatory, where the mean temperature of the external air is $10°.822$ C. ($51°.48$ F.), and of the Caves de l'Observatoire $11°.834$ C. ($53°.3$ F.); the difference being $1°.012$ ($1°.8$ F.) for 28 metres (91.9 English feet). Poisson, Théorie math. de la Chaleur, p. 415

and 462. In the course of the last 17 years, from some cause which has not been completely ascertained, the indications of the lower thermometer have increased 0°.220; but this probably is not due to any actual increase in the general temperature of the caves. Results obtained by means of Artesian wells are, indeed, liable to error, from infiltration, or from penetration from lateral crevices; but currents of cold air, and other circumstances, are still more injurious to the accuracy of the many laborious series of observations that have been made in mines. The general result of Reich's extensive examination into the temperature of the mines in Saxony, gives a somewhat slower increase of the terrestrial heat, or 1° C. to 41.84 metres (1° F. to 76.26 English feet)—(Reich, Beob. über die Temperatur des Gesteins in verschiedenen Tiefen, 1834, S. 134). Phillips, however, found, in the coal mine of Monkwearmouth near Newcastle, in which, as I have already remarked, workings are carried on at a depth of nearly 1500 English feet below the level of the sea, 32.4 metres, or 106.3 English feet to 1° C. (1° F. to 59.06 English feet), a result almost identical with that found by Arago from the Puits de Grenelle (Poggend. Ann. Bd. xxxiv. S. 191).

([139]) p. 165.—Boussingault, Sur la Profondeur à laquelle se trouve la couche de Température invariable entre les Tropiques, in the Annales de Chimie et de Physique, T. iii. 1833, p. 225—247.

[The observations of Mr. Caldecott at Trevandrum in 1842 and 1843 (Quetelet, Climat. de la Belgique, p. 137, et seq.) have shown that the stratum of invariable temperature is not always found within the tropics at so small a depth. At Trevandrum, at the depth of six French feet, the mean temperature of different months, instead of being constant, varies as much as 3° 6′ Fah.; and the curve of annual temperature at that depth exhibits in very marked characters two maxima corresponding to the double passage of the sun over the zenith of Trevandrum.—EDITOR.]

([140]) p. 166.—Laplace, Exp. du Syst. du Monde, p. 229 and 263; Mécanique céleste, T. v. p. 18 and 72. It should be remarked, that the fraction $\frac{1}{110}$th of a centesimal degree of the mercurial thermometer, given in the text as the limit of the permanency of the temperature of the globe since the days of Hipparchus, rests on the supposition that the dilatation from heat of the substances of which the Earth is composed is equal to that of glass, or $\frac{1}{100000}$th for 1° C. Respecting this hypothesis, see Arago, Annuaire pour 1834, p. 177—190.

([141]) p. 167.—William Gilbert, of Colchester, whom Galileo entitled "great to a degree which might be envied," said, "Magnus magnes ipse est globus

408 NOTES.

terrestris." He ridiculed the loadstone mountains which Fracastoro, the great contemporary of Columbus, supposed to constitute the magnetic poles—" rejicienda est vulgaris opinio de montibus magneticis aut rupe aliqua magnetica, aut polo phantastico a polo mundi distante." He assumed the declination of the magnetic needle to be invariable at each point of the surface of the Earth ("variatio uniuscujusque loci constans est") ; and explained the inflexions of the isogonic lines by the form of the continents and the relative positions of the sea-basins, which exercise a less degree of magnetic force than the solid portions which rise above the level of the ocean (Gilbert de Magnete, ed. 1633, p. 42, 98, 152, and 155).

([142]) p. 167.—Gauss, Allgemeine Theorie des Erdmagnetismus, in den Resultaten aus den Beob. des magnet. Vereins, 1838, § 41, S. 56. Translated in "Taylor's Scientific Memoirs," Vol. ii. P. vi. Art. v.

([143]) p. 168.—There are also perturbations which do not extend to such great distances, and of which the causes are more local, and are seated, perhaps, at less depths. I made known, some years ago, a rare instance of this kind, in which the disturbance manifested itself in the Freiberg mines, but not at Berlin. (Lettre de M. de Humboldt à S. A. R. le Duc de Sussex sur les moyens propres à perfectionner la connaissance du magnétisme terrestre, published in Becquerel's Traité expérimental de l'électricité, T. vii. p. 442). Magnetic storms, which were felt simultaneously from Sicily to Upsala, did not extend from Upsala to Alten (Gauss and Weber, Resultate des magnet. Vereins, 1839, S. 128 ; Lloyd, in the Comptes-rendus de l'Académie des Sciences, T. xiii. 1843, Sém. ii. pp. 725 and 827). Amongst the numerous and recent examples of disturbances extending simultaneously over wide portions of the earth's surface, which are assembled in Sabine's important work (Observ. on days of unusual magnetic disturbance, 1843), one of the most memorable is that of September 25, 1841, which was observed at Toronto in Canada, at the Cape of Good Hope, at Prague, and partially at Van Diemen Island. The English observatories suspend their operations on Sunday, and, owing to the difference of longitude, the midnight of Saturday took place at Van Diemen Island soon after the commencement of the disturbance, of which, probably, the greater portion thus escaped observation.

[The disturbance of Sept. 24th and 25th appears also to have extended to Macao (Phil. Trans. 1843, p. 133 ;) and it is worthy of remark that, from Sir Edward Belcher's observations, it may be inferred that the intensity of the horizontal force was increased at Macao at the same hours when it was diminished elsewhere.—EDITOR.]

([144]) p. 168.—I have described, in the Journal de Physique of Lamétherie (1804, T. lix. p. 449), the application of the magnetic Inclination to determinations of latitude along a coast running north and south, and, like that of Chili and Peru, constantly enveloped by mist (*garua*) during part of the year. In the particular locality alluded to, this application is the more practically important, by reason of the strong southerly current, extending to Cape Pariña, which causes great loss of time to navigators, who have inadvertently passed to the north of the latitude of the port to which they are bound. In the Pacific, from Callao to Truxillo, I have found, for a difference of latitude of 3° 57′, a change of Inclination of 8°.1 ; and from Callao to Guayaquil, for a change of latitude of 9° 50′, a change of Inclination of 20°.7 (Rélation historique, T. iii. p. 622). At Guarmey (lat. 10° 4′ S.), Huaura (lat. 11° 3′ S.), and Chancay (lat. 11° 32′), the Inclinations were respectively :—6°.1, 8°.1, and 9°.3. The determination of the ship's place by means of the magnetic Inclination, when her course nearly intersects the isoclinal lines at right angles, is distinguished from all other methods by its independence of the time of the day, and by its not, therefore, requiring the sight of any of the heavenly bodies. I have very recently learned that, as early as the end of the sixteenth century, scarcely twenty years after Robert Norman invented an Inclinatorium, that William Gilbert, in his great work " De Magnete," proposed the determination of latitudes by means of the Inclination of the magnetic needle. In his Physiologia nova de Magnete, Lib. v. cap. 8, p. 200, he extols the advantages of this method in thick weather, " aëre caliginoso." Edward Wright, in the preface which he has added to the work of his illustrious master, says that this proposal is " worth much gold." As he shared Gilbert's error in believing the isoclinal lines to be parallel to the geographical equator, he did not perceive that the method is only applicable in particular localities.

([145]) p. 168.—Gauss and Weber, Resultate des magnetischen Vereins, im J. 1838, § 31, S. 46.

([146]) p. 168.—According to Faraday (London and Edinburgh Philosophical Magazine, 1836, Vol. viii. p. 178), pure cobalt is not magnetic. I am aware that other distinguished chemists (Heinrich Rose and Wöhler) do not consider this a settled point ; but it appears to me that if one of two carefully-purified masses of cobalt, both supposed to be free from any alloy of nickel, is found to be non-magnetic, it is probable that the magnetism shewn in the other mass is due to a want of purity ; and I am inclined, therefore, to believe Faraday's view to be the more correct.

(^{147}) p. 169.—Arago, in the Annales de Chimie, T. xxxii. p. 214; Brewster, Treatise on Magnetism, 1837, p. 111; Baumgartner, in the Zeitschrift für Phys. und Mathem. Bd. ii. S. 419.

(^{148}) p. 169.—Humboldt, Examen critique de l'Hist. de la Géographie, T. iii. p. 86.

(^{149}) p. 169.—Asie cent. T. i. Introduction, p. xxxvii.—xlii. The western nations, the Greeks and the Romans, knew that magnetism could be imparted *permanently* to iron—(" sola hæc materia ferri vires a magnete lapide accipit *retinetque longo tempore*," Plin. xxxiv. 14.) The great discovery of the earth's directive force would, therefore, also have been made in the west, if any one had accidentally observed a fragment of loadstone more long than broad, or a magnetised bar of iron suspended by a thread, or floating freely on the surface of water on a wooden support.

(^{150}) p. 170.—Topographical surveys, made solely by compass, and without any provision for corrections for changes of terrestrial magnetism, cannot fail ultimately to produce great confusion in the boundary lines between different properties, excepting in those parts of the earth where the magnetic declination is either invariable, or at least is subject only to exceedingly small secular changes. Sir John Herschel says, " The whole mass of West India property has been saved from the bottomless pit of endless litigation by the invariability of the magnetic declination in Jamaica, and the surrounding archipelago, during the whole of the last century, all surveys of property there having been conducted solely by the compass. Compare Robertson, in the Phil. Trans. for 1806, Part ii. p. 348, on the permanency of the compass in Jamaica since 1660. In the mother country (England), the magnetic declination has altered during the same period fully 14°.

(^{151}) p. 171.—I have shewn elsewhere that the documents of Columbus's voyages which have come down to us, give, with much certainty, three determinations of points in the Atlantic line of no declination for September 13, 1492, May 21, 1496, and August 16, 1498. The direction of this curve was at that time from N.E. to S.W.; and it touched the continent of South America, a little to the east of Cape Codera, instead of, as at present, in the northern part of Brazil, (Examen critique de l'histoire de la Géographie, T. iii. pp. 44—48). We learn from Gilbert's Physiologia nova de Magnete, Lib. iv. cap. i. the remarkable fact that, in the year 1600, as in the time of Columbus, the magnetic needle had scarcely any declination in the vicinity of the Azores. I think that I have shewn satisfactorily, from documentary evidence (in my Examen critique, T. iii. p. 54), that the famous line of demarcation by which Pope

Alexander VI. divided the western hemisphere between Spain and Portugal, was not drawn through the westernmost of the group of the Azores, because Columbus was desirous of converting a *physical* into a *political* division. He attached great importance to the zone (raya), " in which the compass ceases to shew any variation, where the air and the sea assume a new character, where the sea is covered with weeds, and cool breezes begin to blow, and where the form of the Earth is no longer the same." The last supposition was an infer-ence from erroneous observations of the Pole star.

([152]) p. 171.—It is a question of the highest interest towards the solution of the problem of the physical causes of terrestrial magnetism, whether the two remarkable systems of closed isogonic curves will continue to move forward for centuries to come, still preserving this closed form, or whether they will open out and lose their peculiar character. In the Asiatic system, the declina-tion increases from without inwards ; in the Pacific system, the contrary is the case ; no line of no declination,—indeed, no line below 2° east,—is now known in the Pacific to the east of Kamschatka, (Erman, in Poggend. Annalen, Bd. xxi. S. 129) : but it appears that, in 1616, on Easter-day, Cornelius Schouten found no declination a little to the S.E. of Nukahiva, in 15° S. lat. and 132° W. long. (from Paris), and, therefore, in the middle of the present closed system, (Hansteen, Magnetismus der Erde, 1819, S. 28). In all these con-siderations it must not be forgotten, that we can only trace the lines of magnetic direction and force in their progressive displacements by observations which are confined to their intersection with the surface of the Earth.

([153]) p. 172.—Arago, Annuaire, 1836, p. 284 ; and 1840, pp. 330—338.

([154]) p. 172.—Gauss, Allgemeine Theorie der Erdmagnetismus, § 31.

([155]) p. 172.—Duperrey, de la configuration de l'équateur magnétique, in the Annales de Chimie, T. xlv. pp. 371 and 379 ; also Morlet, in the Mémoires présentés par divers savans à l'Acad. Roy. des Sciences, T. iii. p. 132.

([156]) p. 173.—See, in Sabine's Contributions to Terrestrial Magnetism, 1840, p. 139, the remarkable map of the isoclinal lines in the Atlantic Ocean for 1825 and 1837.

([157]) p. 174.—Humboldt, über die seculäre Veranderung der magnetischen. Inclination, in Poggend. Annalen, Bd. xv. S. 322.

([158]) p. 174.—Gauss, Resultate der Beob. des magn. Vereins im Jahr 1838, § 21 ; Sabine, Report on the Variations of the Magnetic Intensity, p. 63.

[The progress of investigation, since the first volume of " Cosmos" was written, has continued to confirm the opinion expressed by M. de Humboldt in the text (p. 174), that in respect to the theory of terrestrial magnetism, the

isodynamic lines are those from which the most fruitful results are to be expected. Researches into the amount of the magnetic force at different points
of the earth's surface, and graphical representations of the results by lines
drawn through the points where the force has an equal intensity, have shown,
that there are two foci or points of maximum force in each hemisphere, and
consequently four on the whole surface of the globe. The isodynamic lines
which surround each of the two points of maximum in an hemisphere, are not
circles, but are of an ovate form, having the larger axis in a direction which,
if prolonged, would connect the two foci by the shortest line, or nearly so, which
can be drawn between them on the surface of the globe. As the ovals
successively recede from the focus they correspond to weaker and weaker
degrees of force, each in its turn enclosing the ovals of higher intensity.
This continues to be the case until the two systems of ovals encounter in a
point intermediate between the foci: the isodynamic line which corresponds
to the force at this point has consequently the form of a figure of 8, each of
the loops enclosing a focus with its surrounding ovals: this form is called by
geometricians a lemniscate: there is but one such isodynamic line in the
extra-tropical part of each hemisphere; and it separates the isodynamics of
higher intensity than itself which are within the loops, each surrounding a
single point of maximum only, from those which correspond to weaker degrees
of force than that of the lemniscate, and are exterior to it: each of the
exterior isodynamics surrounds both the foci, but without meeting or crossing
in the point between them: their general form is that of parallelism with the
external figure of the lemniscate, but the inflections which produce the double
loop become progressively less marked in the isodynamics of weakest force.

If the two foci in an hemisphere were points of equal force, the ovals surrounding each would be similar in force and area, and the point at which the
two systems would encounter each other would be half way between the foci.
Such, however, does not appear to be the case: the intensity at one of the
foci is greater than at the other: it is so in both hemispheres, and the ratio
of the force at the major and minor focus appears to be nearly the same in
both. The focus of greater intensity in the Northern hemisphere is in North
America, where its position has been ascertained, by a recent survey conducted
by Captain Lefroy of the Royal (British) Artillery, to be in the vicinity of
the S.W. shores of Hudson's Bay in 52° of latitude, (Phil. Trans. 1846.
Art xvii). The weaker focus, of which the approximate position has been determined by MM. Hansteen, Erman, and Due, is in the north of Siberia
about 120° of East longitude from Greenwich. The middle of the lemniscate
which encloses both, or the point where its loops are connected, appears to be

in the Polar Sea, to the North or North east of Bering's Strait, and consequently nearer to the Siberian than to the American focus. The corresponding phænomena of the southern hemisphere are not yet determined with an equal precision, but appear to have the same general characteristics. The two major foci, one in the northern, the other in the southern hemisphere, are not at opposite points of the globe to each other, nor are the two minor foci. Symmetry as regards geographical distribution is also departed from in another respect: the foci in each hemisphere are not separated from each other by an equal number of degrees of geographical longitude; they are nearer to each other in the southern than in the northern hemisphere.

It is well known that in the extra-tropical latitudes of the northern hemisphere, the forces which attract the north end of a magnet (so called, because in the greater part of the accessible portions of the globe it is that end of the magnet which is directed towards the north), and repel the south end, preponderate; and that conversely, in the extra-tropical parts of the southern hemisphere, the forces which attract the south end and repel the north predominate. At both the foci of the northern hemisphere the predominance is of the forces which attract the north end and repel the south; and at both the foci in the southern hemisphere the converse is the case. The line which separates the preponderance of the northern from that of the southern attracting force, is a line drawn through the points in each meridian of the globe where the intensity of the force is weakest on that meridian. Every where on the north side of this line the preponderance of the force attracting the north end of the magnet, and repelling the south, increases; and every where on the south side of the line the preponderance of the force attracting the south end, and repelling the north, increases. This line is not one of those which has been characterised or referred to by M. de Humboldt, but it is an important one in the view which it enables us to take of the magnetical relations of different portions of the globe. Its inflections are various, as will easily be imagined when it is considered that they take place in conformity with forces which produce four points of maximum on the surface of the globe unsymmetrically distributed, and also unsymmetrical in respect to the intensity of the force at each. The most remarkable inflection is the large convexity towards the south, which advances into the southern Atlantic nearly to the latitude of 20° S., in consequence of the wide separation which exists in that quarter between the major and minor foci of the southern hemisphere.

The measurement of the magnetic force in parts of an absolute scale, suggested by M. Poisson, and brought into use by M. Gauss, has added greatly to the value of researches on the terrestrial magnetic force, because the determinations which are now made will be comparable with those which may be made at future epochs. The unit of force in this scale is that amount of magnetic force which, acting on the unit of mass through the unit of time, generates in it the unit of velocity. Taking the units respectively, as a grain, a second, and a foot in British measure, the ratio of the force at the major focus in North America determined by Captain Lefroy's survey is 13·9; at the minor focus in Siberia, from the Observations of MM. Hansteen and Duc, it is 13·3 ; and at St. Helena, which is nearly on the line of least intensity, and at its weakest part, and where consequently the force is nearly a minimum on the surface of the globe, its value is 6·4. The observations made in Sir James Clark Ross's expedition, referred to in page 175 of the text, and which appear to have been made nearly in the centre of the highest isodynamic oval, give 15·6 as the approximate value of the force at the major focus in the southern hemisphere ; whilst the observations of the same voyage render it probable that the value at the minor focus does not much exceed or much fall short of 14·9, which is in the same ratio to 15·6, as 13·3 to 13·9. It has been already noticed that the two foci are nearer to each other in the south than in the north : assuming the general charge of the two magnetic hemispheres to be the same, the greater proximity of the two points of maximum in the south might readily be imagined to augment the force at both, whilst in the opposite geographical longitudes of the hemisphere the intensity would be less than in the analogous positions in the north : the isodynamic maps shew that such is the case, and give reason to believe, that the charge of the two magnetic hemispheres may not differ in the aggregate, although the distribution of the force at the surface of the globe is not precisely similar.

The view which we are now enabled to take of the magnetic system of the globe, by means of the knowledge which we have acquired of the magneto-dynamic variations at its surface, furnishes an explanation of many features in the isoclinal and isogonic lines which had previously occasioned perplexity. We now perceive why the higher isoclinals should be ellipses instead of circles, and why there should be inflections at opposite points of the ellipse nearer to one of its extremities than to the other, causing the isoclinal line to resemble in its general form a curve inclosing a lemniscate, of which one of the loops should be larger than the other. This remarkable conformation of the isoclinal lines has been well described by Erman, in Pogg. Ann. der Phys. vol. xxi. We now see also a confirmation of the conclusion which the sagacity of Halley enabled him to draw more than 150 years ago (Phil. Trans. 1683, No. 148, p. 208), that the complexity and seeming irregularity of the declinations observed in different parts of the earth were due to forces which must produce two points of greatest attraction in each hemisphere.

The interval which has elapsed since the phænomena of the magnetic force have been an object of attention, has been too short to admit of any direct inference being drawn in regard to the effect of secular change on the isodynamic lines: but an attentive consideration of features which are obviously connected with each other in the three systems of elementary lines, confirms—what indeed could scarcely have been doubted—that alterations which are known to have taken place in the configuration and position of the isogonic and isoclinal lines in the last two and a half centuries, have been accompanied by corresponding changes in the lines of equal force. The influence of secular change appears chiefly to affect the portions of the isodynamic lines which are most nearly connected with the two minor foci: from considerations which need not be particularised here, these foci appear to have moved in opposite directions—the Siberian from west to east, and the minor focus in the south from east to west. The remarkable closed systems of isogonic lines in Siberia and in the South Pacific, to which M. de Humboldt has referred in page 171 of the text, and in note 152, appear to have undergone a movement of translation in the same directions as the minor foci. The Ætiology of this science is of so remarkable a character, that it seems not improbable that the secular changes, which now appear to us the most mysterious branch of the phænomena presented to our notice, may eventually prove a means of conducting to a solution of the problem of highest interest—that of the physical causes of terrestrial magnetism.

Although it is scarcely safe to anticipate that any portion of phænomena may be less deserving of attention than another, yet, as M. de Humboldt has remarked that meteorology must seek its foundations and its advance first within the tropics, because the phænomena may there be viewed under an aspect of less complexity than elsewhere,—so the converse may be remarked in respect to magnetism; inasmuch as the characteristics of the magnetic system are less distinctly marked within the tropics than in the higher latitudes, and the influences of the two hemispheres are so blended in the inflexions of the magnetic lines within the tropics, that to understand them it is necessary to have continually present in the mind the phænomena of both hemispheres; and it is not always easy to discriminate, without much patient consideration, to which hemisphere a particular effect is due.—EDITOR.]

([159]) p. 174.—The following is the historical account of the discovery of an important law in terrestrial magnetism, that of the general increase of the intensity of the force with the increase of the magnetic latitude. When, in 1798, I was about to join the expedition of Captain Baudin, on a voyage of circumnavigation, Borda, who took a warm interest in my intended proceedings, proposed to me to observe, in different latitudes and both hemispheres, the oscillations in a vertical plane of a needle moving freely in the magnetic meridian, for the purpose of examining whether the magnetic force varied, or was every where the same. In my subsequent voyage to the equinoctial regions, this investigation formed one of the principal objects which I had in view. I observed the same needle perform in ten minutes, at Paris, 245 oscillations; at Havanna, 246; at Mexico, 242; at San Carlos del Rio Negro (lat. 1° 53′ N., long. 80° 40′ W. from Paris) 216; on the magnetic equator, or on the line where the inclination $= 0$, in Peru (in 7° 1′ S. lat., 80° 40′ W. long. from Paris), only 211; and in Lima (12° 2′ S. lat.) again 219 oscillations. I thus found that at that time, 1799—1803,—if the intensity of the total magnetic force were taken as $= 1,0000$ on the magnetic equator,—in the Peruvian chain of the Andes between Micuipampa and Caxamarca, it should be expressed, in Paris, by 1,3482; in Mexico, by 1,3155; at San Carlos del Rio Negro, by 1,0480; and at Lima, by 1,0773. When, in a Memoir, the mathematical portion of which belonged to M. Biot, I developed to the French Institute in the Seance du 26 Frimaire, An. xiii. de la Republique, this law of the variable intensity of the terrestrial magnetic force, supporting it by numerical values obtained by observations at 104 different points of the Earth's surface, the fact was regarded as perfectly new. It was not until after the reading of this Memoir, as Biot has said in it most distinctly, (Lamétherie, Journal de Physique, T. lix. p. 446, note 2,) and as I have repeated in my "Relation historique," T. i. p. 262, note 1, that M. de Rossel communicated to Biot his six previous observations of the oscillations of a needle made between 1791 and 1794 in Van Diemen's Land, Java, and Amboina. These observations shewed the law of decreasing force to exist also in the Indian Archipelago. It seems reasonable to conjecture that this excellent man had not recognised in his observations the regularity of the increase and decrease of the magnetic intensity, since he had never spoken of this surely not unimportant physical law to our common friends, Laplace, Delambre,

Prony, and Biot, before the reading of my Memoir. It was not until 1808, four years after my return from America, that his observations were published in the Voyage d'Entrecasteaux, T. ii. pp. 287, 291, 321, 480, and 644. The custom of expressing all observations of the magnetic force made in any part of the globe in terms of the force found by me on the magnetic equator in the North of Peru, in which arbitrary scale the force at Paris is taken as = 1·348, has been continued, even up to the present time, in all the tables of magnetic intensity which have been published in Germany, (Hansteen's Magnetismus der Erde, 1819, S. 71 ; Gauss, Beob. des Magnet. Vereins, 1838, S. 36—39; Erman, Physikal. Beob. 1841, S. 529—579): in England, (Sabine's Report on Magnetic Intensity, 1838, pp. 43—62 ; and Contributions to Terrestrial Magnetism, 1840, et seq.): and in France, (Becquerel, Traité d'Electricité et de Magnetisme, T. vii. pp. 354—367. Still earlier than Admiral Rossel's observations were those made by Lamanon, in the unfortunate expedition of La Pérouse, from its stay at Teneriffe, 1785, to its arrival at Macao, 1787, and which were sent to the Académie des Sciences. It is known with certainty, (Becquerel, T. vii. p. 320), that they had reached the hands of Condorcet in July, 1787; but all the attempts which have been hitherto made to discover them have proved fruitless. Captain Duperrey is in possession of a copy of a very important letter, written by Lamanon to the then perpetual secretary of the Academy, and which was omitted to be printed in the "Voyage de La Pérouse." It is expressly said in this letter, " Que la force attractive de l'aimant est moindre dans les tropiques qu'en avançant vers les poles, et que l'intensité magnétique déduite du nombre des oscillations de l'aiguille de la boussole d'inclinaison change et augmente avec la latitude." If the Academy had at that time thought itself justified in anticipating the then hoped-for return of La Pérouse, and in making known a truth which was eventually discovered independently by three travellers unknown to each other, Lamanon, De Rossel, and myself, the theory of terrestrial magnetism would have been advanced eighteen years earlier by the knowledge of a new class of phænomena. This simple narrative of facts may, perhaps, be held to justify the statement contained in the following passage of my " Relation historique," Vol. iii. p. 615 :—
" Les observations sur les variations du magnétisme terrestre auxquelles je me suis livré pendant 32 ans, au moyen d'instrumens comparables entre eux en Amérique, en Europe et en Asie, embrassent, dans les deux hémisphères, depuis les frontières de la Dzoungarie chinoise jusque vers l'ouest à la Mer du Sud qui baigne les côtes du Mexique et du Pérou, un espace de 188° de longitude, depuis les 60° de latitude nord jusqu'aux 12° de latitude sud. J'ai

T 2

regardé la loi du décroissement des forces magnétiques, du pôle à l'équateur, comme le résultat le plus important de mon voyage américain." It is not certain, but it is very probable, that Condorcet read Lamanon's letter of July, 1787, at a meeting of the Academy of Sciences at Paris; and I regard such reading as a sufficient act of publication, (Annuaire du Bureau des Longitudes, 1842, p. 463). The first recognition of the law belongs, therefore, indisputably, to thĕ companion of La Pérouse; but, long unheeded or forgotten, it appears to me that the knowledge of the law of the variation of the magnetic force with the latitude only obtained a real scientific existence by the publication of my observations of 1798—1804. The object and the length of this note will not be surprising to those who are familiar with the recent history of magnetism, and the doubts excited by it, and who know, from their own experience, that one attaches some value to that which has formed a subject of constant occupation during five years of laborious research in tropical climates, and in dangerous mountain journeys.

([160]) p. 175.—The greatest intensity of the magnetic force hitherto observed on the surface of the Earth is 2,071, and the least, 0,706. Both are in the southern hemisphere : the geographical position of the first is, lat. 60° 19′ S., long. 131° 20′ E. from Greenwich, observed by Sir James Ross, where the inclination was 83° 31′, (Sabine, Contributions to Terrestrial Magnetism, No. V. Phil. Trans. 1843, p. 231). The minimum was observed by Erman in 20° 00′ S., and 324° 57′ E. from Greenwich, (Erman, Phys. Beob. 1841, S. 570), at which point the inclination was 7° 56′. These values of the force are in the ratio to each other of 1 to 2·933. It was long supposed that the ratio of the greatest to the least intensity of the force on the surface of the Earth was as 2 to 1, (Sabine, Report on Magnetic Intensity, p. 82).

[This note is slightly altered from the original, with M. de Humboldt's concurrence, in order to bring it into more exact accordance with the facts.— EDITOR].

([161]) p. 176.—Speaking of amber (*succinum, glessum*), Pliny says, xxxvii. 3, " Genera ejus plura.   Attritu digitorum accepta caloris anima trahunt in se paleas ac folia arida quæ levia sunt, ac ut magnes lapis ferri ramenta quoque." Plato, in Timæo, p. 80; Martin, Etudes sur le Timée, T. ii. pp. 343—346 ; Strabo, xv. p. 703, Casaub.; Clemens Alex. Strom. ii. p. 370, where a singular distinction is made between τὸ σούχιον and τὸ ελεκτρον). If Thales, in Aristot. de anima, 1, 2, and Hippias, in Diog. Laertio, 1, 24, ascribe to the magnet and to amber a spirit, it is obvious that the word is meant simply to express a force or a cause of motion.

(^{162}) p. 176.—" The magnet draws iron as amber attracts the smallest grains of mustard seed. It is as if a mysterious breath of air passed through both, and communicated itself with the swiftness of an arrow." Such are the expressions used by Kuopho, a Chinese writer of the beginning of the fourth century, in a speech in praise of the magnet. (Klaproth, Lettre à M. A. de Humboldt, sur l'invention de la boussole, 1834, p. 125).

(^{163}) p. 177.—" The phænomena of periodical variations depend manifestly on the action of solar heat, operating probably through the medium of thermo-electric currents induced on the Earth's surface. Beyond this rude guess, however, nothing is as yet known of the physical cause. It is even still a matter of speculation whether the solar influence be a principal or only a subordinate cause, in the phænomena of terrestrial magnetism," (Observ. to be made in the Antarctic Expedition, 1840, p. 35).

(^{164}) p. 177.—Barlow, in the Phil. Trans. for 1822, Part i. p. 117 ; Sir David Brewster, Treatise on Magnetism, p. 129. The influence of heat in diminishing the directive force of the magnetic needle had been taught in the Chinese work, Ou-thsa-tsou, long before the time of Gilbert and Hooke, (Klaproth, Lettre à M. A. de Humboldt, sur l'invention de la boussole, p. 96).

(^{165}) p. 178.—See the Memoir on Terrestrial Magnetism in the Quarterly Review, 1840, Vol. lxvi. pp. 271—312.

(^{166}) p. 178.—When the first proposal to establish a system of observatories, forming a net-work of stations, all provided with similar instruments, was made by myself, I could hardly entertain the hope that I should actually live to see the time when, thanks to the united activity of excellent physicists and astronomers, and especially to the munificent and persevering support of two governments, the Russian and the British, both hemispheres should be covered with magnetic observatories. In 1806 and 1807, my friend, M. Oltmanns, and myself, frequently observed the march of the declination needle at Berlin, for five or six days and nights consecutively, from hour to hour, and often from half hour to half hour, particularly at the equinoxes and solstices. I was persuaded that continuous uninterrupted observations (*observatio perpetua*), during several days and nights, were preferable to detached observations continued during an interval of many months. The apparatus employed was Prony's magnetic telescope, placed in a glass case, and suspended by a thread without torsion, by which angles of 7 or 8 seconds could be read on a finely-divided scale, placed at a distance, and illuminated at night by a lamp Magnetic perturbations, or storms recurring sometimes at the same hours for several successive nights, led me even then to express an earnest wish for the employment of similar instruments, to the east and west of Berlin, to enable us to distinguish between general phænomena and those which might belong to local perturbations from the unequally heated crust of the Earth, or the atmosphere in which clouds are formed. My departure for Paris, and the long political troubles of Europe, prevented the accomplishment of my wishes at that time. The light which, in 1820, the great discovery of Oersted threw on the intimate connection between electricity and magnetism, excited, after a long slumber, a general and lively interest in the periodical variations in the Earth's electro-magnetic charge. Arago, who had began several years before at the observatory at Paris, with an excellent declination apparatus of Gambey, the longest uninterrupted series of hourly observations existing in Europe, shewed, by the comparison with simultaneous perturbations at Kasan, the great advantage which might be derived from such comparisons. When I returned to Berlin, after a residence of eighteen years in France, I had a small magnetic observatory erected during the autumn of 1828, not only for the purpose of continuing the work which I had begun in 1806, but also and especially for the purpose of instituting a series of simultaneous observations, at concerted hours, at Berlin, Paris, and at Freiberg, at the depth of 216 feet below the surface. The simultaneity of the perturbations, and the parallelism of the movements for October and December, 1829, were represented graphically at the time in Pogg. Ann. Bd. xix. S. 357, Tafel i.—iii. An expedition, undertaken under the orders of the Emperor of Russia, in 1829, to Northern Asia, gave me an opportunity of carrying out my plan on a more extensive scale. It was developed in the Report of a Commission specially appointed by the Imperial Academy of Sciences ; and under the protection of Count Cancrin, Chef du Corps des Mines, and the able direction of Professor Kupffer, magnetic stations were established throughout Northern Asia, from Nicolaieff by Catharinenburg, Barnaul, and Nertchinsk, to Pekin. The year 1832, (Göttinger gelehrte Anzeigen, S. 206), forms an epoch in the history of the science ; for it was then that the illustrious founder of a general theory of terrestrial magnetism, Friedrich Gauss, established in the Göttingen Astronomical Observatory apparatus constructed on new principles. The magnetic observatory of Göttingen was completed in 1834 ; and in the same year, (Resultate der Beob. des magnetischen Vereins im Jahr 1838, S. 135, und Poggend. Annalen, Bd. xxxiii. S. 426), by the zealous and active assistance of an ingenious physicist, Wilhelm Weber, Gauss's instruments and methods were made known and brought into use throughout Italy, Sweden, and a large part of

Germany. A "magnetic union," of which Göttingen was the centre, was thus established, making simultaneous observations four times a year, beginning from 1836, for periods of 24 hours. The appointed days, which were called "magnetic term days," did not coincide with the epochs which I had adopted and proposed in 1830, viz. those of the equinoxes and solstices. Great Britain, though the nation which possesses the greatest trade and most extensive navigation of the world, had hitherto taken no part in the movement which, since 1828, had begun to promise such important results towards the establishment of the science of terrestrial magnetism on a sure basis. In April, 1836, by a request, addressed in a public manner directly to the then President of the Royal Society of London, the Duke of Sussex, (Lettre de M. de Humboldt à S. A. R. le Duc de Sussex sur les moyens propres à perfectionner la connaissance du magnétisme terrestre par l'établissement de stations magnétiques et d'observations correspondantes), I had the happiness of exciting in those who had so much in their power a feeling of interest in an undertaking the enlargement of which had long been the object of my warmest wishes. In this letter I urged upon the Duke of Sussex the establishment of permanent stations in Canada, St. Helena, at the Cape of Good Hope, in the Isle of France, in Ceylon, and in New Holland,—all localities which, five years previously, I had pointed out as desirable. A "joint physical and meteorological committee, appointed by the Royal Society from among its Fellows, proposed to the government the establishment of fixed magnetic observatories in both hemispheres, and, in addition, the equipment of a naval expedition for magnetic observations in the Antarctic Seas." I need not here dwell on how deeply science is indebted to the great and zealous exertions, on this occasion, of Herschel, Sabine, Airy, and Lloyd, and to the powerful support of the British Association for the Advancement of Science at its meeting at Newcastle in 1838. In June, 1839, the magnetic Antarctic expedition was determined on, and placed under the command of Captain James Clark Ross. It has now returned, crowned with success and honour; having enriched science with most important geographical discoveries in the vicinity of the Southern Pole, with simultaneous observations on eight or ten magnetic term days,—[and with a determination of the lines of equal Declination, equal Inclination, and equal Force, over three-fourths of the accessible portion of the high latitudes of the southern hemisphere.—EDITOR.]

([167]) p. 179.—Instead of attributing the internal heat of the Earth to the transition from a nebulous to a solid state of the matter of which it is formed, Ampère proposes what appears to me a very improbable hypothesis, in which the heat is regarded as resulting from the long-continued chemical action of a nucleus, composed of the metals of the earths and alkalies, on an already oxydized external crust. In his great work, "Théorie des phénomènes électro-dynamiques," 1826, p. 66, he says, " On ne peut douter qu'il existe dans l'intérieur du globe des courants électro-magnétiques, et que ces courants sont les causes de la chaleur qui lui est propre. Ils naissent d'un noyau métallique central composé des métaux que Davy nous a fait connaître, agissant sur la couche oxidée qui entoure le noyau."

([168]) p. 179.—The remarkable resemblance between the magnetic and the isothermal lines was first pointed out by Sir David Brewster in the Transactions of the Royal Society of Edinburgh, Vol. ix. 1821, p. 318, and in his "Treatise on Magnetism," 1837, pp. 42, 44, 47, and 268. This distinguished physicist supposes the existence of two poles of maximum cold in the northern hemisphere; one American (lat. 73°, long. 100° W., near Cape Walker); and one Asiatic (lat. 73°, long. 80° E.); and he considers that these occasion two meridians of maximum heat, and two meridians of maximum cold. In the sixteenth century, however, Acosta taught the existence of four lines of no declination, which he inferred from the observations of a very experienced Portuguese pilot, (Historia natural de las Indias, 1589, Lib. i. Cap. 17). This opinion would seem to have had some influence on Halley's theory of four magnetic poles, if we may judge from some discussions between Henry Bond, Author of " The Longitude Found," 1676, and Beckborrow. See my " Examen critique de l'hist. de la Géographie," T. iii. p. 60.

([169]) p. 179.—Halley, in the Phil. Trans. Vol. xxix. (for 1714—1716), No. 341.

([170]) p. 180.—Dove, in Poggendorff's Annalen, Bd. xx. S. 341, Bd. xix. S. 388 : " The declination needle is acted upon nearly like an atmospheric electrometer, of which the divergence increases until a spark (lightning) is produced." See also the ingenious comparisons of Professor Kämtz, in his Lehrbuch der Meteorologie, Bd. iii. S. 511—519; Sir David Brewster, Treatise on Magnetism, p. 280. See also Casselmann's Observations (Marburg, 1844), S. 56—62, on the magnetic properties of the galvanic flame, or luminous arch, in a Bunsen's carbon and zinc battery.

([171]) p. 181.—Argelander, in the important Memoir on the Aurora which he has incorporated in the " Vortragen, gehalten in der physikalisch-ökonomischen Gesellschaft zu Königsberg," Bd. i. 1834. S. 257—264.

([172]) p. 181.—On the results of the observations of Lottin, Bravais, and Siljerström, who passed a winter at Bosckop on the coast of Lapland (Lat.

70°), and saw 160 Auroras in 210 nights, see Comptes-rendus de l'Acad. des Sciences, T. x. p. 289, and Martins' Météorologie, 1843, p. 453; also Argelander, in the "Vortragen, gehalten in der Königsberg Gesellschaft," Bd. i. S. 259.

(¹⁷³) p. 183.—Captain John Franklin, "Narrative of a Journey to the Shores of the Polar Seas in the years 1819—1822," pp. 552 and 597; Thienemann, in the Edinburgh Philos. Journal, Vol. xx. p. 366; Farquharson, in the same journal, Vol. vi. p. 392; Wrangel, Phys. Beob. S. 59. Parry saw the Aurora borealis in plain day-light, (Journal of a Voyage performed in 1821—1823, p. 156). A nearly similar observation was made in England on the 9th of September, 1827, (Journal of the Royal Institution, 1828, Jan. p. 489).

(¹⁷⁴) p. 183.—On my return from America, I described, under the name of "polar bands," cirro-cumuli, in which the small detached masses were distributed at very regular intervals, as if by the action of repulsive forces. I made use of the expression "polar bands," because their perspective points of convergence usually appeared at first in the prolongation of the dipping needle, so that the parallel lines of cirrus corresponded with the magnetic meridian. Sometimes the point of convergence appeared to move first in one direction, and then in the opposite; and, at other times, to advance gradually in one direction. These bands usually shew themselves completely formed only in one quarter of the sky; and their movement is first directed from south to north, and then gradually from east to west. I do not think the movements can be explained by variations in the currents of air in the higher regions of the atmosphere. The bands are seen when the air is extremely calm, and the sky very serene; and they are much more frequent within the tropics than in the temperate and frigid zones. I have seen the phænomenon develope itself in a manner so strikingly similar on the Andes, at an elevation of 14000 French feet, almost under the equator, and in Northern Asia, in the plains of Krasnojarski, to the south of Buchtarminsk, that it seems difficult not to regard it as a process depending on very general and widely-diffused natural forces. See the important remarks of Kämtz, (Vorlesungen über Meteorologie, 1840, S. 146; and the more recent ones of Martins and Bravais, (Météorologie, 1843, p. 117). Arago observed at Paris, June 23, 1844, in the day-time, south polar bands, consisting of extremely light clouds, and saw dark rays shoot upwards from an arc having an east and west direction. In page 181 of the text, mention is made of the appearance of black smoke-like rays in brilliant nocturnal auroras.

([175]) p. 184.—In the Shetland Islands, the auroral streamers are called " the merry dancers," (Kendal, in the Quarterly Journal of Science, new series, Vol. iv. p. 395).

([176]) p. 184.—See the excellent article of Muncke, in the last edition of Gehler's Physikalisches Wörterbuch, Bd. vii. 1, p. 113—268 ; and particularly p. 158.

([177]) p. 184.—Farquharson, in Edinb. Philos. Journal, Vol. xvi. p. 304 ; Phil. Trans. for 1829, p. 113.

([178]) p. 187.—Kämtz, Lehrbuch der Meteorologie, Bd. iii. S. 498 and 501.

([179]) p. 189.—On the dry fogs of 1783 and 1831, which appeared luminous at night, see Arago, in the Annuaire du Bureau des Longitudes, 1832, pp. 246 and 250 ; and on some singular phænomena of light from clouds not being storm or thunder clouds, in his " Notices sur le Tonnerre," in the Annuaire pour l'an 1838, pp. 279—285.

[Being at Loch Scavig, in the Island of Skye, in a friend's yacht, in August 1836, the summit of the mountain which rises on the east side of the harbour to the height of 2000 feet or thereabouts, was enveloped during the day by a thin veil of mist, extending 3 or 400 feet below the summit, and so thin as to permit the outline of the hill, as well as the rocky inequalities of the surface, to be seen through it : as night came on the mist remained, but became distinctly and decidedly luminous, still continuing so thin that the hill was seen through it : towards 8 or 9 o'clock in the evening, streamers of the Aurora ascended from it for about 10° or 15° towards the zenith, and continued to do so for an hour or thereabouts.— EDITOR.]

([180]) p. 191.—Herod. iv. 28.   The ancients were prepossessed with an idea that Egypt is exempt from earthquakes, (Plin. ii. 80) ; although the necessity of restoring the statue of Memnon is, in some degree, evidence to the contrary, (Letronne, La Statue vocale de Memnon, 1833, pp. 25—26).   It is, however, true that the valley of the Nile is situated outside the earthquake district of Byzantium, the Archipelago, and Syria, (Ideler ad Aristol. Meteor. p. 584).

([181]) p. 191.—Saint-Martin, in the learned notes which he has appended to Lebeau's Histoire du Bas Empire, T. ix. p. 401.

([182]) p. 191.—Humboldt, Asie Centrale, T. ii. pp. 110—118.   On the difference between the agitation of the surface and that of the inferior strata, see Gay-Lussac, in the Annales de Chimie et de Physique, T. xxii. p. 429.

([183]) p. 192.—Tutissimum est cum vibrat crispante ædificiorum crepitu ;

et cum intumescit assurgens alternoque motu residet, innoxium et cum concurrentia tecta contrario ictu arietant; quoniam alter motus alteri renititur. Undantis inclinatio et fluctus more quædam volutatio infesta est, aut cum in unam partem totus se motus impellit (Plin. ii. 82).

(184) p. 193.—The absence of any connection between earthquakes, and the state of the weather or appearance of the sky immediately preceding their occurrence, begins to be recognised even in Italy. The numerical data obtained by Friedrich Hoffmann on this subject agree perfectly with the experience of the Abbate Scina of Palermo. See F. Hoffmann's "hinterlassene Werke," Bd. ii. S. 366—375. I have indeed myself observed more than once the appearance of reddish clouds a short time before earthquake shocks; and on the 4th of November, 1799, I experienced two strong shocks simultaneously with a violent clap of thunder (Relation historique, Liv. iv. chap. 10); and a physicist of Turin, Vasalli Eandi, observed great disturbance in Volta's electrometer during the long-continued earthquakes at Pignerol, which lasted from April 2 to May 17, 1808 (Journal de Phys. T. lxvii. p. 291). But we have no reason for regarding any of these phænomena, such as haze or clouds, disturbances in the electric state of the atmosphere, or calms, as having any general and necessary connection with earthquakes; for in Quito, Peru, and Chili, as well as in Canada and in Italy, shocks have been felt when the sky was most serene and free from clouds or haze, and when the freshest land or sea breezes were blowing. But although we have no reason to believe that earthquakes are preceded or accompanied by any particular meteorological indications, yet we ought not peremptorily to reject altogether the popular belief of the influence of particular seasons, the vernal and autumnal equinoxes, the setting in of tropical rains after long-continued drought, and the change of the monsoons, solely because we do not at present understand the causal connection which may exist between meteorological and subterranean phænomena. Numerical data, collected with great care by M. de Hoff, Peter Merian, and Friedrich Hoffmann, for the purpose of elucidating the comparative frequency of earthquakes at the different seasons of the year, indicate a maximum about the times of the equinoxes. It deserves notice, that Pliny, at the end of his fanciful theory of earthquakes, terms these awful phænomena "subterranean storms," not so much on account of the noise resembling thunder, which often accompanies them, as from a notion that the elastic forces which by their increasing tension thus agitate the ground, accumulate beneath the surface of the earth when they are wanting in the atmosphere:

Ventos in causa esse non dubium reor. Neque enim unquam intremiscunt terræ, nisi sopito mari cœloque adeo tranquillo, ut volatus avium non pendeant, subtracto omni spiritu qui vehit; nec unquam nisi post ventos conditos, scilicet in venas, et cavernas ejus occulto afflatu. Neque aliud est in terra tremor, quam in nube tonitruum; nec hiatus aliud quam cum fulmen erumpit, incluso spiritu luctante et ad libertatem exire nitente (Plin. ii. 79). We find in Seneca (Nat. Quæst. vi. 4—31), the germ of all that has been observed or imagined up to very recent times regarding the causes of earthquakes.

([185]) p. 193.—In my Relation historique, T. i. p. 311 and 513, I have given evidence that the horary march of the barometer continues undisturbed both before and after the occurrence of earthquake shocks.

([186]) p. 194.—Humboldt, Rel. hist. T. i. p. 515—517.

([187]) p. 196.—On the " bramidos" of Guanaxuato, see my Essai polit. sur la Nouv. Espagne, T. i. p. 303. The town of Guanaxuato is situated 6420 French feet above the level of the sea. The subterranean thunder was not accompanied by any sensible shock either in the deep mines or at the surface: it was not heard on the neighbouring plateau, but only in the mountainous part of the Sierra, from the Cuesta de los Aguilares, not far from Marfil, to the north of Santa Rosa. The waves of sound did not reach detached portions of the Sierra situated 24 to 28 miles north-west of Guanaxuato, in the neighbourhood of the thermal spring of San Jose de Comangillas beyond Chichimequillo. Measures of extraordinary severity were adopted by the magistrates of the city, when the panic caused by the subterranean thunder was at the highest. The flight of a family was punished by a fine of 1000 piastres or by two months' imprisonment, and the militia were commanded to arrest and bring back the fugitives. The most curious circumstance, however, was the confidence which the authorities,—" el cabildo,"—seem to have reposed in their own superior knowledge, saying, in a proclamation of the period, which I had an opportunity of seeing—" the magistracy in their wisdom (en su sabiduria) will be well aware of the period of real and imminent danger, should it arise, and they will then give orders for flight; but for the present it is sufficient to continue the processions." A famine ensued, as the inhabitants of the table lands were prevented by their fears from bringing corn to the town. The ancients were acquainted with the phænomenon of subterranean noises unaccompanied by earthquake: see Aristot. Meteor. ii. p. 802; and Plin. ii. 80. The singular noise which was heard in the Dalmatian island of Meleda, 16 miles from Ragusa, from March 1822 to September 1824, was occasionally accompanied by shocks. Much light has been thrown on this phænomenon by Partsch.

([188]) p. 198.—Drake, Nat. and Statist. View of Cincinnati, p. 232—238; Mitchell, in the Trans. of the Literary and Philosophical Society of New York. In the Piedmontese county of Pignerol, glasses filled to the brim with water exhibited agitation for several hours.

([189]) p. 199.—The Spanish expression is, " rocas que hacen puente." These local interruptions to the transmission of the shock through the *upper* strata, seem analogous to the remarkable phænomenon which took place in the deep silver mines of Marienberg, in Saxony, at the beginning of the present century, when earthquake shocks drove the miners in alarm to the surface, where, meanwhile, nothing of the kind had been experienced. The converse phænomenon was observed in November 1823, when the workmen in the mines of Falun and Persberg felt no movement whatsoever, whilst above their heads a violent shock of earthquake spread terror among the inhabitants at the surface.

([190]) p. 200.—Sir Alex. Burnes, " Travels into Bokhara," Vol. i. p. 18; Wathen, " Mem. on the Usbek State," in the Journal of the Asiatic Soc. of Bengal, Vol. iii. p. 337.

([191]) p. 201.—Phil. Trans. Vol. xlix. p. 414.

([192]) p. 202.—On the frequency of earthquakes in Kashmir, see Troyer's translation of the ancient Radjatarangini, Vol. ii. p. 297; and Carl von Hugel's Travels, Bd. ii. S. 184.

([193]) p. 203.—Strabo, Lib. i. p. 100, Casaub. It is evident from another passage (Strabo, Lib. vi. p. 412) that the expression πηλὸν διαπύθον ποταμὸν signifies lava, and not erupted mud. Compare Walter, " über Abnahme der vulkanischen Thätigkeit in historischen Zeiten," 1844, S. 25.

([194]) p. 205.—Bischof's valuable Memoir, " Wärmelehre des inneren Erdkörpers."

([195]) p. 205.—On the artesian fire-wells (Ho-tsing) of China, and the antiquity of the use of portable gas (carried in tubes of bamboo) in the city of Khiung-tscheu: see Klaproth, in my Asie centrale, T. ii. p. 519—530.

([196]) p. 205.—Boussingault (Annales de Chimie, T. lii. p. 181) did not observe any hydrochloric acid in the exhalations from the volcanoes of New Grenada; whereas it was found in immense quantities by Monticelli, during the eruption of Vesuvius in 1813.

([197]) p. 206.—Humboldt, Recueil d'Observ. astronomiques, T. i. p. 311 (Nivellement barométrique de la Cordillère des Andes, No. 206.)

([198]) p. 206.—Adolphe Brongniart, in the Annales des Sciences Naturelles, T. xv. p. 225.

([199]) p. 207.—Bischof, "Wärmelehre des inneren Erdkörpers," S. 324, Anm. 2.

([200]) p. 207.—Humboldt, Asie Centrale, T. i. p. 43.

([201]) p. 208.—On the theory of the isogeothermal lines (chthonisothermals), see the ingenious discussions of Kupffer, in Poggend. Ann. Bd. xv. S. 184, and Bd. xxxii. S. 270; in his "Voyage dans l'Oural," p. 382—398; and in the Edinburgh Journal of Sciences (New Series, Vol. iv. p. 355). Compare also Kämtz, "Lehrb. der Meteor. Bd. ii. S. 217; and on the rising up of the isogeothermals in mountainous countries, see Bischof, S. 174—198.

([202]) p. 208.—Leopold von Buch, in Poggend. Ann. Bd. xii. S. 405.

([203]) p. 208.—See my Rel. hist. T. ii. p. 22. The temperature of drops of rain at Cumana was 22°·3 C. (72°·2 Fah.), when that of the air was 30° and 31° C. (86° to 88° Fah.); and during the fall of rain the temperature of the air sunk to 23·4 C. (about 74° Fah.) The initial temperature of the drops depends on the height of the cloud in which they are formed, and on the heat which its upper surface may have received from the direct effect of the sun's rays; but this temperature alters during their fall. The latent heat, set free in their formation, causes them to have at first a temperature rather above that of the surrounding medium; and as they pass through lower and warmer strata of air, this is somewhat raised, while at the same time their size is increased by the condensation of the aqueous vapour contained in those strata. (Bischof, Wärmelehre des inneren Erdkörpers, S. 73). The increase of temperature from the cause described is, however, compensated by the loss of heat which the drops undergo from evaporation from their surface. Apart from the effect probably due to atmospheric electricity in thunder showers, the cooling influence of rain may be ascribed, first to the low initial temperature which the drops have acquired in the higher regions of the atmosphere, and to the colder air of the higher strata which they bring down with them; and, lastly, to evaporation from the wet ground. This is the ordinary course of the phænomenon; but in some rare instances (Humboldt, Rel. hist. T. iii. p. 513) the drops are warmer than the lower strata of the atmosphere through which they fall: this may possibly be caused either by warm upper currents, or by the heating of an extensive surface of thin cloud by the sun's rays. Arago, in the "Annuaire for 1836," p. 300, has shewn the connection between the phænomenon of "supplementary rainbows," which are explained by the interferences of luminous rays, and the size and increasing bulk of falling drops of rain. This ingenious discussion also shews how, under certain conditions, an accurately observed optical phænomenon may throw light on difficult meteorological phænomena.

(204) p. 208.—Boussingault's researches appear to me to leave no doubt on the fact, that within the tropics the temperature of the ground, at a very small depth below the surface, corresponds with the mean atmospheric temperature. I subjoin a few examples :—

Stations.	Temperature at the depth of one foot.	Mean temperature of the air.	Elevation.
Guayaquil . . .	78·80° Fah.	78·08° Fah.	0 Eng. ft.
Anserma nuevo . .	74·66	74·84	3443
Zupia	70·70	70·70	4018
Popayan	64·76	65·66	5930
Quito	59·90	59·90	9559

The doubt respecting the temperature of the earth within the tropics, to which I may myself have given occasion by my observations in the caves of Guaripe Cueva del Guacharo (Rel. hist. T. iii. p. 191—196), is removed by considering that I compared the presumed mean temperature of the air at the convent of Caripe (18°·5 C., 65°·3 Fah.) not with the temperature of the air in the cavern (which was 18°·7 C., or 65°·6 Fah.), but with that of the subterranean rivulet (16°·8 C., or 62°·2 Fah.), although I had before remarked (Vol. iii. p. 164 and 194) that the waters of the cavern might very probably be affected by waters coming from greater heights. [See addition to note 139. —EDITOR.]

(205) p. 209.—Boussingault, in the Annales de Chemie, T. lii. p. 181. The temperature of the spring of Chaudes-Aigues, in Auvergne, is not more than 80° C. (176° Fah.) It may also be remarked, that while the aguas calientes de las Trincheras, south of Porto Cabello (in Venezuela), issuing from granite divided in beds, show a temperature of 97° (206°·6 F.), all the springs on the still active volcanoes of Pasto, Cotopaxi, and Tunguragua, have temperatures of only 36° to 54° C. (96°·8 to 129°·2 Fah.)

(206) p. 210.—See the descriptions of the Kassotis (the fountain of St. Nicholas), and, of the Castalian spring at the foot of the Phedriades, in Pausanias, x. 24—5, and x. 8—9 ; of the Pirenean spring (Acrocorinthus), in Strabo, p. 379 ; of that of Erasinos (on the Chaon, south of Argos), in

Herodotus, vi. 67, and Pausanias, ii. 24, 7 ; of those of Ædepsos in Eubæa (of which some have a temperature of 31° C. (87°·1 Fah.), and others of 62° to 75° C. (or 143°·6 to 167°·0 F.), in Strabo, p. 60, 447, and Athenæus, ii. 3, 73 ; the hot springs of Thermopylæ, at the foot of Mount Oeta, having a temperature of 65° C. (149° Fah.), are described by Pausanias, x. 21, 2.— These references are taken from manuscript notes of Professor Curtius, the learned travelling companion of Otfried Müller.

(207) p. 210.—Plin. ii. 106 ; Seneca, Epist. 79, § 3, ed. Ruhkopf; Beaufort, Survey of the Coast of Karamania, 1820, art. Yanar, near Deliktash, the ancient Phaselis, p. 24. Compare likewise Ctesias, Fragm. cap. x. p. 250, ed. Bähr ; and Strabo, lib. xiv. p. 665, Casaub.

(208) p. 211.—Arago, Annuaire, 1835, p. 234.

(209) p. 211.—Acta S. Patricii, p. 555, ed. Ruinart, T. ii. p. 385, Mazochi. Dureau de la Malle first called attention to this remarkable passage, in the " Recherches sur la Topographie de Carthage," 1835, p. 276. (Compare Seneca, Nat. Quæst. iii. 24).

(210) p. 213.—Humboldt, Rel. hist. T. iii. pp. 562—567 ; Asie centrale, T. i. p. 43, T. ii. pp. 505—515; Vues des Cordillères, Pl. xli. On the Macalubi (from the Arabic Makhlub, overturned, from the root Khalaba), and on " fluid Earth issuing from the Earth," see Solinus, cap. v. ; idem ager Agrigentinus eructat limosas scaturigines, et ut venæ fontium sufficiunt rivis subministrandi, it ain hac Siciliæ parte solo nunquam deficiente, æterna rejectatione terram terra evomit.

(211) p. 214.—See the excellent little map of the island of Nisyros, in Ross, Reisen auf den griechischen Inseln, Bd. ii. 1843, S. 69.

(212) p. 215.—Leopold von Buch, Phys. Beschreibung der Canarischen Inseln, S. 326 ; and the same author, über Erhebungscratere und Vulcane, in Poggend. Ann. Bd. xxxvii. S. 169. Strabo distinguishes extremely well between two modes of formation of islands, when he is speaking of the separation of Sicily from Calabria, Lib. vi. p. 258. Casaub. He says, " Some islands are detached parts of the main land ; others have been raised from the bottom of the sea, as we still see. The islands of the open sea, i. e. those far from the shore, have probably been raised from the bed of the sea ; and those situated near promontories have been detached from the main land."

(213) p. 215.—Ocre Fisove (Mons Vesuvius). The word ocre is true Umbrian, and, according to Festus himself, signifies " mountain " in the ancient Umbrian language, (Lassen, Deutung der Eugubinischen Tafeln, im Rhein.

Museum, 1832, S. 387). If, according to Voss, Αιτνη is of Grecian origin, and connected with αιθω or αιθινος, Ætna would signify a burning and shining mountain; but the learned Parthey doubts the Grecian origin, both from etymological considerations, and also because Ætna could not have been a beacon light for Greek navigators and travellers, like the continually-active volcano, Stromboli (Strongyle), which Homer appears to refer to in the Odyssey (xii. 68, 202, and 219), although the geographical position is only very vaguely indicated. I imagine that the word Ætna would more probably be found to belong to the language of the ancient Siculi, if any considerable portions of it were in our possession. According to Diodorus (v. 6), the Sicani, who were the aboriginal inhabitants of Sicily before the arrival of the Siculi, were forced by the eruptions of Ætna, which lasted several years, to take refuge in the western part of the country. The oldest recorded eruption of Ætna is that mentioned by Pindar and Æschylus as having taken place in the reign of Hiero, in the second year of the 75th Olympiad. Hesiod was probably aware of devastating eruptions of Ætna having occurred previous to the Greek settlements. There are, however, some doubts respecting the word Αιτνη, which I have discussed at some length in my " Examen. critique de la Géographie," T. i. p. 168.

(^{214}) p. 215.—Seneca, Epist. 79.

(^{215}) p. 215.—Ælian. Var. hist. viii. 11.

(^{216}) p. 218.—Petri Bembi Opuscula (Ætna Dialogus), Basil. 1556, p. 63 ; " Quicquid in Ætnæ matris utero coalescit, nunquam exit ex cratere superiore, quod vel eo inscondere gravis materia non queat, vel, quia inferius alia spiramenta sunt, non fit opus. Despumant flammis urgentibus ignei rivi pigro fluxu totas delambentes plagas, et in lapidem indurescunt."

(^{217}) p. 218.—See my drawings of the volcano of Jorullo, of its " Hornitos," and of the upraised Malpays, in the " Vues des Cordillères," Pl. xliii. p. 239.

(^{218}) p. 219.—Humboldt, Essai sur la Géographie des Plantes et Tableau physique des Régions équinoxiales, 1807, p. 130, and Essai géognostique sur le gisement des roches, p. 321. A consideration of the greater part of the volcanos of the Island of Java is sufficient to shew that the entire absence of lava currents, during a period of uninterrupted activity, cannot be ascribed to the form, situation, and absolute elevation of the mountains. (*Vide* Leop. von Buch, Descr. phys. des Iles Canaries, p. 419 ; Reinwardt and Hoffmann, in Poggend. Ann. Bd. xii. S. 607).

(^{219}) p. 221.—See the comparison of my measurements with those of

Saussure and Earl Minto, in the Abhandlungen der Academie der Wiss. zu Berlin aus den J. 1822 and 1823, S. 30.

([220]) p. 222.—Pimelodes cyclopum, *vide* Humboldt, Recueil d'Observations de Zoologie et d'Anatomie comparée, T. i. pp. 21—25.

([221]) p. 224.—Leop. von Buch, in Poggend. Ann. Bd. xxxvii. S. 179.

([222]) p. 224.—On the chemical origin of specular iron in volcanic masses, see Mitscherlich, in Poggend. Ann. Bd. xv. S. 630; and on the disengagement of hydrochloric acid gas in craters, see Gay-Lussac, in the Annales de Chimie et de Physique, T. xxii. p. 423.

([223]) p. 226.—See the fine experiments on the cooling of masses of stone, in Bischof's " Wärmelehre," S. 384, 443, 500—512.

([224]) p. 226.—Berzelius and Wöhler, in Poggend. Annalen, Bd. i. S. 221, and Bd. xi. S. 146; Gay-Lussac, in the Annales de Chimie, T. xxii. p. 422; Bischof, Reasons against the Chemical Theory of Volcanoes, in the English edition of his Wärmelehre, pp. 297—309.

([225]) p. 227.—In Plato's geognostical ideas, as developed in the Phædon, the Pyriphlegethon plays nearly the same part, in respect to the activity of volcanos, as that which we now assign to the internal heat of the globe, increasing with increasing depth, and to the state of fusion of the deeper strata. (Phædon, ed. Ast. pp. 603 and 607, Annot. pp. 808 and 817). " Within and around the Earth there are subterranean channels of various magnitudes. Water flows through them abundantly, as do currents of fire, and streams of liquid mud, more or less impure, like the flow of mud which, in Sicily, precedes the issuing forth of torrents of fire, both alike overwhelming every thing in their path. The Pyriphlegethon pours itself into a wide space, where a strong fire burns fiercely, and it there forms a lake larger than our sea, in which the water and mud are always boiling; and, re-issuing thence, it rolls its troubled and muddy waves around the earth." This river of molten earth and mud is so far the *general* source of volcanic phenomena, that Plato says expressly, in his continuation, " Such is the Pyriphlegethon, of which the fiery currents (οι ρυακες), wherever they are found on the earth (οπη αν τυχσωι της γης), are a part. Volcanic scoriæ and currents of lava are therefore regarded as parts of the Pyriphlegethon itself, or of a subterranean molten mass, in continued motion, in the interior of the earth. That οι ρυακες signifies lava currents, and not " fire ejecting mountains," as Schneider, Passow, and Schleiermacher would suppose, is evident from many passages, of which part have been brought together by Ukert, (Geogr. der Griechen und

Römer, Th. ii. 1, S. 200). Ρυαξ is the volcanic phenomenon taken in its most striking feature, namely, the lava current : hence the expression, the ρυακες of Ætna. Aristol. Mirab. T. ii. p. 833, sect. 38, Bekker; Thucyd. iii. 116; Theophr. de Lap. 22, p. 427, Schneider; Diod. v. 6, and xiv. 59, where there are the remarkable words : " Many towns near the sea, and not far from Ætna, have been destroyed," υπο του καλουμενου ρυακος; Strabo, vi. p. 269, xiii. p. 628 ; on the celebrated " burning mud" of the Lelantine plains in Eubæa, see i. p. 58, Casaub.; and, lastly, Appian. de bello civili, v. 114. The censure which Aristotle (Meteor. ii. 2, 19), passes on the geognostical fancies in Phædon only applies, strictly, to the origin of the rivers which flow on the surface of the Earth. We cannot but be struck with Plato's distinctly expressed assertion, that, in Sicily, " humid emissions of mud *precede* the burning streams," or currents of lava. May we suppose that rapilli and ashes, formed into a paste by melted snow and water during a volcano-electric storm over the crater of eruption, may have been regarded as erupted mud ? or is it not more probable that Plato's streams of liquid mud are mere reminiscences of the salses (mud-volcanoes) of Agrigentum, which eject mud with a loud noise, and have been noticed in Note 210. We have to regret, on this sub-ject, among the many lost writings of Theophrastus, the loss of one " on the volcanic current in Sicily," (περι του ρυακος εν Σικελια), mentioned by Diog. Laert. v. 39.

(226) p. 228.—Von Buch, Physikal. Beschreib. der Canarischen Inseln, S. 326—407. I doubt the correctness of the view to which Darwin appears to incline, (Geological Observations on the Volcanic Islands, 1844, p. 127) ; according to which, central volcanoes would be regarded generally as short volcanic chains over parallel fissures. Hoffmann had already supposed that in the group of the Lipari Islands, which he has so well described, and in the two fissures of eruption which intersect near Panaria, he had found an intermediate link between Von Buch's central volcanoes and volcanic chains, (Poggend. Ann. der Physik. Bd. xxvi. S. 81—88),

(227) p. 229.—Humboldt, Geognost. Beob. über die Vulkane des Hochlandes von Quito, in Poggend. Annalen, Bd. xliv. S. 194.

(228) p. 229.—Seneca, in speaking very pertinently of the problematical lower-ing of Mount Ætna, says, in his 79th letter : " Potest hoc accidere, non quia montis altitudo desedit, sed quia ignis evanuit et minus vehemens ac largus effertur : ob eandem causam, fumo quoque per diem segniore. Neutrum autem incredibile est, nec montem qui devoretur quotidie minui, nec ignem non manere eundem ; quia non ipse ex se est, sed in aliqua inferna valle conceptus exæstuat

et alibi pascitur : in ipso monte non alimentum habet sed viam," (Ed. Ruhkopfiana, T. iii. p. 32). Strabo distinctly recognises the probable existence of a subterranean communication between the volcanoes of Sicily and those of Lipari, Pithecusa (Ischia), and Vesuvius, " which we may suppose had once been a fiery crater." He speaks of the whole district as undermined by fire" (Lib. i. pp. 247 and 248).

(²²⁹) p. 229.—Humboldt, Essai politique sur la Nouvelle Espagne, T. ii. pp. 173—175.

(²³⁰) p. 230.—Ovid's Description of the Elevation of Methone, (Metamorph. xv. 296—306), runs thus :—

> " Est prope Pittheam tumulus Trœzena sine ullis
> Arduus arboribus, quondam planissima campi
> Area, nunc tumulus ; nam—res horrenda relatu—
> Vis fera ventorum, cæcis inclusa cavernis,
> Exspirare aliqua cupiens, luctataque frustra
> Liberiore frui cœlo, cum carcere rima
> Nulla foret, toto nec pervia flatibus esset,
> Extentam tumefecit humum ; ceu spiritus oris
> Tendere vesicam solet, aut direpta bicorni
> Terga capro.   Tumor ille loci permansit, et alti
> Collis habet speciem, longoque induruit ævo."

This geologically important description of the upheaving of a bell or dome-shaped elevation on the mainland, agrees remarkably with that given by Aristotle of the upheaving of an island of eruption, (Meteor. ii. 8, 17—19). " The Earth continues to tremble until the wind (ανεμος) which caused the trembling has made its way through and escaped from the ground. This is what happened lately at Heraclea in Pontus, and formerly in Hiera, one of the Æolian islands. At Hiera, part of the ground was inflated, and, with a loud noise, rose into a hill, until the " strong breath" (πνευμα) found an outlet. It then threw out sparks and ashes, which covered the neighbouring town of the Liparians, and even reached some of the cities of Italy." This description distinguishes very well between the eruption itself, and the inflation and upheaving of the ground by which it was preceded. Strabo describes the phenomenon of Methone, (Lib. i. p. 59, Casaub.) as " an eruption of flames, in which a volcano was raised to a height of seven stadia (?). In the day it was inaccessible from the heat and the smell of sulphur ; but in the night it had a fragrant odour (?). The heat was so great that the sea boiled for a distance of five stadia ; and twenty stadia off, the waters were disturbed and muddied by the fall of ejected fragments of rock." Respecting the present mineralogical constitution of the peninsula of Methana, see Fiedler, Reise durch Griechenland, Th. i. S. 257—263.

(²³¹) p. 230.—See Leop. von Buch's Physik. Beschr. der Canar. Inseln, S. 356—358, and particularly the French translation of this excellent work, p. 402; and Von Buch, in Poggendorff's Annalen, Bd. xxxvii. S. 183. In very modern times, a submarine island has been again in process of formation within the crater of Santorin. In 1810, it was still 15 fathoms below the surface of the sea; and in 1830, had risen to within 3 or 4 fathoms of the surface. Its sides are nearly perpendicular. The continued activity of the submarine crater is manifested by the mixture of sulphureous gases in the waters of the eastern bay of Neo-Kammeni, as at Vromolimni near Methana. Copper-bottomed ships cast anchor in the bay for the purpose of cleansing their copper sheathing by this natural (or volcanic) process, and rendering it bright.

(²³²) p. 230.—The appearances of the new island near St. Michael, in the Azores, took place—June 11, 1638; December 31, 1719; June 13, 1811.

(²³³) p. 231.—Prévost, in the Bulletin de la Société géologique, T. ii. p. 34; Hoffmann, Hinterlassene Werke, Bd. ii. S. 451—456.

(²³⁴) p. 231.—"Accedunt vicini et perpetui Ætnæ montis ignes et insularum Æolidum, veluti ipsis undis alatur incendium; neque enim aliter durare tot seculis tantus ignis potuisset, nisi humoris nutrimentis aleretur," (Justin, Hist. Philipp. iv. 1). This physical description of Sicily commences with a very complicated volcanic theory. Deep-seated beds of sulphur and resin; a very thin soil, full of cavities, and very subject to fissures; a great agitation caused by the waves of the sea, which, as they beat against the shore, draw down with them the air, causing a wind, which blows the fire, are the elements of Trogus's theory. The ancients probably connected the idea of the wind being forced into the interior of the Earth, there to act on the volcanic fire, with the influence which they ascribed to the direction of particular winds on the degree of volcanic activity of Ætna, Hiera, and Stromboli, (see a remarkable passage in Strabo, Lib. vi. pp. 275 and 276). The island of Stromboli passed for the dwelling of Æolus, "the regulator of the winds," because mariners predicted the weather from the degree of violence of the eruptions of the volcano. Some connection between the eruptions of small volcanoes and the direction of the wind is generally admitted at the present time, although our knowledge of volcanic phenomena, and the small variations of atmospheric pressure which accompany different winds, are far from enabling us to assign, as yet, any satisfactory explanation. Bembo, who was educated in Sicily by Greek refugees, has given us a pleasing relation of his youthful wanderings in his " Ætna Dialogus," written in the middle of the sixteenth century, in which he propounds the theory of the introduction of sea water to the focus of volcanic activity, and of the necessity of the proximity of the sea to the existence of active volcanoes. During the ascent of Ætna, the following questions were proposed:—" Explana potius nobis quæ petimus, ea incendia unde oriantur et orta quomodo perdurent? In omni tellure nuspiam majores fistulæ aut meatus ampliores sunt quam in locis, quæ vel mari vicina sunt, vel a mari protinus alluuntur : mare erodit illa facillime pergitque in viscera terræ. Itaque cum in aliena regna sibi viam faciat, ventis etiam facit ; ex quo fit, ut loca quæque maritima maxime terræ motibus subjecta sint, parum mediter- ranea. Habes quum in sulfuris venas venti furentes inciderint, unde incendia oriantur Ætnæ tuæ. Vides, quæ mare in radicibus habeat, quæ sulfurea sit, quæ cavernosa, quæ a mari aliquando perforata ventos admiserit æstuantes, per quos idonea flammæ materies incenderetur.

(235) p. 232,—Compare Gay-Lussac, " sur les Volcans," in the Annales de Chimie, T. xxii, p, 427 ; and Bischof, Wärmelehre, S. 272. The eruptions of smoke and of aqueous vapour which have been seen at different times round Lancerote, Iceland, and the Kurile islands during eruptions of the neighbouring volcanoes, show a reaction of the volcanic foci, in which the hydrostatic pressure of the waters of the adjacent sea is overcome by the greater expansive force of the vapours or gases.

(236) p. 232.—Abel-Rémusat, Lettre à M, Cordier, in the Annales des Mines, T. v. p. 137.

(237) p. 232,—Humboldt, Asie centrale, T. ii. pp. 30—33, 38—52, 70—80, and 426—428. The existence of active volcanoes in Kordofan, 540 miles from the Red Sea, has been recently contradicted by Rüppell, (Reisen in Nubien, 1829, S. 151).

(238) p. 234.—Dufrénoy et Elie de Beaumont, Explication de la Carte géologique de la France, T. i. p. 89.

(239) p. 234.—Sophocl. Philoct. v. 971 and 972. On the supposed epoch of the extinction of the " Lemnian fire" in the time of Alexander, compare Buttmann in Museum der Alterthumswissenschaft, Bd. i. 1807, S. 295 ; Dureau de la Malle in Malte-Brun, Annales des Voyages, T. ix. 1809, p. 5 ; Ukert, in Bertuch, Geogr. Ephemeriden, Bd. xxxix. 1812, S. 361 ; Rhode, Res Lemnicæ, 1829, p. 8, and Walter über Abnahme der vulkanischen Thätigkeit in historischen Zeiten, 1844, S. 24. Choiseul's Chart of Lemnos

makes it appear very probable that both the extinct crater of Mosychlos and the island of Chryse, the desert habitation of Philoctetes, (Otfried Müller, Minyer, S. 300), have been long swallowed up by the sea. Reefs and shoals to the north-east of Lemnos still indicate the place where the Ægean Sea once possessed an active volcano similar to Ætna, Vesuvius, Stromboli, and Volcano. [The last-named mountain is of the Lipari group of islands].

(²⁴⁰) p. 234.—Compare Reinwardt and Hoffmann, in Poggend. Ann. Bd. xii. S. 607 ; Leop. von Buch, Descr. des Iles Canaries, pp. 424—426. The eruptions of argillaceous mud from Carguairazo, when that volcano was destroyed in 1698, the Lodazales of Igualata, and the Moya of Pelileo on the table and of Quito, are volcanic phenomena of the same kind.

(²⁴¹) p. 236.—In a profile of the district round Tezcuco, Totonilco, and Moran, (Atlas géogr. et phys. Pl. vii.), which I designed originally (1803) for an inedited work, (Pasigraphia geognostica destinada al uso de los Jovenes del Colegio de Mineria de Mexico), I applied, at a later period (1832), the term *endogenous* (generated in the interior) to erupted plutonic and volcanic rocks, and *exogenous* (generated externally on the surface of the globe) to the sedimentary rocks. In the graphical system which I adopted, the first mentioned class of rocks were indicated by an arrow directed upwards $\uparrow$, and the exogenous by an arrow directed downwards $\downarrow$. These marks appear to me to be at least less unsightly, than the mode in which the eruption of masses of basalt, porphyry, and syenite, and the penetration of sedimentary strata by them, are often represented by figures of ascending veins, which are drawn in an arbitrary manner, and with little conformity to nature. The names proposed in the pasigraphico-geognostical section were taken from Decandolle's botanical nomenclature—endogenous for monocotyledonous, and exogenous for dycotyledonous plants ; Mohl's more accurate analysis of plants has, however, shewn that it is not strictly and generally true, that monocotyledones increase from within, and dycotyledones from without. (Link, Elementa philosophiæ Botanicæ, T. i. 1837, p. 287 ; Endlicher and Unger, Grundzüge der Botanik, 1843, S. 89 ; and Jussieu, Traité de Botanique, T. i. p. 85.) The rocks which I have called endogenous, are designated by Lyell, in his Principles of Geology, 1833, V. iii. p. 374, by the characteristic expression of "netherformed" or "hypogene rocks."

(²⁴²) p. 236.—Compare Von Buch, über Dolomit als Gebirgsart, 1823, S. 36 ; and the same writer in the Abhandl. der Akad. der Wissensch. zu Berlin, 1842, pp. 58 and 63 ; and in the Jahrb. für wissenschaftliche Kritik, 1840, p. 195 ; on the degree of fluidity to be attributed to plutonic rocks at the time of their eruption; as well as on the transformation of schist into gneiss by the action of granite, and of the substances accompanying its eruption.

([243]) p. 237.—Darwin, Volcanic Islands, 1844, pp. 49 and 154.

([244]) p. 238.—Moreau de Jonnès, Hist. phys. des Antilles, T. i. pp. 136, 138, and 543.   Humboldt, Relation historique, T. iii. p. 367.

([245]) p. 238.—Near Teguiza.   Leop. von Buch, Canarische Inseln, S. 301.

([246]) p. 238.—Leop. von Buch, Can. Inseln. S. 9.

([247]) p. 239.—Bernhard Cotta, Geognosie, 1839, S. 273.

([248]) p. 239.—Leop. von Buch, über Granit und Gneuss in den Abhandl. der Berl. Akad. aus dem J. 1842, S. 60.

([249]) p. 239.—The mural masses of granite near Lake Kolivan, divided into numerous parallel beds, contain few crystals of titanium; feldspar and albite predominate.   Humboldt, Asie centrale, T. i. p. 295; Gustav Rose, Reise nach dem Ural, Bd. i. S. 524.

([250]) p. 239.—Humboldt, Relat. hist. T. ii. p. 99.

([251]) p. 239.—See in Rose's work above cited, Vol. i. p. 584, the plan of Biri-tau, which I drew from the south side, where the Kirghis tents stood.— On spheroids of granite which separate into concentric layers, see my Relat. hist. T. ii. p. 597; and Essai géogn. sur le gisement des roches, p. 78.

([252]) p. 240.—See Humboldt, Asie centrale, T. i. pp. 299—311; and the drawings in Rose's Reise, Bd. i. S. 611: the latter shew the curvature in the layers of granite which Von Buch has pointed out as characteristic.

([253]) p. 240.—This remarkable superposition was first described by Weiss, in Karsten's Archiv für Bergbau und Huttenwesen, Bd. xvi. 1827, S. 5.

([254]) p. 240.—Dufrénoy and Elie de Beaumont, Géologie de la France, T. i. p. 130.

([255]) p. 240.—These intercalated beds of diorite form an important feature in the mining district of Naila, near Steben,—where I studied mining during the latter part of the last century, and with which my happiest youthful recollections are associated.   Compare Hoffmann, in Poggendorf's Annalen, Bd. xvi. S. 558.

([256]) p. 241.—In the southern and Bashkirian portion of the Ural.   Rose, Reise, Bd. ii. S. 171.

([257]) p. 241.—Gustav Rose, Reise nach dem Uural, Bd. ii. S. 47—52. On the identity of eleolithe and nepheline (the latter containing rather more lime), see Scheerer in Poggend. Ann. Bd. xlix. S. 359—381.

([258]) p. 245.—See Mitscherlich's admirable researches, in the Abhandl.

der Berl. Akad. for the years 1822 and 1823, S. 25—41; in Poggendorff's Annalen, Bd. x. S. 137—152; Bd. xi. S. 323—332; Bd. xli. S. 213—216 (Gustav Rose, über Bildung des Kalkspaths und Aragonits, in Poggend. Ann. Bd. xlii. S. 353—366; Haidinger, in the Transactions of the Royal Society of Edinburgh, 1827, p. 148).

(^{259}) p. 245.—Lyell, Principles of Geology, Vol. iii. pp. 353 and 359.

(^{260}) p. 247.—The view here given of the relations of position under which granite is found, expresses its general or leading character. But its aspect in some localities gives reason to believe that it was occasionally more fluid at the time of eruption: see p. 239, as well as the description of part of the Narym chain, near the frontiers of the Chinese territories, in Rose's Reise nach dem Ural, Bd. i. S. 599. Similar exceptional indications have been observed in trachyte (Dufrénoy et Elie de Beaumont, Description géologique de la France, T. i. p. 70). Having before spoken in the text of the narrow apertures through which basalts have sometimes issued, I would mention the large fissures which have afforded a passage to melaphyres, which must not be confounded with basalts. See Murchison's interesting description of a fissure 480 feet wide, through which the melaphyre has been ejected in the coal mine at Cornbrook, Hoar-Edge (Silurian System, p. 126).

(^{261}) p. 247.—Sir James Hall, in the Edinb. Trans. Vol. v. p. 43; Vol. vi. p. 71; Gregory Watt, in the Phil. Trans. for 1804, Pt. ii. p. 279; Dartigues and Fleuriau de Bellevue, in the Journ. de Phys. T. lx. p. 456; Bischoff, Wärmelehre, S. 313 and 443.

(^{262}) p. 248.—Gustav Rose, in Poggendorff's Annalen der Physik, Bd. xlii. S. 364.

(^{263}) p. 248.—On the dimorphism of sulphur, see § 55—63, in Mitscherlich's Lehrbuch der Chemie.

(^{264}) p. 248.—On gypsum considered as a uniaxal crystal, and on the sulphate of magnesia and oxides of zinc and nickel, see Mitscherlich, in Poggend. Ann. Bd. xi. S. 328.

(^{265}) p. 248.—Coste, Versuche, im Creusot über das brüchig werden des Stabeisens, in Elie de Beaumont, Mém. géol. T. ii. p. 411.

(^{266}) p. 248.—Mitscherlich, über die Ausdehnung der krystallisirten Körper durch die Wärme in Poggend. Ann. Bd. x. S. 151.

(^{267}) p. 249.—On the double system of divisional planes, see Elie de Beaumont, Géologie de la France, p. 41; Credner, Geognosie Thüringens und des Harzes, S. 40; Römer, das Rheinische Uebergangsgebirge, 1844, S. 5 und 9.

(^{268}) p. 249.—The silex is not merely coloured by oxide of iron, but is accompanied by clay, lime, and potash: Rose, Reise, Bd. ii. S. 187. On the formation of jasper by the action of dioritic porphyry, augite, and hypersthene rock, see Rose, Reise, Bd. ii. S. 169, 187, and 192. Compare also Vol. i. p. 427, containing drawings of the globes of porphyry between which jasper presents itself in the calcareous *grauwacke* of Bogoslowsk, as produced by the plutonic influence of the augitic rock, Bd. ii. S. 545. Humboldt, Asie Centrale, T. i. p. 486.

([269]) p. 249.—Rose, Reise nach dem Ural, Bd. i. S. 586—588.

([270]) p. 249.—In respect to the volcanic origin of mica, it is important to notice that crystals of mica are found, in the basalt of the Bohemian Mittel-gebirge,—in the lava of Vesuvius of 1822 (Monticelli, Storia del Vesuvio negli anni 1821 e 1822, § 99),—and in the fragments of argillaceous schist imbedded in scoriaceous basalt on the Hohenfels, not far from Gerolstein, in the Eifel (see Mitscherlich, in Leonhard, Basalt-Gebilde, S. 244). On the production of feldspar in argillaceous schist by the contact of porphyry between Urval and Poïet (Forez), see Dufrénoy, in Géol. de la France, t. i. p. 137. It is probably owing to a similar cause that certain schists, which I had an opportunity of seeing near Paimpol, in Brittany, during a geological pedestrian excursion with Professor Kunth through that interesting country, derive their amygdaloïdal and cellular character (T. i. p. 234, of the same work).

([271]) p. 249.—Leopold von Buch, in the Abhandlungen der Akad. der Wissensch. zu Berlin aus dem J. 1842, S. 63 ; and in the Jahrbüchern für wissenschaftliche Kritik, Jahrg. 1840, S. 196.

([272]) p. 249.—Elie de Beaumont, in the Annales des Sciences naturelles, T. xv. p. 362—372: " En se rapprochant des masses primitives du Mont Rose et des montagnes situées à l'ouest de Coni, on voit les couches secondaires perdre de plus en plus les caractères inhérents à leur mode de dépôt. Souvent alors elles en prennent qui semblent provenir d'une toute autre cause, sans perdre pour cela leur stratification, rappelant par cette disposition la structure physique d'un tison à moitié charbonné dans lequel on peut suivre les traces des fibres ligneuses, bien au-delà des points qui présentent encore les caractères naturels du bois." Compare also Annales des Sciences naturelles, T. xiv. p. 118—122, and H. von Dechen, Geognosie, S. 553. We may reckon among the most striking proofs of the transformation of rocks by plutonic action the belemnites in the schists of Nuffenen, (in the Alpine valley of Eginen, and near the Gries-glacier), and in the belemnites which M. Charpentier found in what was called primitive limestone on the western descent of the Col de la Seigne, between the Enclove de Montjovet and the *chálet* of La Lanchette, aud which he shewed to me at Bex, in the autumn of 1822. (Annales de Chimie, T. xxiii. p. 262.)

(273) p. 250.—Hoffman, in Poggend. Annalen, Bd. xvi. S. 552. "Strata of transition argillaceous schist in the Fichtelgebirge, which can be traced for a distance of sixteen miles, are altered into gneiss only at the two extremities, where they come into contact with granite. We are there able to trace the gradual formation of the gneiss, and the development of the mica and of the feldspathic amygdaloids, in the interior of the mass of schist, which, indeed, contains in itself almost all the elements of those minerals.

(274) p. 250.—Among the works of art which have come down to us from Greek and Roman antiquity, we remark the absence of columns or large vases of jasper, and, at the present time, large blocks of jasper are obtained only from the Ural mountains. The material worked under the name of jasper in part of the Altai mountains (Revennaia Sopka), is a superb ribboned porphyry. The word jasper belongs to the Semitic languages, and, according to the confused descriptions of Theophrastus (De Lap. 23 and 27) and of Pliny (xxxvii. 8 and 9), who enumerate jasper among "opaque gems," the name appears to have been given to fragments of *Jaspachat*, and to a sub-stance which the ancients called *jasponyx*, and which we call opal-jasper. Pliny considered a piece of jasper of the size of eleven inches so remarkable as to deserve his mentioning that he had himself actually seen so great a rarity:—"Magnitudinem jaspidis undecim unciarum vidimus, formatamque inde effigiem Neronis thoracatam." According to Theophrastus, the stone which he calls *smaragd*, or emerald, and from which large obelisks were cut, would be merely an imperfect jasper.

(275) p. 250.—Humboldt, Lettre à M. Brochant de Villiers, in the Annales de Chimie et de Physique, T. xxiii. p. 261; Leop. von Buch, Geogn. Briefe über das südliche Tyrol, S. 101, 105, and 273.

(276) p. 250.—On the transformation of compact into granular limestone by the action of granite in the Pyrenees, at the Montagnes de Rancie, see Dufrénoy, in the Mémoires géologiques, T. ii. p. 440;—in the Montagnes de l'Oisans, Elie de Beaumont, Mem. Géol. T. ii. p. 130: on a similar effect by the action of dioritic and pyroxenic porphyry (*ophyte*; Elie de Beaumont, Géol. de la France, T. i. p. 72) between Tolosa and St. Sebastian, Dufrénoy, in the Mém. géol. t. ii. p. 130;—and by syenite in the Isle of Skye, the fossils in the altered limestone being still distinguishable (Von

Dechen, Geognosie, S. 573). In the alteration of chalk by contact with basalt, the displacement of the particles in the process of crystallization or granulation is the more remarkable, because Ehrenberg's microscopic investigations have shewn that, in the unaltered state, the particles of chalk form an infinity of small separate rings (Poggend. Ann. Bd. xxxix. S. 105.) See also on the rings in arragonite deposited by solutions, Gustav Rose, Ann. Bd. xlii. S. 354.

(²⁷⁷) p. 251.—Beds of granular limestone in the granite at the Port d'Oo and on the Mont de Labourd, Charpentier, Constitution géol. des Pyrénées, pp. 144, 146.

(²⁷⁸) p. 251.—Leop. von Buch, Desc. des Canaries, p. 394; Fiedler, Reise durch das Königreich Griechenland, Th. II. S. 181, 190, and 516.

(²⁷⁹) p. 251.—I have before alluded to this remarkable passage in Origen's Philosophumena, cap. 14 (Opera, ed. Delarue, T. i. p. 893.) The whole context renders it extremely improbable that Xenophanes meant an impression of a laurel (τύπον δάφνης), instead of an impression of a fish (τύπον ἀφύης). Delarue appears to me to be wrong in blaming Jacob Gronovius for substituting the latter reading.

(²⁸⁰) p. 251.—On the geological character of the environs of Carrara ("the city of the moon," Strabo, lib. v. p. 222), see Savi, Osservazioni sui terreni antichi Toscani, in the Nuovo Giornale de' Letterati di Pisa, No. 63; and Hoffmann, in Karsten's Archiv für Mineralogie, Bd. vi. S. 258—263, and also his Geogn. Reise durch Italien, S. 244—265.

(²⁸¹) p. 252. — This hypothesis is that of an excellent and experienced observer, Karl von Leonhard: see his Jahrbuch für Mineralogie, 1834, S. 329; and Bernhard Cotta, Geognosie, S. 310.

(²⁸²) p. 252.—Leop. von Buch, Geognostische Briefe an Alex. von Humboldt, 1824, S. 36 and 82; also in the Annales de Chimie, T. xxiii. p. 276, and in the Abhandl. der Berliner Akad. aus den J. 1822 und 1823, S. 83—136; H. von Dechen, Geognosie, S. 574—576.

(²⁸³) p. 254.—Hoffmann, Geogn. Reise bearbeitet von H. von Dechen, S. 113—119, 380—386; Poggend. Ann. der Physik, Bd. xxvi. S. 41.

(²⁸⁴) p. 254.—Dufrénoy, in Mémoires géologiques, T. ii. pp. 145 and 179.

(²⁸⁵) p. 254.—Humboldt, Essai geogn. sur le Gisement des Roches, p. 93; Asie centrale, T. iii. p. 532.

(²⁸⁶) p. 255.—Elie de Beaumont, in the Annales des Sciences naturelles, T. xv. p. 362; Murchison, Silurian System, p. 286.

[287] p. 255.—Rose, Reise nach dem Ural, Bd. i. S. 364 and 367.

[288] p. 255.—Leop. von Buch, Briefe, S. 109—129. Compare also Elie de Beaumont on the contact of granite with the beds of the Jura (Mém. géol. T. ii. p. 408).

[289] p. 255.—Hoffmann, Reise, S. 30 and 37.

[290] p. 255.—On the chemical process in the formation of specular iron, see Gay-Lussac, in the Annales de Chemie, T. xxii. p. 415 ; and Mitscherlich, in Poggend. Ann. Bd. xv. p. 630. Crystals of olivine have been formed (probably from sublimation) in the cavities of the obsidian which I brought from the Cerro del Jacal, in Mexico (Gustav Rose, in Poggend. Ann. Bd. x. S. 323). We thus find olivine in basalt, in lava, in obsidian, in artificial scoriæ, in meteoric stones, in the syenite of Elfdale, and (under the name of hyalosiderite) in the *wacke* of the Kaiserstuhl.

[291] p. 256.—Constantin von Beust über die Porphyrgebilde, 1835, S. 89 —96 ; his Beleuchtung der Werner'schen Gangtheorie, 1840, S. 6 ; C. von. Weissenbach, Abbildungen merkwürdiger Gangverhältnisse, 1836, fig. 12. The structure in narrow bands is not however general, nor does the order of succession of the different members of these masses necessarily indicate their relative age : see Freiesleben, über die sächsischen Erzgänge, 1843, S. 10—12.

[292] p. 256.—Mitscherlich über die kunstliche Darstellung der Mineralien, in the Abhandl. der Akad. der Wiss. zu Berlin, 1822—23, S. 25—41.

[293] p. 257.—Of minerals accidentally produced in the scoriæ of artificial works, crystals of feldspar have been discovered by Heine in a furnace for fusing copper, and have been analysed by Kersten (Poggend. Ann. Bd. xxxiii. S. 337) ; crystals of augite in scoriæ at Sahl (Mitscherlich, in the Abhandl. der Akad. zu Berlin, 1822—23, S. 40) ; crystals of olivine under similar circumstances (Sefström, in Leonhard's "Basalt Gebilde," Bd. ii. S. 495) ; of mica in old scoriæ of Schloss Garpenberg (Mitscherlich, in Leonhard, S. 506) ; crystals of magnetic oxide of iron in the scoriæ of Chatillon sur Seine (Leonhard, S. 441) ; specular iron produced in potters' clay (Mitscherlich, in Leonhard, S. 234).

[294] p. 257.—Of minerals purposely produced, there have been idocrase and garnet (Mitscherlich, in Poggendorff's Annalen der Physik, Bd. xxxiii. S. 340) ; ruby (Gaudin, in the Comptes rendus de l'Académie des Sciences, T. iv. P. 1, p. 999) ; olivine and augite (Mitscherlich and Berthier, in the Annales de Chimie et de Physique, T. xxiv. p. 376). Although augite and hornblende present, according to G. Rose, the greatest similarity in the form of their crystals, and are almost identical in their chemical composition, yet

hornblende has never been met with in scoriæ accompanied by augite, nor have chemists ever succeeded in reproducing either hornblende or feldspar (Mitscherlich, Poggend. Ann. Bd. xxxiii. p. 340, and Rose, Reise nach dem Ural, Bd. ii. S. 358 and 363).   Compare also Beudant, in the Mem. de l'Acad. des Sciences, T. viii. p. 221, and Becquerel's ingenious investigations in his Traité de l'Electricité, T. i. p. 334; T. iii. p. 218; T. v. 1, p. 148 and 185.

(295) p. 257.—D'Aubuisson, in the Journal de Physique, T. lxviii. p. 128.

(296) p. 258.—Leop. von Buch, Geognost. Briefe, S. 75—82; where it is also shewn why the red sandstone (the todtliegende of the floetz strata of Thuringia) and the coal measures should be regarded as produced by the eruption of porphyritic rocks.

(297) p. 260.—This discovery was made by Miss Mary Anning, who was also the discoverer of the coprolites of fishes, which, as well as those of the Ichthyosaurus, have been found in such abundance in England (near Lyme Regis particularly), that Buckland compares them to beds of pototoes. (Buckland's Geology with reference to Natural Theology, Vol. i. p. 188—202, and 305).   Respecting Hooke's hope " to raise a chronology" from the study of broken and fossilised shells, "and to state the intervals of time wherein such or such catastrophes and mutations have happened," see his Posthumous Works, Lecture, Feb. 29, 1688.

(298) p. 261.—Leop. von Buch, in Abhandlungen der Akad. der Wiss. zu Berlin aus dem J. 1837, S. 64.

(299) p. 262.—The same author's Gebirgsformationen von Russland, 1840, S. 24—40).

(300) p. 262.—Agassiz, Monographie des Poissons fossiles du vieux Grès Rouge, p. vi. and 4.

(301) p. 262.—Leop. von Buch, in Abhandl. der Berl. Akad. 1838, S. 149—168; Beyrich, Beitr. zur Kenntniss des Rheinischen Uebergangs-gebirges, 1837, S. 45.

(302) p. 262.—Agassiz, Recherches sur les Poissons fossiles, T. i. Introd. p. 18 (Davy, Consolations in Travel, Dial. iii.)

(303) p. 263.—According to Hermann von Meyer, it would be a Proto-saurus.   The case of the rib of a saurian, supposed to have been found in the mountain limestone of Northumberland, is considered by Lyell extremely doubtful (Lyell's Geology, 1832, Vol. i. p. 148).   The discoverer himself referred it to the alluvial strata which cover the mountain limestone.

(304) p. 263.—F. von Alberti, Monographie des Bunten Sandsteins, Mus-chelkalks und Keupers, 1834, S. 119 und 314.

(305) p. 263.—See Hermann von Meyer's ingenious considerations on the organisation of the flying saurians (Palæologica, S. 228—252). In the fossil Pterodactylus crassirostris, which, as well as the longer known P. longirostris (Ornithocephalus of Sömmering), was found at Solenhofen, in the lithographic slate of the upper jurassic formation or oolite, Professor Goldfuss has even discovered traces of the membranous wing "with the impressions of tufts of reversed hair, some of the hair being even an inch long."

(306) p. 264.—Cuvier, Recherches sur les Ossemens fossiles, T. i. p. 52—57. See also the geological scale of epochs in Phillips's Geology, 1837, p.166—185.

(307) p. 265.—Agassiz, Poissons fossiles, T. i. p. 30, and T. iii. p. 1—52; Buckland, Geology, Vol. i. p. 273—277.

(308) p. 265.—Ehrenberg, über noch jetzt lebende Thierarten der Kreide-bildung in den Abhandl. der Berliner Akad. aus dem J. 1839, S. 164.

(309) p. 265.—Valenciennes, in the Comptes rendus de l'Acad. des Sciences, T. vii. 1838, pt. ii. p. 580.

(310) p. 265.—In the Weald-Clay ; Beudant, Géologie, p. 173. The number of ornitholites increases in the gypsum of the tertiary formations. (Cuvier, Ossemens fossiles, T. iii. p. 302—328).

(311) p. 266.—Leop. von Buch, in the Abhandl. der Berl. Akad. 1830, S. 135—187.

(312) p. 266.—Quenstedt, Flözgebirge Wurtembergs, 1843, S. 135.

(313) p. 267.—The same work, S. 13.

(314) p. 267.—Murchison makes two divisions of the *bunter sandstein*,— the upper being the trias of Alberti, whilst of the lower division (to which Elie de Beaumont's Vosges sandstone belongs) of the zechstein, and of the todtliegende, he forms his Permian system. He makes the secondary formations begin with the upper trias, *i. e.* with the upper division of the German bunten sandstein. The Permian system, the carboniferous or mountain lime-stone, and the devonian and silurian strata, constitute his *palæozoic* strata. In this system the chalk and the oolites are the upper, and the keuper, the muschelkalk, and the bunter sandstein, are the lower secondary formations ; the Permian system and the carboniferous limestone are the upper, and the devonian and silurian strata are the lower palæozoic formations. The bases of this general classification are developed in the great work in which the indefatigable British geologist designs to describe the geology of a large portion of Eastern Europe.

(315) p. 268.—Cuvier, Ossemens fossiles, 1821, T. i. p. 157, 261, and 264 ; Humboldt, über die Hochebene von Bogota in der Deutschen Viertcljahrs-Schrift, 1839, Bd. i. S. 117.

(³¹⁶) p. 268.—Journal of the Asiatic Society, 1844, No. xv. p. 109.

(³¹⁷) p. 268.—Beyrich, in Karsten's Archiv für Mineralogie, 1844, Bd. xviii. S. 218.

(³¹⁸) p. 269.—By the valuable labours of Count Sternberg, Adolphe Brongniart, Göppert, and Lindley.

(³¹⁹) p. 269.—R. Brown's Botany of Congo, p. 42; and D'Urville, in the memoir entitled De la distribution des Fougères sur la surface du globe terrestre.

(³²⁰) p. 269.—Such are the cycadeæ discovered by Count Sternberg in the old carboniferous formation at Radnitz in Bohemia, and described by Corda. (Two species of *cycadites* and *zamites Cordai*, see Göppert, fossile Cycadeen in den Arbeiten der Schles. Gesellschaft, für vaterl. Cultur im J. 1843, S. 33, 37, 40, and 50). A cycadea (Pterophyllum gonorrhachis, Göp.) has also been found in a coal bed at Königshütte, in Upper Silesia.

(³²¹) p. 269.—Lindley, Fossil Flora, No. xv. p. 163.

(³²²) p. 269.—Fossil Coniferæ, in Buckland, Geology, p. 483—490. Witham has the great merit of having been the first to recognise the existence of coniferæ in the primitive vegetation of the carboniferous period; before his time, almost all the trunks of trees found in this formation were considered to be palms. The species of the genus Araucaria are not peculiar to the coal formations of the British islands; they are also found in Upper Silesia.

(³²³) p. 270.—Adolphe Brongniart, Prodrome d'une Hist. des Végétaux fossiles, p. 179; Buckland, Geology, p. 479; Endlicher and Unger, Grundzüge der Botanik, 1843, S. 455.

(³²⁴) p. 270.—" By means of Lepidodendron a better passage is established from Flowering to Flowerless Plants, than by either Equisetum or Cycas, or any other known genus."—Lindley and Hutton, Fossil Flora, Vol. ii. p. 53.

(³²⁵) p. 270.—Kunth, Anordnung der Pflanzenfamilien, Handb.der Botanik, S. 307 and 314.

(³²⁶) p. 270.—A very striking proof of coal having been formed (not by the action of fire in charring vegetable vessels, but more probably) by decomposition under the influence of sulphuric acid, is afforded, as Göppert has acutely remarked (Karsten, Archiv für Mineralogie, Bd. xviii. S. 530) by the conversion of a piece of the amber tree into black coal; the coal and the unaltered amber being found close together. On the part which the smaller vegetation may have had in the formation of beds of coal, see Link, in the Abhandl. der Berliner Akademie der Wissenschaften, 1838, S. 38.

(³²⁷) p. 271.—See the accurate investigations of Chevandier, in the Comptes rendus de l'Acad. des Sciences, 1844, T. xviii. P. i. p. 285. In comparing this bed of carbonaceous matter, 0·6 in. in thickness, with beds of coal, the enormous pressure to which the latter have been subjected, and which shews itself in the flattened form of the trunks of trees which they contain, should also be remembered. The "wood-hills" seen on the south shore of the island of New Siberia, discovered in 1806 by Sirowatskoi, were described by Hedenström as 30 fathoms in height, consisting of alternate layers of sandstone and bituminous trunks of trees, of which those at the summit were in a vertical position. The bed of driftwood is visible for seven wersts. (Wrangel, Reise längs der Nordküste von Siberien in den Jahren 1820—1824, Th. i. S. 102; or p. 486 of the second edit. of the English translation).

([328]) p. 272.—This *corypha* is the soyate (in aztec zoyatl), or the *Palma dulce* of the natives; see Humboldt and Bonpland, Synopsis Plant. Æquinoct. Orbis Novi, T. i. p. 302. A writer deeply versed in the American languages, Professor Buschmann, remarks, that the Palma soyate is so designated in Yepe's "Vocabulario de la Lengua Othomi," and that the aztec word zoyatl (Molina's Vocabulario en Lengua Mexicana y Castellana, p. 25) recurs in names of places, such as Zoyatitlan and Zoyapanco, near Chiapa.

([329]) p. 272.—At Baracoa and Cayos de Moya; see the Admiral's journal of November 25 and 27, 1492, and Humboldt, Examen critique de l'Hist. de la Geogr. du Nouveau Continent, T. ii. p. 252, and T. iii. p. 23. Columbus's attention was so unremittingly alive to all natural objects, that he even distinguished (and was indeed the first who did so) between *Podocarpus* and *Pinus*: "I find," said he, "en la tierra aspera del Cibao pinos que no llevan pinas (fir-cones), pero portal orden compuestos por naturaleza, que (los frutos) parecen azeytunas del Axarafe de Sevilla." When the great botanist Richard published his admirable memoir on Cycadeæ and Coniferæ, he could scarcely have imagined that before the time of L'Heritier, and even before the end of the 15th century, Podocarpus had been separated from the Abietineæ by a navigator of the 15th century.

([330]) p. 272.—Charles Darwin, Journal of the Voyages of the Adventure and Beagle, 1839, p. 271.

([331]) p. 273.—Göppert describes three other cycadeæ found in schistose clay with brown coal of Altsattel and Commotau in Bohemia. They may possibly belong to the Eocene period. (Page 61 of the work quoted in Note 320).

([332]) p. 273.—Buckland, Geology, p. 509.

([333]) p. 274.—Leopold von Buch, in the Abhandl. der Akad. der Wiss. zu Berlin aus den J. 1814—1815, S. 161, and in Poggendorff's Annalen, Bd. ix. S. 575; Elie de Beaumont, in the Annales des Sciences Nat. T. xix. p. 60.

(³³⁴) p. 275.—Compare Elie de Beaumont, Descr. géol. de la France, T. i. p. 65; Beudant, Géologie, 1844, p. 209.

(³³⁵) p. 279.—Transactions of the Cambridge Philosophical Society, Vol. vi. Pt. 2, 1837, p. 297. According to other writers, the ratio is 100 : 284.

(³³⁶) p. 280.—It was a prevalent opinion in the middle ages, that only a seventh part of the surface of the earth was covered by sea. Cardinal d'Ailly grounded this opinion on the 4th Apocryphal Book of Esdras, Imago Muudi, cap. viii. Columbus, who took his cosmological views from the works of the Cardinal, was greatly interested in maintaining this supposition of the smallness of the sea, to which the misunderstood expression of "ocean river" also contributed. Compare Humboldt, Examen critique de l'Hist. de la Géographie, T. i. p. 186.

(³³⁷) p. 281.—Agathemeros, in Hudson, Geographi minores, T. ii. p. 4 ; Humboldt, Asie centr. T. i. p. 120—125.

(³³⁸) p. 281.—Strabo, lib. i. p. 65, Casaub. Compare Humboldt, Examen crit. T. i. p. 152.

(³³⁹) p. 282.—On the mean latitude of the north coast of Asia, and on the true denomination of Cape Taimura (Cape Siewero-Wostotschnoi), and Cape North-East (Schalagskoi Mys), see Humboldt, Asie centr. T. iii. p. 35 and 37.

(³⁴⁰) p. 282.—T. i. p. 198—200 of the same work. The southern point of America, and the Archipelago which we term Terra del Fuego, are in the meridian of the most northern part of Baffin's Bay, and of the great polar land, the limits of which are still undetermined, and which possibly forms a part of West Greenland.

(³⁴¹) p. 283.—Strabo, lib. ii. p. 92 and 108, Casaub.

(³⁴²) p. 283.—Humboldt, Asie centrale, T. iii. p. 25. As early as 1817, in my work entitled "De distributione geographica plantarum secundum cœli temperiem et altitudinem montium," I called attention to the important distinction between compact and intersected continents, as bearing on climatology and on human civilisation : " Regiones vel per sinus lunatos in longa cornua porrectæ, angulosis littorum recessibus quasi membratim discerptæ, vel spatia patentia in immensum, quorum littora nullis incisa angulis ambit sine anfractu Oceanus" (p. 81 and 182). On the proportion of the length of coast line to the area of a continent (in some degree as a measure of the accessibility of the interior), see Berghaus, Annalen der Erdkunde, Bd. xii. 1835, S. 490, and Physikal. Atlas, 1839, No. iii S. 69.

(³⁴³) p. 283.—Strabo, lib. ii. p. 126, Casaub.

(³⁴⁴) p. 283.—Pliny, in speaking of Africa, said (v. 1), " Nec alia pars terrarum pauciores recipit sinus." The Indian peninsula on this side the Ganges, presents a third, smaller but very similar, triangular form. The idea was very prevalent amongst the Greeks of the existence of a certain *regularity* in the configuration of the shape of the dry land. They considered that there were four great gulphs, among which the Persian gulph was the opposite or counterpart of the Caspian or Hyrcanian Sea, (Arrian, vii. 16 ; Plut. in vita Alexandri, cap. 44 ; Dionys. Perieg. v. 48 and 630, pp. 11 and 38, Bernh.) ; and, according to the fantastical conception of Agesianax, the four gulphs and isthmuses were even supposed to recur on the Moon's disk, (Plut. de Facie in Orbe Lunæ, p. 921, 19), as a sort of reflection of the great outlines of the terrestrial surface. Respecting the division of the Earth into four quarters or continents, two north and two south of the Equator, see Macrobius, Comm. in Somnium Scipionis, ii. 9. I have submitted this part of ancient geography, respecting which great confusion of ideas has prevailed, to a new and careful investigation in my Examen crit. de l'hist. de la Géogr. T. i. pp. 119, 145, 180—185, as well as in my Asie Centr. T. ii. pp. 172—178.

(³⁴⁵) p. 283.—Fleurieu, in the " Voyage de Marchand autour du Monde," T. iv. pp. 38—42.

(³⁴⁶) p. 283.—Humboldt, in the Journal de Physique, T. liii. 1799, p. 33 ; and Rel. hist. T. ii. p. 19, T. iii. pp. 189 and 198.

(³⁴⁷) p. 284.—Humboldt, in Poggendorff's Annalen der Physik, Bd. xl. S. 171. On the labyrinth of fiords at the south-east end of the American continent, see Darwin's Journal, (Narrative of the Voyages of the Adventure and Beagle, Vol. iii.) 1839, p. 266. The parallelism of the two chains of mountains is maintained from 5° N. to 5° S. lat. The change of direction of the coast near Arica appears to be the consequence of an analogous change in the direction of the immense fissure over which the Cordillera of the Andes has been upheaved.

(³⁴⁸) p. 286.—De la Beche, Sections and Views illustrative of Geological Phenomena, 1830, Tab. 40 ; Charles Babbage, Observations on the Temple of Serapis at Pozzuoli, near Naples, and on certain causes which may produce Geological Cycles of great extent, 1834. " If a stratum of sandstone, five English miles in thickness, should have its temperature raised 100° Fah., its surface would rise 25 feet ; heated beds of clay would, on the contrary, cause a sinking of the ground by their contraction." Compare, in Bischof's Wärmelehre des Innern unseres Erdkörpers, S. 303, his calculations on the secular elevation of Sweden, on the supposition of the small increase of 3° of

Reaumur (6·°8 Fah.) in a stratum 140000 French feet in thickness, and heated to a state of fusion.

([349]) p. 286.—"The assumption, which has hitherto appeared so secure, of the invariability of the force of gravity at any given point of the Earth's surface, has become subject to some degree of uncertainty, since we have become aware of the slow elevation of large districts," (Bessel über Maass und Gewicht, in Schumacher's Jahrbuch für 1840, S. 134).

([350]) p. 287.—Th. ii. (1810), S. 389. Compare Hallström, in Kongl. Vetenskaps-Academiens Handlingar (Stockh.), 1823, p. 30; Lyell, in the Phil. Trans. for 1835, p. 1; Blom (Amtmann in Budskerud), Stat. Beschr. von Norwegen, 1843, S. 89—116. Previous to von Buch's publication of his Scandinavian journey, but not previous to the journey itself, Playfair surmised, in 1802, in the Illustrations of the Huttonian Theory, § 393, that it was the mainland of Sweden that was rising, not the sea that was sinking; and Keilhau has remarked (Om Lanjordens Stigning in Norge in dem Nyt Magazin for Naturvidenskaberne), that the Dane Jessen had expressed such a conjecture at a still earlier period. Their writings were, however, entirely unknown to the great German geologist; nor have they, so far as I am aware, influenced the progress of physical geography on this point. In a work entitled Kongeriget Norge fremstillet efter dets naturlige og borgerlige Tilstand, Kjöbenh. 1763, Jessen has attempted to investigate the changes which have taken place in the relative level of the coast and the sea, taking for bases the early determinations of Celsius, Kalm, and Dalin. He shews some confusion of ideas as to the possibility of increase by internal growth of the rocks which form the coast, but at last declares himself in favour of the elevation of the land as a consequence of earthquakes. " Although," he says, " immediately after the earthquake at Egersund no such elevation was perceived, yet the earthquake may have opened the way for the action of other causes."

([351]) p. 287.—Berzelius, Jahresbericht über die Fortschritte der physischen Wiss. No. 18, S. 686. The islands of Bornholm, and of Saltholm opposite to Copenhagen, rise, however, very little. Bornholm hardly rises one foot in a century. See Forchhammer, Phil. Mag. Series iii. Vol. ii. p. 309.

([352]) p. 287.—Keilhau, in Nyt Mag. for Naturvid. 1832, Bd. i. pp. 105—254, Bd. ii. p. 57; Bravais, " sur les lignes d'ancien niveau de la Mer," 1843, pp. 15—40. Compare also Darwin on the Parallel Roads of Glen-Roy and Lochaber, Phil. Trans. 1839, p. 60.

([353]) p. 288.—Humboldt, Asie centrale, T. ii. pp. 319—324, T. iii. pp. 549—551. The depression of the Dead Sea has been successively investigated by the barometric measurements of Count Bertou, the far more careful ones of Russegger, and by the trigonometrical measurements of Lieutenant Symond, of the British Navy. By a letter from Mr. Alderson to the Geographical Society of London, communicated to me by my friend Captain Washington, the measurement of Lieutenant Symond gave the level of the Dead Sea at 1506 French feet (1605 English feet) below the highest house in Jaffa. Mr. Alderson, at that time (November 28, 1841), considered the Dead Sea to be about 1314 feet (1400 English) below the level of the Mediterranean. In a more recent communication of Lieutenant Symond, (Jameson's Edin. New Philos. Journal, Vol. xxxiv. 1843, p. 178), he gives 1231 feet (1312 English) as the final result of two very accordant trigonometric operations.

[354] p. 288.—Sur la Mobilité du fond de la Mer Caspienne, in my Asie centr. T. ii. pp. 283—294. At my request, the Imperial Academy of Sciences at St. Petersburg, charged the learned physicist, Lenz, with fixing solid and well-secured marks on the peninsula of Abscheron, near Baku, for the purpose of shewing the mean level of the water at a given epoch. In a similar manner I requested and obtained the insertion, in the supplementary instructions given to Captain James Ross for the Antarctic expedition, of a direction to fix marks, at suitable places, on the cliffs and rocks of the Antarctic Seas, as has been done in Sweden and on the Caspian. If similar measures had been taken in Cook's and Bougainville's earliest voyages, we should be now in possession of the necessary data for determining whether secular variation in the relative level of land and sea is a general or a merely local phænomenon, and whether any law is discoverable in the direction of the points which rise or sink simultaneously.

[355] p. 288.—On the sinking and rising of the bottom of the sea in the Pacific, and the different areas of alternate movements, see Darwin's Journal, pp. 557 and 561—566.

[356] p. 291.—Humboldt, Rel. hist. T. iii. pp. 232—234. Compare also some ingenious remarks on the form of the Earth and the position of its lines of elevation, in Albrechts von Roon Grundzügen der Erd-Völker-und Staaten-kunde, Abth. i. 1837, S. 158, 270, and 276.

[357] p. 292—Leop. von Buch über die geognostischen Systeme von Deutschland in his Geogn. Briefen an Alexander von Humboldt, 1824, pp. 265—271; and Elie de Beaumont, Recherches sur les Révolutions de la Surface du Globe, 1829, pp. 297—307.

[358] p. 292.—Humboldt, Asie centrale, T. i. pp. 277—283; Essai sur le Gisement des Roches, 1822, p. 57; and Relat. hist. T. iii. pp. 244—250.

(359) p. 292.—Asie centrale, T. i. pp. 284—286. The Adriatic also follows a south-east and a north-west direction.

(360) p. 293.—De la hauteur moyenne des continents in Asie centrale, T. i. pp. 82—90, and 165—189. The results which I have obtained are to be regarded as the extreme values or "nombres-limites." Laplace's estimation of 3078 French feet as the mean height of continents, is at least three times too great. The illustrious geometer was conducted to this erroneous result by hypotheses as to the mean depth of the sea, (Mécanique Céleste, T. v. p. 14). I have shown, in my Asie centrale, T. i. p. 93, that the mathematicians of the Alexandrian school supposed the depth of the sea to be determined by the height of the mountains, (Plut. in Æmilio Paulo, cap. 15). The height of the centre of gravity of the volume of the continental masses probably undergoes small alterations in the course of many centuries.

(361) p. 294.—Zweiter geologischer Brief von Elie de Beaumont and Alexander von Humboldt, in Poggendorff's Annalen, Bd. xxv. S. 1—58.

(362) p. 295.—Humboldt, Relation hist. T. iii. chap. xxix. pp. 514—530.

(363) p. 296.—See the series of observations made by me in the Pacific, from 0° 5′ to 13° 16′ N. lat. (Asie centr. T. iii. p. 354).

(364) p. 296.—" On pourra (par la température de l'Océan sous les tropiques) attaquer avec succès une question capitale restée jusqu'ici indécise, la question de la constance des températures terrestres, sans avoir à s'inquiéter des influences locales naturellement fort circonscrites, provenant du déboisement des plaines et des montagnes, du dessêchement des lacs et des marais. Chaque siècle, en léguant aux siècles futurs quelques chiffres bien faciles à obtenir, leur donnera le moyen peut-être le plus simple, le plus exact et le plus direct, de décider si le soleil, aujourd'hui source première, à peu près exclusive, de la chaleur de notre globe, change de constitution physique et d'éclat, comme la plupart des étoiles, ou si au contraire cet astre est arrivé à un état permanent," (Arago, in the Comptes rendus des séances de l'Acad. des Sciences, T. xi. P. 2, p. 309).

(365) p. 297.—Humboldt, Asie centr. T. ii. pp. 321 and 327.

(366) p. 297.—See the numerical results in pp. 328—333 of the volume just named. The geodesical levelling which, at my request, my friend, General Bolivar, caused to be executed in 1828 and 1829, by Lloyd and Falmarc, has shewn that the Pacific is, at the utmost, 3·4 French feet higher than the Caribbean Sea, and even that, at different hours of the day, each of the two seas is in turn the highest, according to their respective hours of ebb and flood. As the levelling extended over a distance of 64 miles, and

comprised 933 stations, an error of three feet would not be at all surprising ; and we may consider this result as rather tending to confirm the equilibrium of the waters which communicate round Cape Horn, (Arago, Annuaire du Bureau des Longitudes pour 1831, p. 319). I had inferred, from my barometric observations in 1799 and 1804, that, if there were any difference of level between the two oceans, it could not exceed 3 metres, (Rel. hist. T. iii. pp. 555—557 ; and Annales de Chimie, T. i. pp. 55—64.) The measurements which seem to establish a higher level for the waters of the gulph of Mexico, and for those of the Adriatic, (in the latter case, by the combination of the trigonometrical operations of Delcros and Choppin with those of the Swiss and Austrian engineers), appear to me to be subject to many doubts. Notwithstanding the form of the Adriatic, it is scarcely probable that the level of its waters at the upper end should be almost 26 French feet higher than that of the Mediterranean at Marseilles, or 23.4 French feet higher than the level of the Atlantic. (See my Asie centr. T. ii. p. 332).

[367] p. 298.—Bessel über Fluth und Ebbe, in Schumacher's Jahrbuch für 1838, S. 225.

[368] p. 299.—The density of sea water depends concurrently on the temperature and on the degree of saltness—a consideration which has not been sufficiently attended to in investigations into the cause of currents. The submarine current which brings towards the Equator the cold water of the poles, would follow an exactly opposite course, if difference in respect to saltness were alone concerned. In this point of view, the geographical distribution of temperature and of density in the waters of the ocean is of great importance. The numerous observations obtained by Lenz, (Poggendorff's Annalen, Bd. xx. 1830, S. 129), and by Captain Beechey, (Voyage to the Pacific, Vol. ii. p. 727), are deserving of particular attention. Compare also Humboldt, Relat. hist. T. i. p. 74 ; and Asie centrale, T. iii. p. 356.

[369] p. 300.—Humboldt, Relat. hist. T. i. p. 64 ; Nouvelles Annales des Voyages, 1839, p. 255.

[370] p. 300.—Humboldt, Examen crit. de l'hist. de la Géogr. T. iii. p. 100. Columbus adds shortly after, (Navarrete, Coleccion de los viages y descubrimientos de los Españoles, T. i. p. 260), " that the movement is strongest in the Caribbean Sea." In fact, Rennell terms this region, " not a current, but a sea in motion."

[371] p. 300.—Humboldt, Examen crit. T. ii. p. 250 ; Relat. hist. T. i. pp. 66—74.

[372] p. 300.—Martyr de Angleria, De Rebus Oceanicis et Orbe Novo, Bas.

1523, Dec. iii. Lib. vi. p. 57.   See Humboldt, Examen critique, T. ii. pp. 254—257, and T. iii. p. 108.

(373) p. 301.—Humboldt, Examen crit. T. iii. pp. 64—109.

[M. de Humboldt has not noticed the important and philosophical classification of currents, according to their origin, into *drift* and *stream* currents, introduced by the late Major Rennell, in his Investigation of the Currents of the Atlantic Ocean, p. 21, Lond. 1832; a distinction most necessary to be borne in mind, when inquiring into or discussing the history of any particular current.   According to this classification, a *drift* current is the effect of a constant or of a very prevalent wind on the surface water of the ocean, impelling it to leeward until it meets with some obstacle which stops it, and occasions an accumulation; the accumulation giving rise to a *stream* current: the obstacle may be either land or banks, or a stream current already formed.   A *stream* current is the flowing off of the accumulated waters of a drift current, caused by the effort of the water to restore the equilibrium of the general level surface of the ocean.   A stream current may be of any bulk, or depth, or velocity; a drift current is shallow, and rarely exceeds in velocity the rate of half a mile an hour.   When drift currents are opposed by a stream already formed, they either fall into it and augment it, if the angle which their direction makes with that of the stream current be less than a right angle; or if it be greater, the drift current itself forms a stream current, which takes a parallel but opposite course to that of the stream by which its progress has been stopped. Thus the equatorial current referred to in the text, which flows from the accumulated waters in the bend of the western coast of Africa near the Bight of Benin, is banked up on its southern side, in its passage across the Atlantic, by the drift current impelled by the S.E. trade-wind, which current falls into and continually augments the equatorial stream, maintaining it in such strength that, on its arrival on the Brazilian coast, it is able to furnish the two great branches into which it is divided by Cape St. Roque, the N.E. promontory of South America, one branch pursuing its course to the Caribbean Sea, and the other running southward along the coast of South America to Cape Horn. It appears to require a further investigation, to decide whether the stream current, referred to in the text, which flows along the coast of Norway and round the North Cape of Europe, and is, at least, mainly supplied from the accumulated waters of the drift impelled by the west and south-west winds which prevail to the northward of the trades, derive any portion whatsoever of its force from the original impulse given to the waters of the gulf-stream at its outlet from the Gulf of Mexico, in the Bahama Channel.   The transport of West Indian seeds to the coast of Norway is undoubted; and even parts of the cargoes of vessels wrecked on the coast of Africa have reached the Norwegian coast, after having made the circuit of the West Indian Islands :—[such an instance occurred when the Editor was at Hammerfest, near the North Cape of Europe, in 1823 ; casks of palm oil were thrown on shore belonging to a vessel which had been wrecked at Cape Lopez, on the African coast, near the Equator, under circumstances which made her loss the subject of discus . sion when the Editor was in that quarter of the globe, the year preceding his visit to Hammerfest :]—but it is quite conceivable that objects conveyed a certain distance by the gulf-stream, and thrown off on its north side into the waters which do not participate in its movement, may be subsequently drifted by the prevailing westerly and south-westerly winds, in accompaniment with the surface water of the sea, across the remaining portion of the Atlantic. The stream current which terminates in ordinary years at the Azores, and which in rare instances extends to the coasts of Europe, is unquestionably traceable the whole way to the outlet of the Gulf of Mexico, by a continuous strength of current and warmth of water; but with respect to a northern branch of the gulf-stream, supposed to detach itself to the N.E., and to convey the waters which have issued through the Bahama Channel in a continuous stream to the North Cape of Europe, positive information is greatly wanting. It may be hoped that the importance to navigation, as well as the interest to physical geography, of a full and complete knowledge of all the details of so remarkable a stream as the gulf-stream, will cause it to become ere long the subject of a systematic examination and survey.

Amongst the sources which contribute to produce the currents which are met with in navigating the ocean, M. de Humboldt has not mentioned the discharge of large bodies of fresh water by the great rivers of the globe, which, nevertheless, deserves to be included in such an enumeration, because the river water is sometimes found to preserve its original direction, and to flow with a very slowly-diminishing velocity over the surface of the ocean, for several hundred miles from its first entrance into the sea. Thus, the current occasioned by the discharge of the River Plate preserves an easterly direction, and is still found to have a velocity of a mile an hour, and a breadth of more than 300 miles, at a distance of not less than 600 miles from the mouth of the river (Rennell, Investigation, p. 65). The current produced by the waters of the River Amazon is another example of the same kind. This current was crossed by the Editor in the "Pheasant" sloop of war, in the year 1822, at a distance of upwards of 300 miles from the mouth of the river, still preserving a velocity of nearly three miles an hour, its original direction being but little altered, and the fresh water but partially mixed with that of the ocean. In both the cases referred to, the river current crosses nearly at a right angle a stream current of the ocean, viz. the two branches into which the equatorial current divides at Cape St. Roque. From the less specific gravity of its water, the river current flows over as well as across the ocean current, which reappears on either side of it. On the side where the ocean current impinges on the river stream, the line of separation of the waters was very distinctly marked by a difference of colour in the case of the Amazon, and the river water was nearly a degree of temperature warmer than that of the ocean. On the opposite side of the river stream the distinction between its water and that of the sea was gradually and insensibly lost, but was clearly determinable for a breadth of above 100 miles (Sabine, Pendulum and other Experiments, London, 1825, pp. 445 to 448).—EDITOR.]

(374) p. 304.—The unknown voice said to him, "Maravillosamente Dios hizo sonar tu nombre en la tierra; de los atamientos de la mar Oceana, que estaban cerrados con cadenas tan fuertes, te dió las llaves." The dream of Columbus is related in the letter to the Catholic monarchs of July 7, 1503, (Humboldt, Examen critique, T. iii. p. 234).

(375) p. 305.—Boussingault, Recherches sur la composition de l'Atmosphère, Annales de Chimie et de Physique, T. lvii. 1834, pp. 171—173; ib. T. lxxi. 1839, p. 116. According to Boussingault and Lewy, the proportion of carbonic acid in the atmosphere at Andilly, at a distance, therefore, from the exhalations of cities, varied only between 0·00028 and 0·00031 in volume.

(376) p. 305.—Liebig, in his important work, entitled Die organische Chemie in ihrer anwendung auf Agricultur und Physiologie, 1840, S. 64—72. On the influence of atmospheric electricity in the production of nitrate of ammonia, which is changed into carbonic acid by contact with lime, see Boussingault's Economie rurale considérée dans ses rapports avec la Chimie et la Météorologie, 1844, T. ii. pp. 247 and 697. Compare also T. i. p. 84.

(377) p. 306.—Lewy, in the Comptes rendus de l'Acad. des Sciences, T. xvii. P. 2, pp. 235—248.

(378) p. 306.—J. Dumas, in the Annales de Chimie, 3ᵉ Série, T. iii. 1841, p. 257.

(379) p. 306.—I have not included in this enumeration the exhalation of carbonic acid gas by plants during the night, when they inhale oxygen, because this source of addition to the quantity of carbonic acid in the atmosphere is fully compensated by the respiration of plants during the day. Compare

Boussingault's "Econ. rurale," T. i. pp. 53—68 ; and Liebig's "Organische Chemie," S. 16 and 21.

([380]) p. 307.—Gay-Lussac, in the "Annales de Chimie," T. liii. p. 120 ; Payen, "Mém. sur la composition chimique des Végétaux," pp. 36 and 42 ; Liebig, Org. Chemie, S. 299—345 ; Boussingault, Econ. rurale, T. i. pp. 142—153.

([381]) p. 307.—By applying the formulæ which Laplace communicated to the Board of Longitude a short time before his death, Bouvard found, in 1827, that the portion of the horary oscillations of the pressure of the atmosphere which results from the attraction of the Moon cannot raise the column of mercury in the barometer at Paris more than 0·018 of a millimètre ; while the mean oscillation of the barometer, derived from 11 years' observation at Paris, is 0·756 of a millimètre from 9 A.M. to 3 P.M., and 0·373 of a millimètre from 3 P.M. to 9 P.M. See Mémoires de l'Acad. des Sciences, T. vii. 1827, p. 267.

([382]) p. 308.—Observations faites pour constater la marche des variations horaires du baromètre sous les tropiques, in my Relation historique du Voyage aux Régions Equinoxiales, T. iii. pp. 270—313.

[The impulse and systematic direction which has been given to meteorological observations, by the establishment of meteorological observatories in different parts of the globe, has already thrown a new light on the diurnal variations of the barometer. It has become known that at stations situated in the interior of great continents, very distant from the ocean or from large bodies of water from whence supplies of aqueous vapour may be derived, and where the air consequently is at all times extremely dry, the double maximum and minimum of the diurnal variation of the barometer either wholly or almost wholly disappear, and the variation consists in a single maximum and minimum, which occur respectively nearly at the coldest and at the hottest hours of the day ; the greatest height of the mercury being at or near the coldest hour, and the least height at or near the warmest hour. For this simple state of the phænomenon an equally simple explanation presents itself : the surface of the earth becoming warmed by the sun's rays, imparts heat to the strata of the atmosphere in contact with it ; the superincumbent air thus rarefied, the column, extending in height, overflows laterally in the higher regions of the atmosphere, and the statical pressure at the base of the column is diminished. In the afternoon the converse takes place : the column of air cools, condenses, and, contracting in height, receives the overflow from the adjacent air, which in its turn becomes heated ; and thus the statical pressure is increased. The stations at which this more simple form of the barometric diurnal curve has hitherto been found to take place are Catherinenbourg, Nertchinsk, and Barnaoul, all in the Russian territories, on the confines of Europe and Asia.

Whenever a free communication with a sufficiently extensive evaporating surface exists, a diurnal variation is also found to take place in the tension of the aqueous vapour in the atmosphere, proportioned in amount to the diurnal variation of the temperature. If the sources of vapour be ample to supply the drain of the ascending current of the air which carries vapour with it, as well as to furnish the increasing tension which the increasing temperature demands, the diurnal variation of the vapour tension has its maximum at or near the hottest hour of the day, and its minimum at or near the coldest hour, being the converse of the diurnal variation of the dry air (or of the gaseous portion of the atmosphere), which, as before stated, has its maximum at or near the coldest hour, and its minimum at or near the warmest hour. The combination of the diurnal variations of the vapour and of the dry air, which conjointly produce the diurnal variation in the pressure on the barometer, occasions the double maximum and minimum of the barometric curve in the temperate zone, at stations not so far removed as the Russian ones from the sources from whence vapour can be supplied. If the elastic force of the vapour be observed by means of an hygrometer with the same care that the barometer is observed, and if the respective pressures of the elastic forces of the air and of the vapour upon the mercury of the barometer be separated from each other, the diurnal variation of the dry air exhibits at all stations in the temperate zone at which observations have hitherto been made, a similar curve to that which the whole barometric pressure produces at the Russian stations where the air is naturally dry.

At Prague in the interior of Europe, at Toronto in the interior of America but situated near extensive lakes, and at Greenwich in the vicinity of the ocean, the diurnal variation of the dry air has but one maximum and one minimum, and these coincide, or nearly coincide, with the coldest and the warmest hours (Sabine, Reports of the Brit. Assoc. 1844, pp. 42—62). Hence it has been inferred, that the normal state of the diurnal variation of the gaseous portion of the atmosphere in the temperate zone is that of a single progression, having a maximum at or near the coldest hour of the day, and a minimum at or near the warmest hour; that at stations of remarkable natural dryness, this curve is given directly by the barometer; and that at other stations, where aqueous vapour mixes in the atmosphere in a greater degree, the same curve is deducible from that of the barometer, by separating the elastic force of the vapour from the whole pressure shewn by that instrument.

As a concomitant phænomenon with the diurnal variation of the gaseous pressure, and due originally to the same cause, (viz. the alternate heating and cooling of the earth's surface by radiation, although an immediate consequence of the ascending current,) we find a diurnal variation taking place in the force of the wind at or near the base of the column. At Greenwich and Toronto, where the force of the wind is continually recorded by self-registering instruments, it is found to undergo a diurnal variation consisting of a single progression, having a minimum at or near the coldest hour, and a maximum at or near the warmest hour of the day. Soon after the ascending current has commenced, a lateral influx is produced to supply the drain which it occasions; the cooler air as it arrives becomes heated in its turn and ascends; as the upward current gathers strength with the increasing temperature of the day, the lateral influx augments also, and attains its maximum about the hour when the phænomena of increasing temperature give place generally to those of decreasing temperature.

The insight which has been obtained into the mutual relations of these meteorological phænomena, and into the sequence of natural causes and effects by which their connection is explained, is a further illustration of the progress which is made in the physical sciences by the aid of *mean numerical values.* The connection which the knowledge of these values has established between the diurnal phænomena of the different elements, and the dependence which it has shown them to have on a common cause, (viz. on the rotation of the earth on its axis, whereby each portion of its surface is successively turned towards the sun, and each meteorological element is thus subjected to a fluctuation of which the period is measured by a day,) adds another beautiful instance to those which have been cited by M. de Humboldt in the text, of the simplicity with which general results in the physical sciences can be presented, when the links of mutual relation are discovered between phænomena, which, when looked at singly and superficially, appear unconnected, and when a deeper insight is obtained into the intricate play of natural forces, by pursuing the strictly inductive method of investigation, resting on the secure basis of correct quantitative determinations. The *annual* variations of the meteorological elements afford a not less striking example of the more or less immediate dependence of several apparently unconnected phænomena on the variations of the temperature occasioned by the earth's annual revolution in its orbit.

The method and systematic direction which characterises the meteorological researches of the present period promises in a peculiar degree to reveal the
" constant amid change," the " stable amid the flow of phænomena."

M. Dove, (to whom is primarily due the new aspect which this beautiful
branch of physical investigation has assumed by the separation of the pressures
of the aqueous and gaseous portions of the atmosphere,) has very recently
shewn, in a memoir read to the Academy of Sciences at Berlin (March 1846),
that the same single progression of the diurnal variation of the dry air extends
also into the intertropical regions. At Buitenzorg in Java the dry air is found
to have a single maximum and minimum, the epochs of which coincide respectively with the coldest and warmest hours. It was previously known
(Sabine, Report of the British Association, 1845, pages 78—82), that at
Bombay, also within the tropics, a less simple law prevailed, the gaseous
atmosphere having there a double maximum and minimum in the twenty-four
hours, accompanied by a corresponding double progression in the force of the
wind. The phænomena at Bombay are by no means, however, to be viewed as
a contradiction to the principles on which the more simple progression prevailing elsewhere has been explained : on the contrary, an extension of the same
principles to the more complicated relations produced by the juxtaposition of
surfaces of land and sea, and by the different affections of these surfaces by
temperature, had led to the expectation that such exceptional cases would be
found within the tropics ; the mutual relations of the diurnal variations of the
gaseous pressure and the force of the wind have received a further and very
striking exemplification in this exceptional case ; one minimum of the pressure
is found to coincide with the greatest strength of the land-breeze, the other
minimum with the greatest strength of the sea-breeze, and the epochs of the
two maxima of pressure correspond respectively with those of the two minima
of the force of the wind.—EDITOR.]

(³⁸³) p. 309.—Bravais, in " Kaemtz et Martins, Météorologie," p. 263.
At Halle (lat. 51° 29′) the oscillation still amounts to 0·28 French lines.
Apparently a very great number of observations will be requisite to obtain a
good determination of the hours of maximum and minimum on mountains in
the temperate zone. Compare observations made in 1832, 1841, and 1842,
on the summit of the Faulhorn, (Martins, Météorologie, p. 254).

(³⁸⁴) p. 309.—Humboldt, Essai sur la Géographie des Plantes, 1807, p. 90;
and in Rel. hist. T. iii. p. 313. On the diminution of the atmospheric
pressure in the tropical portion of the Atlantic Ocean, (Humboldt, in Poggend.
Annalen der Physik, Bd. xxxvii. S. 245—258, and S. 468—486).

(385) p. 310.—Daussy, in Comptes rendus, T. iii. p. 136.

(386) p. 310.—Dove über die Stürme, in Poggend. Ann. Bd. lii. S. 1.

(387) p. 310.—Leopold von Buch, barometrische Windrose, in Abhandl. der Akad. der Wiss. zu Berlin aus den J. 1818—1819, S. 187.

(388) p. 310.—Dove, Meteorologische Untersuchungen, 1837, S. 99—343 ; and Kämtz's ingenious remarks on the descent in the higher latitudes of the upper current, and on the general phænomena of the direction of the wind, in his "Vorlesungen über Meteorologie," 1840, S. 58—66, 196—200, 327—336, 353—364. Also in Schumacher's Jahrbuch für 1838, S. 291—302. Dove has given a very happy and lively representation of meteorological phenomena systematically viewed, in a short memoir entitled Witterungs-verhältnisse von Berlin," 1842. Concerning the early knowledge of the rotation of the winds possessed by navigators, compare Churruca, "Viage al Magellanes," 1793, p. 15 ; and respecting a remarkable expression made use of by Columbus, and preserved by his son Don Fernando Colon in the " Vida del Almirante," cap. 55, see Humboldt's Examen critique de l'hist. de la Géographie, T. iv. p. 253.

(389) p. 311.—" Monsun," (in Malay, musim, the hippalus of the Greeks,) is derived from the Arabic word " *mausim*," fixed time or epoch, season, time of the assembling of the pilgrims at Mecca. The word has been trans-ferred to the season of the regular or periodical winds, which are named from the countries from whence they blow,—as " the Mausim of Aden," " of Guzerat," " of Malabar," &c. (Lassen, Indische Alterthumskunde, Bd. i. 1843, S. 211). On the different relations which prevail where the atmosphere has a solid and where it has a liquid base (land and sea), see Dove, in the Abhandl. der Akad. der Wiss. zu Berlin aus dem J. 1842, S. 239.

(390) p. 316.—Humboldt, Recherches sur les causes des Inflexions des Lignes isothermes, in Asie centr. T. iii. pp. 103—114, 118, 122, 188.

(391) p. 317.—Georg. Forster, kleine Schriften, Th. iii. 1794, S. 87; Dove, in Schumacher's Jahrbuch für 1841, S. 289 ; Kämtz, Meteorologie, Bd. ii. S. 41, 43, 67, and 96 ; Arago, in Comptes rendus, T. i. p. 268.

(392) p. 319.—Dante, Divina Commedia, Purgatorio, canto iii.

(393) p. 320.—Humboldt sur les Lignes isothermes, in the " Mémoires de physique et de chimie de la Société d'Arcueil," T. iii. Paris, 1817, p. 143—165 ; Knight, in Transactions of the Horticultural Society of London, Vol. i. p. 32; Watson, Remarks on the Geographical Distribution of British Plants, 1835, p. 60; Trevelyan, in Jameson's Edinburgh New Phil. Journal, No. 18,

p. 154; Mahlmann, in his excellent German translation and completion of my Asie centrale, Th. ii. S. 60.

([394]) p. 321.—" Hæc de temperie æris, qui terram late circumfundit, ac in quo, longe a solo, instrumenta nostra meteorologica suspensa habemus. Sed alia est caloris vis, quem radii solis nullis nubibus velati, in foliis ipsis et fructibus maturescentibus, magis minusve coloratis, gignunt, quemque, ut egregia demonstrant experimenta amicissimorum Gay-Lussacii et Thenardi de combustione chlori et hydrogenis, ope thermometri metiri nequis. Etenim locis planis et montanis, vento libe spirante, circumfusi æris temperies eadem esse potest cœlo sudo vel nebuloso ; ideoque ex observationibus solis thermometricis, nullo adhibito Photometro, haud cognosces, quam ob causam Galliæ septentrionalis tractus Armoricanus et Nervicus, versus littora, cœlo temperato sed sole raro utentia, Vitem fere non tolerant. Egent enim stirpes non solum caloris stimulo, sed et lucis, quæ magis intensa locis excelsis quam planis, duplici modo plantas movet, vi sua tum propria, tum calorem in superficie earum excitante." (Humboldt, De distributione geographica plantarum, 1817, pp. 163—164).

([395]) p. 321.—The work last quoted, pp. 156—161 ; Meyen, in his " Grundriss der Pflanzengeographie," 1836, S. 379—467 ; Boussingault, " Economie rurale," T. ii. p. 675.

([396]) p. 322.—I subjoin a Table, exhibiting, in a descending scale, the capability of different places in Europe for the production of wine. See my " Asie centrale," T. iii. p. 159. The numerical data for the banks of the Rhine and the Main are given in addition to those for the places referred to in the text of Cosmos. A comparison of Cherbourg and Dublin with places in the interior of Europe shews that, with but little difference of temperature, so far as the indications of the thermometer in the shade are concerned, the question of maturity or immaturity of fruit is determined by the habitual serenity or cloudiness of the sky.

The great accordance in the distribution of the annual temperature throughout the different seasons of the year in the valleys of the Rhine and the Main, tends to confirm the accuracy of the observations. The months of December, January, and February, are taken as winter months, as is both the usual and the most advantageous arrangement in meteorological tables. When we compare the qualities of the wines of Franconia and Berlin, and the mean summer and autumn temperatures of Würzburg and Berlin, we are almost surprised to find that the temperatures differ only 1° or 1°·2

PLACES.	Latitude.	Elevation.	Mean of Year.	Winter.	Spring.	Summer.	Autumn.	Number of Years of Observations.
		In Toises.	*Cent.*	*Cent.*	*Cent.*	*Cent.*	*Cent.*	
Bordeaux . . .	44° 50′	4	13·9°	6·1	13·4	21·7	14·4°	10
Strasburg . .	48 35	75	9·8	1·2	10·0	18·1	10·0	35
Heidelberg . .	49 24	52	9·7	1·1	10·0	17·9	9·9	20
Manheim . .	49 29	47	10·3	1·5	10·4	19·5	9·8	12
Würzburg . .	49 48	88	10·1	1·6	10·2	18·7	9·7	27
Frankfort-on-Main .	50 7	60	9·6	0·8	10·0	18·0	9·7	19
Berlin . . .	52 31	16	8·6	—0·6	8·1	17·5	8·6	22
Cherbourg (no wine)	49 39	0	11·2	5·2	10·4	16·5	12·5	3
Dublin . . .	53 23	0	9·5	4·6	8·4	15·3	9·8	13
		In Eng. ft.	*Fah.*	*Fah.*	*Fah.*	*Fah.*	*Fah.*	
Bordeaux . . .	44° 50′	25·6	57°	43·0°	56·1°	71·0°	58·0°	
Strasburg . .	48 35	479·6	49·6	34·2	50·0	64·6	50·0	
Heidelberg . .	49 24	333·5	49·4	34·0	50·0	64·2	49·8	
Manheim . .	49 29	300·5	50·6	34·9	50·8	67·1	49·6	
Würzburg . .	49 48	562·7	50·2	35·0	50·4	65·7	49·4	
Frankfort . .	50 07	383·7	49·4	33·3	50·0	64·4	49·4	
Berlin . . .	52 31	102·3	47·7	31·0	46·7	63·5	47·5	
Cherbourg . .	49 39	0	52·1	41·4	50·8	61·7	54·3	
Dublin . . .	53 23	0	49·1	40·2	47·1	59·6	49·7	

464   NOTES.

of the cent. thermometer, or about 2° of Fah. The spring difference is greater, being about 2° cent., or nearly 4° of Fah. The influence of late May frosts on the flowering season of the vine, after a winter of correspondingly lower temperature, is an element of no less importance than the late season of the ripening of the grapes, and the influence of direct, not diffused, solar light unobscured by clouds. The difference, alluded to in the text, between the true temperature of the surface of the ground, and the indications of a thermometer placed in the shade and protected from extraneous influences, has been investigated by Dove, from the observations collected during fifteen years at the garden of the Horticultural Society at Chiswick, near London, in Bericht über die Verhandl. der Berl. Akad. der Wiss. August, 1844, S. 285.

(397) p. 323. — Compare my memoir, " über die Haupt-ursachen der Temperaturverschiedenheit auf der Erdoberfläche," in Abhandl. der Akad. der Wissensch. zu Berlin aus dem Jahre, 1827, S. 311.

(398) p. 323.—The general level of Siberia, between Tobolsk, Tomsk, and Barnaul, from the Altai to the icy sea, is not so high as Manheim and Dresden; and even Irkutsk, far to the east of the Jenissei, is only 208 toises (1330 English feet) above the level of the sea, or about one-third lower than Munich.

(399) p. 325.—Humboldt, Recueil d'Observations astronomiques, T. i. pp. 126—140; Relation historique, T. i. pp. 119, 141, and 227; Biot, in Connaissance des temps pour l'an 1841, pp. 90—109.

(400) p. 327.—Anglerius de Rebus Oceanicis, Dec. 11, Lib. ii. p. 140 (ed. Col. 1574). In the Sierra de Santa Marta, the highest summits of which appear to exceed 18000 (above 19000 English) feet, (Humboldt, Relat. hist. T. ii. p. 214), one peak is still called Pico de Gaira.

(401) p. 328.—Compare the table of the heights of perpetual snow in both hemispheres, from 71° 15′ of North to 53° 54′ South latitude, in my " Asie centrale," T. iii. p. 360.

(402) p. 329.—Darwin, " Journal of the Voyages of the Adventure and Beagle," p. 297. As the volcano of Aconcagua was not then in eruption, the remarkable phænomenon of the absence of snow cannot have been caused, as it sometimes is on Cotopaxi, by the rapid heating of the interior of the crater, or by the emission of heated gases through fissures, (Gillies, in the " Journal of Nat. Sciences," 1830, p. 316).

(403) p. 330.—See my " Second Mémoire sur les Montagnes de l'Inde," in the " Annales de Chimie et de Physique," T. xiv. pp. 5—55, and " Asie centrale," T. iii. pp. 281—327. The fact of the greater elevation of the snow-line on the Thibetian side of the Himalaya was supported by the most

experienced and best informed Indian travellers—Colebrooke, Webb and Hodgson, Victor Jacquemont, Forbes Royle, Carl von Hügel, and Vigne, all of whom knew the mountains by personal examination. It was, however, treated as doubtful by John Gerard; the geologist Mac Clelland, editor of the "Calcutta Journal;" and Lieutenant Thomas Hutton, assistant-surveyor of the Agra division. The publication of my work on Central Asia occasioned the renewal of the discussion. A recent number of a journal published in India (Mac Clelland and Griffiths' "Calcutta Journal of Natural History," Vol. iv. January, 1844), contains a remarkable notice, which seems nearly conclusive as to the limits of snow on the Himalaya. Mr. Batten, of the Bengal service, writes from the camp of Semulka, on the river of Cosillah, in the province of Kumaoon :—"I have only recently read, and with surprise, the statements of Lieut. Thomas Hutton respecting the limits of perpetual snow. I feel it a duty towards science to contradict these assertions, because Mr. Mac Clelland goes so far as to speak of the ' service which Lieut. Hutton has rendered to science by dispelling a widely-prevailing error,' (Journal of the Asiatic Society of Bengal, Vol. ix. Calcutta, 1840, pp. 575, 578, and 580). It is an erroneous assertion that every traveller in the Himalaya must participate in Hutton's doubts. I am one of those by whom the western portion of our great chain of mountains has been most visited. I have gone through the Borendo pass into the Buspa valley and the lower Kunawer, and have returned by the lofty pass of Rupin in the mountains of Gurwal. I visited the sources of the Jumna at Jumnotri, and from thence the tributaries of the Ganges, from Mundakni and Vischnu-Aluknunda to Kadarnath and the celebrated snowy peak of Nundidevi. I have repeatedly crossed the Niti pass to the highlands of Thibet; and the settlement of Bhote-Mehals was established by me. My residence in the very midst of the mountains has for the last six years brought me constantly into intercourse with travellers, both European and native, who I carefully interrogated, and from whom I was able to derive the best information respecting the aspect of the country. All the knowledge gained in these different ways, by personal observation, and by the relations of others, has led me to a conviction which I feel prepared to support on all occasions—that, in the Himalaya, the limit of perpetual snow is higher on the northern declivity towards Thibet, than on the southern declivity towards India. Mr. Hutton changes the question at issue; for whilst he thinks he is attacking M. de Humboldt's view of the phænomenon taken in its generality, he is really only combating an imaginary point of difference. He tries to prove what we are quite willing to admit ;

namely, that, on particular mountains belonging to the Himalaya range, the snow lies longer on the northern than on the southern declivity :" (compare also Note 5, p. 363). If the mean of the plateau of Thibet be 1800 toises (11510 English feet), it may be justly compared with the lovely and fertile Peruvian plateau of Caxamarca; but it would still be 1200 French, or about 1300 English, feet lower than the plateau of Bolivia round the lake of Titiaca, and than the pavement of the streets of Potosi. Ladak, according to Vigne's determination by means of the boiling point of water, is situated at an altitude of 1563 toises (9994 English feet). Probably this is also about the altitude of H'Lassa (Yul-sung), a monastic city, surrounded by vineyards, and called by Chinese writers " the kingdom of joy." The vineyards may possibly be situated in deeply-cleft valleys.

([404]) p. 331.—Compare Dove, Meteorologische Vergleichung von Nordamerika und Europa, in Schumacher's Jahrbuch für 1841, and his Meteorologische Untersuchungen, S. 140.

([405]) p. 331.—The mean annual quantity of rain in Paris, from 1805 to 1822, was, according to Arago, 18 inches 9 lines ; in London (from 1812 to 1827), according to Howard, 23 inches 4 lines; and in Geneva, by a mean of 32 years, 28 inches and 8 lines. On the coast of Hindostan, the quantity of rain is from 108 to 120 inches ; and in the island of Cuba, there fell, in 1821, fully 133 inches. (The above quantities are in French measure, corresponding in English inches to 20 inches at Paris ; 24·9 in London ; 30·5 at Geneva, about 115 to 128 in Hindostan ; and 141·7 in Cuba). On the distribution of the annual fall of rain into the different seasons of the year in middle Europe, see the excellent observations of Gasparin, Schouw, and Bravais, in the Bibliothèque Universelle, T. xxxviii. p. 54 and 264; Tableau du Climat de l'Italie, p. 76 ; and the Notes with which Martins has enriched his French translation of Kämtz's Vorlesungen über Meteorologie (Leçons de Météorologie), p. 142.

([406]) p. 331.—According to Boussingault (Economie rurale, T. ii. p. 693), 60 inches and 2 lines of rain fell, on the mean of the years 1833 and 1834, at Marmato, in lat. 5° 27′, at an altitude of 731 toises (4674 E. feet), and with a mean temperature of 20°·4 C. (68°·7 Fah.) ; whilst at Bogotá, in lat. 4°·36′, at an elevation of 1358 toises (8684 E. feet), and with a mean temperature of 14°·5 C. (58° Fah.), the annual fall of rain is only 37 inches and 1 line. [In English measure, the above quantities are—at Marmato, 64·1 inches, and at Bogotá, 39·4 inches of rain.]

([407]) p. 332.—For the details of this observation, see my Asie centrale,

T. iii. p. 85—89 and 567 ; and on the hygrometric state of the atmosphere over the lowlands of tropical South America, see my Relation hist. T. i. p. 242—248, and T. ii. p. 45—164.

(408) p. 332.—Kämtz, Vorlesungen über Meteorologie, S. 117.

(409) p. 333.—On electricity from evaporation at a high temperature, see Peltier, in the Annales de Chimie, T. lxxv. p. 330.

(410) p. 333.—Pouillet, in the Annales de Chimie, T. xxxv. p. 405.

(411) p. 333.—De la Rive, in his admirable Essai historique sur l'Electricité, p. 140.

(412) p. 333.—Peltier, in the Comptes-rendus de l'Acad. des Sciences, T. xii. p. 307 ; Becquerel, Traité de l'Electricité et du Magnétisme, T. iv. p. 107.

(413) p. 334.—Duprez, sur l'Electricité de l'air (Bruxelles, 1844), p. 56—61.

(414) p. 334.—Humboldt, Relation historique, T. iii. p. 318. I would refer, however, exclusively to those experiments which were made with Saussure's electrometer, with a metallic conductor of a metre in length, and in which the electrometer was not moved either upwards or downwards, nor the conductor armed, according to Volta's proposal, with a sponge dipped in burning alcohol. Those of my readers who are acquainted with the points at present in discussion with reference to the subject of atmospheric electricity, will understand the reason of this limitation. On the formation of thunderstorms in the tropics, see my Rel. hist. T. ii. p. 45, and 202—209.

(415) p. 334.—Gay-Lussac, Annales de Chimie et de Physique, T. viii. p. 167. The discordant views of Lamé, Becquerel, and Peltier, render it difficult to arrive at present at any conclusion respecting the cause of the specific distribution of electricity in clouds, some of which have a positive, and others a negative tension. The negative electricity, which, near lofty cascades, is developed in the air by the finely-divided particles of water, is a very striking phænomenon : it was first observed by Tralles, and since then by myself in many latitudes. With a sensitive electrometer the effect can be distinctly recognised at a distance of three or four hundred feet.

(416) p. 335.—Arago, in the Annuaire du Bureau des Longitudes pour 1838, p. 246.

(417) p. 335.—Arago, p. 249—266 of the above-named volume. Compare p. 268—279.

(418) p. 336.—Arago, p. 388—391 of the same volume. Von Baer, who has rendered such great services to the meteorology of the north of Asia, has not

discussed the rare occurrence of thunderstorms in Iceland and Greenland; he has only noticed that thunder has sometimes been heard in Nova Zembla and Spitzbergen. (Bulletin de l'Acad. de St.-Pétersbourg, 1839, Mai).

([419]) p. 337.—Kämtz, in Schumacher's Jahrbuch für 1838, S. 285. On the comparison of the laws of the distribution of heat to the East and to the West, in Europe and North America, see Dove's Repertorium der Physik, Bd. iii. S. 392—395.

([420]) p. 339.—The "natural history of plants," which has been ably and briefly sketched by Endlicher and Unger (Grundzüge der Botanik, 1843, S. 449—468), was distinguished by myself from the "geography of plants" more than half a century ago, in a passage of the aphorisms appended to my Subterranean Flora :—" Geognosia naturam animantem et inanimam vel, ut vocabulo minus apto, ex antiquitate saltem haud petito, utar, corpora organica æque ac inorganica considerat. Sunt enim tria quibus absolvitur capita : Geographia oryctologica quam simpliciter Geognosiam vel Geologiam dicunt, virque acutissimus Wernerus egregie digessit; Geographia zoologica, cujus doctrinæ fundamenta Zimmermannus et Treviranus jecerunt ; et Geographia plantarum quam æquales nostri diu intactam reliquerunt. Geographia plantarum vincula et cognationem tradit, quibus omnia vegetabilia inter se connexa sint, terræ tractus quos teneant, in ærem atmosphæricum quæ sit eorum vis ostendit, saxa atque rupes quibus potissimum algarum primordiis radicibusque destruantur docet, et quo pacto in telluris superficie humus nascatur, commemorat. Est itaque quod differat inter Geognosiam et Physiographiam, historia naturalis perperam nuncupatam, quum Zoognosia, Phytognosia et Oryctognosia, quæ quidem omnes in naturæ investigatione versantur, non nisi singulorum animalium, plantarum, rerum metallicarum vel (venia sit verbo) fossilium formas, anatomen, vires scrutantur. Historia Telluris, Geognosiæ magis quam Physiographiæ affinis, nemini adhuc tentata, plantarum animaliumque genera orbem inhabitantia primævum, migrationes eorum compluriumque interitum, ortum quem montes, valles, saxorum strata et venæ metalliferæ ducunt, ærem, mutatis temporum vicibus, modo purum, modo vitiatum, terræ superficiem humo plantisque paulatim obtectam, fluminum inundantium impetu denuo nudatam, iterumque siccatam et gramine vestitam commemorat. Igitur Historia zoologica, Historia plantarum et Historia oryctologica, quæ non nisi pristinum orbis terræ statum indicant, a Geognosia probe distinguendæ."—(Humboldt, Flora Fribergensis subterranea, cui accedunt aphorismi ex Physiologia chemica plantarum, 1793, p. ix.—x.) Respecting the "spontaneous motion" which is spoken of farther on in the text, see the remarkable passage in Aristotle de Cœlo, ii. 2, p. 284, Bekker, where the distinction between animate and inanimate bodies is based on whether their movements are determined from within or from without. The Stagirite says, " The life of vegetables produces no movement, because it is plunged in a profound slumber from which nothing arouses it" (Aristot. de generat. animal. V. i. p. 778, Bekker) ; and " plants have no desires which incite them to spontaneous motion" (Aristot. de somno et vigil. cap. i. p. 455, Bekker).

[421] p. 342.—Ehrenberg's memoir, über das kleinste Leben im Ocean, read to the Academy of Sciences at Berlin, May 9, 1844.

[422] p. 343.—Humboldt, Ansichten der Natur (2te Ausgabe, 1826), Bd. ii. S. 21.

[423] p. 343.—On multiplication by spontaneous division and intercalation of new substance, see Ehrenberg, von den jetzt lebenden Thierarten der Kreidebildung, in den Abhandl. der Berliner Akad. der Wiss. 1839, S. 94. The most powerful productive faculty in nature is that of the Vorticellæ. Estimations of the possible increase of masses composed of these animals, are given in Ehrenberg's great work, Die Infusionsthierchen als vollkommne Organismen, 1838, S. xiii. xix. and 244. " The milky way of these organisms is formed of the genera Monas, Vibrio, Bacterium, and Bodo." Life is distributed in nature with such profusion, that small infusoria live parasitically on larger, and are themselves inhabited by smaller (S. 194, 211, and 512).

[424] p. 344.—Aristot. Hist. Animal, vol. xix. p. 552, Bekker.

[425] p. 345.—Ehrenberg, S. xiv. 122 and 493 of the work last quoted. The rapid multiplication of microscopic animalcula is in some species accompanied by an astonishing tenacity of life : for example, in Wheat-eels, Wheel-animalcules, and Water-bears or tardigrade animalcula. They have been seen to come to life from a state of apparent death after being dried during 28 days in a vacuum with chloride of lime and sulphuric acid, and exposed to a heat of 120° Cent. (248° Fah.) See M. Doyère's experiments, in his Mémoire sur les Tardigrades, et sur leur propriété de revenir à la vie, 1842, pp. 119, 1£., 131, and 133. On the revival of infusoria which had been for years in a state of desiccation, compare generally Ehrenberg, p. 492—496.

[426] p. 345.—On the supposed " primitive transformation" of organized or unorganized matter into plants and animals, compare Ehrenberg, in Poggendorff's Annalen der Physik, B. xxiv. S. 1—48, and the same author's Infusionsthierchen, S. 121 and 525, with Joh. Müller, Physiologie des Menschen, (4the Aufl.) B. i. S. 8—17. It seems to me particularly deserving of notice, that St. Augustine, in treating the question, " how islands may have

been provided with new plants and animals after the Deluge," shows himself in no respect disinclined to the hypothesis of what has been called "spontaneous generation" (generatio æquivoca, spontanea aut primaria). He says, "If animals have not been conveyed to the remoter islands by angels, or possibly by inhabitants of continents addicted to the chase, they must have sprung directly from the earth; though in this case the question arises, why all kinds of animals should have been assembled in the Ark." "Si e terra exortæ sunt (bestiæ) secundum originem primam, quando dixit Deus: producat terra animam vivam! multo clarius apparet, non tam reparandorum animalium causa, quam figurandarum variarum gentium (?) propter ecclesiæ sacramentum in Arca fuisse omnia genera, si in insulis, quo transire non possent, multa animalia terra produxit."—Augustinus de Civitate Dei, lib. xvi. cap. 7 (Opera, ed. Monach. Ordinis S. Benedicti, T. vii. Venet. 1732, p. 422). Two centuries prior to the Bishop of Hippo, we find by extracts from Trogus Pompeius that he had established a similar connection between the "generatio primaria," the drying of the ancient world, and the high table-land of Asia, as is found in the hypothesis of the great Linnæus, between the terraces of Paradise and the reveries of the eighteenth century respecting the fabled Atlantis. "Quod si omnes quondam terræ submersæ profundo fuerunt, profecto editissimam quamque partem decurrentibus aquis primum detectam; humillimo autem solo eandem aquam diutissime immoratum et quanto prior quæque pars terrarum siccata sit, tanto prius animalia generare cœpisse. Porro Scythiam adeo editiorem omnibus terris esse ut cuncta flumina ibi nata in Mæotium, tum deinde in Ponticum et Ægyptium mare decurrant."—Justinus, lib. ii. cap. 1. The erroneous supposition of Scythia being an elevated table-land is so ancient, that we find it very distinctly expressed in Hippocrates (De Aere et Aquis, cap. vi. § 96, Coray): he says, "Scythia consists of elevated barren plains, which, without being crowned with mountains, rise higher and higher towards the north."

(427) p. 345.—Humboldt, Aphorismi ex Physiologia chemica plantarum, in the Flora Fribergensis subterranea, 1793, p. 178.

(428) p. 346.—Ueber die Physiognomik der Gewächse, in Humboldt, Ansichten der Natur, Bd. ii. S. 1—125.

(429) p. 346.—Ætna Dialogus. Opuscula, Basil, 1556, p. 53—54. A finely executed geography of the plants of Etna has been recently published by Philippi. (See Linnæa, 1832, p. 733.)

(430) p. 348.—Ehrenberg, in the Annales des Sciences naturelles, T. xxi. p. 387—412; Humboldt, Asie centrale, T. i. p. 339—342; T. iii. p. 96—101.

(431) p. 349.—Schleiden, über die Entwicklungsweise der Pflanzenzellen, in Müller's Archiv für Anatomie und Physiologie, 1838, S. 137—176; also his Grundzüge der wissenschaftlichen Botanik, Th. i. S. 191, Th. ii. S. 11, Schwann, Mikroskopische Untersuchungen über die Uebereinstimmung in der Struktur und dem Wachsthum der Thiere und Pflanzen, 1839, S. 45 und 220. On hereditary form, see Joh. Müller, Physiologie des Menschen, 1840. Th. ii. S. 614.

(432) p. 349.—Schleiden, Grundzüge der wissenschaftlichen Botanik, 1842, Th. i. S. 192—197.

(433) p. 351.—Tacitus, in his speculations on the population of Britain, (Agricola, cap. ii.) distinguishes finely between that which is due to the influence of climate, and that which in the immigrating tribes is hereditary in race, and not susceptible of change. "Britanniam qui mortales initio coluerunt, indigenæ an advecti, ut inter barbaros, parum compertum. Habitus corporis varii, atque ex eo argumenta; namque rutilæ Caledoniam habitantium comæ, magni artus Germanicam originem adseverant. Silurum colorati vultus et torti plerumque crines, et posita contra Hispania, Iberos veteres trajecisse, easque cedes occupasse fidem faciunt: proximi Gallis, et similes sunt: seu durante originis vi; seu, procurrentibus in diversa terris, positio cœli corporibus habitum dedit." Respecting the persistency of types and configuration in' the warm and cold regions of the earth, and in the mountainous districts of the New Continent, see my Relation historique, T. i. p. 498—503; T. ii. p. 572—574.

(434) p. 351.—On the American race generally, see the magnificent work of Samuel George Morton, entitled, Crania Americana, 1839, p. 62—86; on the skulls brought by Pentland from the highlands of Titiaca, see the Dublin Journal of Medical and Chemical Science, vol. v. 1834, p. 475; and Alcide d'Orbigny, l'Homme américain considéré sous ses rapports physiologiques et moraux, 1839, p. 221. See also Prince Maximilian of Wied's Reise in das Innere von Nordamerika, which is rich in refined ethnographical remarks.

(435) p. 351.—Rudolph Wagner, über Blendlinge und Bastarderzeugung, in his notes to Naturgesch. des Menschengeschlects, Th. i. S. 174—188, translated from Prichard's Natural History of Man.

(436) p. 352.—Prichard (in Wagner's German translation), Th. i. S. 431. Th. ii. S. 363—369.

(437) p. 352.—Onesicritus, in Strabo, xv. p. 690 and 695, Causab. Welcker (Griechische Tragödien, Abth. iii. S. 1078) supposes that the verses of Theo-

dectes, which Strabo has quoted, are taken from a lost tragedy, which, perhaps, bore the title of "Memnon."

(438) p. 353.—Joh. Müller, Physiologie des Menschen, Bd. ii. S. 768, 772—774.

(439) p. 353.—Prichard (in the German translation), Th. i. S. 295; T. iii. S. 11.

(440) p. 353.—The late arrival of the Turkish and Mongolian tribes, both on the Oxus and on the Kirghis Steppes, is opposed to the hypothesis of Niebuhr, which makes the Scythians of Herodotus and Hippocrates *Mongolians*. It seems to me far more probable that the Scythians, Scoloti, should be referred to the Indo-germanic Massagetæ (the Alani). The Mongolians, true Tatars, (a name long afterwards improperly given to purely Turkish tribes in Russia and Siberia,) then dwelt far in the eastern part of Asia. Compare my Asie centr. T. i. p. 239 and 400; Examen critique de l'hist. de la Géog. T. ii. p. 320. A distinguished linguist, Professor Buschmann, remarks that Firdusi, in a half mythic history with which he begins the Shahnamah, mentions a "fortress of the Alani," on the seashore, in which Selm, the eldest son of king Feridun, who certainly lived two centuries before Cyrus, sought shelter. The Kirghis of the steppe, called Scythian, are originally a Finnish race: their three hordes probably constitute the most numerous nomade nation of the present time, and in the sixth century they lived on the same steppe where I have myself seen them. The Byzantine Menander (p. 380—382, ed. Nieb.) relates expressly that the Chakan of the Turks (Thu-kiu), in 569, made Zemarchus, the ambassador of the emperor Justinus II., a present of a Kirghis female slave: he calls her χερχίς, and also in Abulgasi (Historia Mongolorum et Tatarorum), the Kirghis are called *Kirkiz*. Similarity of manners and customs, where the nature of the country determines their principal characters, is a very uncertain evidence of identity of race. The life of the steppes produces among the Turks (Ti Tu-kiu), the Baschkirs (Fins), the Kirghis, the Torgodi, and the Dsungari (Mongolians), the customs common to a nomade life, and the same use of felt tents, carried on waggons, and pitched among the herds of cattle.

(441) p. 354.—Wilhelm von Humboldt, über die Verschiedenheit des menschlichen Sprachbaues, in his great work, Ueber die Kawi-Sprache auf der Insel Java, Bd. i. S. xxi. xlviii. and ccxiv.

(442) p. 355.—The cheerless and since often repeated doctrine of inequality in men's right to freedom, and of slavery as an institution in conformity with nature, is found, alas! systematically developed in Aristotle's Politica, i. 3, 5, 6.

(443) p. 357.—Wilhelm von Humboldt, über die Kawi-Sprache, Bd. iii. S. 426. I here subjoin another extract from the same work:—" The impetuous conquests of Alexander, the more politic and deliberate extension of the Roman dominion, the savagely cruel wars of the Mexicans, and the despotic territorial acquisitions of the Incas of Peru, have contributed in both hemispheres to terminate the segregation of nations, and to form more extensive societies. Men of great and powerful minds, as well as whole nations, acted under the dominion of an idea, to which, nevertheless, in its moral purity, they were entire strangers. Christianity first made known its true significancy and profound charity: and even from her voice it has obtained but a slow and gradual reception. Until that voice had spoken, a few solitary accents alone foreshadowed this great truth. In modern times an increased impulse has been given to the idea of civilization; and the desire of extending more widely friendly relations between nations, as well as the benefits of intellectual and moral culture, is increasingly felt. Even selfishness begins to perceive that its interests are thus better served, than by forcibly maintaining a constrained and hostile isolation. Language, more than any other faculty, binds mankind together. Diversities of idiom produce, indeed, to a certain extent, separation between nations; but the necessity of mutual understanding occasions the acquirement of foreign languages, and reunites men without destroying national peculiarity."

END OF VOL. I.

Printed in the United States
By Bookmasters